AF356605

Biomass and Waste Conversion and Valorization to Chemicals, Energy and Fuels

Biomass and Waste Conversion and Valorization to Chemicals, Energy and Fuels

José A.P. Coelho
Roumiana P. Stateva

Basel • Beijing • Wuhan • Barcelona • Belgrade • Novi Sad • Cluj • Manchester

José A.P. Coelho
Chemical Engineering
Department
Instituto Superior de
Engenharia e Lisboa, IPL
Lisboa
Portugal

Roumiana P. Stateva
Institute of Chemical Engineering
Bulgarian Academy of Sciences
Sofia
Bulgaria

Editorial Office
MDPI AG
Grosspeteranlage 5
4052 Basel, Switzerland

This is a reprint of articles from the Special Issue published online in the open access journal *Molecules* (ISSN 1420-3049) (available at: www.mdpi.com/journal/molecules/special_issues/05KH605Q02).

For citation purposes, cite each article independently as indicated on the article page online and using the guide below:

Lastname, A.A.; Lastname, B.B. Article Title. *Journal Name* **Year**, *Volume Number*, Page Range.

ISBN 978-3-7258-1836-5 (Hbk)
ISBN 978-3-7258-1835-8 (PDF)
https://doi.org/10.3390/books978-3-7258-1835-8

Contents

About the Editors

José A.P. Coelho

José Augusto Paixão Coelho earned his Ph.D. in Chemical Engineering in February 1996. Since then, he has specialized in Chemical Engineering and has developed expertise in using greener technologies such as supercritical fluids and microwaves, thermodynamics, the extraction of natural products, the characterization of extracts, and the reuse of biomass from agricultural and food residues.

He is currently a Coordinating Professor at IPL-ISEL and Deputy vice president of the Polytechnic Institute of Lisbon (IPL)'s research, development, and innovation. He was president of the Scientific Board of Instituto Superior de Engenharia de Lisboa (IPL-ISEL) from 2008 to 2010 and president of the Chemical Engineering Department from 2002 to 2005.

Since April 2023, he has been an evaluator for the European Commission, Research Executive Agency, Central Finance; a member and secretary of the Executive Board of ISASF; and a member of ACS and SPQ. He has published 60 papers, 2 books, 3 chapters of books, and a patent. He has enaged in more than 150 communications, both oral and as a poster. He is a member of the Scientific Committee and Chairperson in 13 international conferences, and he is a researcher and coordinator for 30 national and international projects in extraction, characterization, modeling, and extraction with green technologies of plants and biomasses. The aim is to develop alternative extraction techniques to conventional methods.

Roumiana P. Stateva

Roumiana P. Stateva is a full professor at the Institute of Chemical Engineering, Bulgarian Academy of Sciences, Sofia, Bulgaria. She has an MSc in Chemical Engineering and Chemical Cybernetics and holds a PhD in Chemical Engineering, both from Dmitry Mendeleev University of Chemical Technology, Faculty of Physical Chemistry, Moscow, The Russian Federation. Her main fields of scientific research are chemical engineering thermodynamics, predictive methods for estimating thermophysical properties, advanced innovative methods for biomass valorization, green chemistry, modelling, simulation and the design of processes. Dr. Stateva was a visiting professor at the Departments of Chemical Engineering of the University of Valencia, Spain, Imperial College London, UK, and Ben-Gurion University of Negev, Israel. Twice, she was a visiting fellow at the Department of Chemical Engineering, Carnegie-Mellon University, Pittsburgh, Pa. USA. She has co-authored over 90 scientific publications in high-impact international journals, one book, and several book chapters. Prof. Stateva has participated as a principal investigator in a number of international research projects financed by the Royal Society, UK; EC (COST and HORIZON 2020 MSCA-RISE), and is the agreement coordinator of seven Inter-Institutional ERASMUS+ Programme 2022–2027 Agreements.

Preface

Biomass and waste conversion and the valorization of resulting chemicals, energies, and fuels are important areas of research and development. In general, biomass and bio-waste are underutilized and, in some cases, even neglected. However, they have the potential to replace depleting fossil resources as alternative, sustainable feedstocks, which aligns with the circular bioeconomy concept. By utilizing biomass, our reliance on non-renewable resources will be decreased, and some of the severe environmental issues caused by the careless exploitation of fossil fuels will be avoided. Furthermore, valorizing waste involves converting potential pollutants into valuable products, thus lessening the environmental impact of waste disposal.

Biomass and wastes are a rich storehouse of various high-value compounds and secondary metabolites that can be transformed into bioproducts, biofuels, and various platform chemicals by applying different innovative advanced techniques. Hence, effective waste management and conversion are scientific endeavors of major importance and are crucial in reducing our environmental impact. The use of biomass, with its potential to significantly decrease our carbon footprint, is a key part of this effort. Since biomass absorbs CO_2 during its growth, it can offset the CO_2 released during its conversion to energy-related compounds and/or bioactives. This could lead to a lower net carbon footprint compared to fossil fuels. Furthermore, effective waste management and conversion are crucial in reducing landfill usage, lowering methane emissions, and minimizing soil and water pollution.

Minimizing carbon footprint and greenhouse gas emissions by extracting the maximum value from materials before recovery and regeneration address environmental concerns, supports economic growth, enhances energy security, and promotes technological advances. This approach is based on the "cradle to cradle" concept and contributes significantly to the circular bio-economy.

In view of the foregoing, we were privileged to compile this Special Issue, titled *Biomass and Waste Conversion and Valorization to Chemicals, Energy and Fuels*, comprising 10 papers (8 research articles and 2 reviews).

This Special Issue is a result of the consorted efforts of scientists with extensive expertise in biomass conversion and valorization. It has been an honor for the Guest Editors to work with them, and they thank each and every one of the authors, experts from all around the world, for their outstanding contributions.

José A.P. Coelho and Roumiana P. Stateva
Editors

Article

A Network of Processes for Biorefining Burdock Seeds and Roots

Luigi di Bitonto [1], Enrico Scelsi [1], Massimiliano Errico [2], Hilda Elizabeth Reynel-Ávila [3,4],
Didilia Ileana Mendoza-Castillo [3,4], Adrián Bonilla-Petriciolet [4], Marcos Lucio Corazza [5],
Luis Ricardo Shigueyuki Kanda [5], Martin Hájek [6], Roumiana P. Stateva [7] and Carlo Pastore [1,*]

1 Water Research Institute (IRSA), National Research Council (CNR), Viale De Blasio 5, 70132 Bari, Italy;
luigi.dibitonto@ba.irsa.cnr.it (L.d.B.); enrico.scelsi@ba.irsa.cnr.it (E.S.)
2 Department of Green Technology, Faculty of Engineering, University of Southern Denmark, Campusvej 55,
5230 Odense, Denmark; maer@igt.sdu.dk
3 Consejo Nacional de Humanidades, Ciencias y Tecnologías (CONAHCYT), Ciudad de México 03940, Mexico;
helizabeth_00@hotmail.com (H.E.R.-Á.); didi_men@hotmail.com (D.I.M.-C.)
4 Department of Chemical Engineering, Instituto Tecnológico de Aguascalientes, Aguascalientes 20256, Mexico;
petriciolet@hotmail.com
5 Department of Chemical Engineering, Universidade Federal do Paraná (UFPR), P.O. Box 19011,
Curitiba 81531-980, PR, Brazil; corazza.marcos@gmail.com (M.L.C.); kanda@ufpr.br (L.R.S.K.)
6 Department of Physical Chemistry, Faculty of Chemical Technology, University of Pardubice, Studentská 95,
532 10 Pardubice, Czech Republic; martin.hajek@upce.cz
7 Institute of Chemical Engineering, Bulgarian Academy of Sciences, Acad. G. Bonchev Str. 103,
1113 Sofia, Bulgaria; thermod@bas.bg
* Correspondence: carlo.pastore@ba.irsa.cnr.it

Citation: di Bitonto, L.; Scelsi, E.;
Errico, M.; Reynel-Ávila, H.E.;
Mendoza-Castillo, D.I.;
Bonilla-Petriciolet, A.; Corazza, M.L.;
Shigueyuki Kanda, L.R.; Hájek, M.;
Stateva, R.P.; et al. A Network of
Processes for Biorefining Burdock
Seeds and Roots. *Molecules* **2024**, *29*,
937. https://doi.org/10.3390/
molecules29050937

Academic Editor: Alejandro
Rodríguez Pascual

Received: 13 January 2024
Revised: 17 February 2024
Accepted: 19 February 2024
Published: 21 February 2024

Abstract: In this work, a novel sustainable approach was proposed for the integral valorisation of
Arctium lappa (burdock) seeds and roots. Firstly, a preliminary recovery of bioactive compounds,
including unsaturated fatty acids, was performed. Then, simple sugars (i.e., fructose and sucrose)
and phenolic compounds were extracted by using compressed fluids (supercritical CO_2 and propane).
Consequently, a complete characterisation of raw biomass and extraction residues was carried out
to determine the starting chemical composition in terms of residual lipids, proteins, hemicellulose,
cellulose, lignin, and ash content. Subsequently, three alternative ways to utilise extraction residues
were proposed and successfully tested: (i) enzymatic hydrolysis operated by Cellulases (*Thricoderma
resei*) of raw and residual biomass to glucose, (ii) direct ethanolysis to produce ethyl levulinate;
and (iii) pyrolysis to obtain biochar to be used as supports for the synthesis of sulfonated magnetic
iron-carbon catalysts (Fe-SMCC) to be applied in the dehydration of fructose for the synthesis of
5-hydroxymethylfurfural (5-HMF). The development of these advanced approaches enabled the full
utilisation of this resource through the production of fine chemicals and value-added compounds in
line with the principles of the circular economy.

Keywords: *Arctium lappa*; burdock seeds; burdock roots; biomass valorization; enzymatic hydrolysis;
ethyl levulinate; 5-hydroxymethylfurfural; circular economy

1. Introduction

Arctium lappa, also known as burdock, is a perennial edible plant belonging to the
family *Asteraceae*. It was originally cultivated in Asia and Europe for its pharmaceutical
and therapeutic properties [1–3] and was also used as a nutritious food. Nowadays, it
has become an invasive weed on soils characterised by high nitrogen content, mainly
in North American countries and Australia [4,5]. In recent years, several studies have
demonstrated the beneficial effects of burdock due to the presence of various chemical
compounds, including polyunsaturated fatty acids, phenolic acids, polyphenols, aldehydes,
terpenoids, phytosterols, and mono- and sesquiterpenes [6–8]. Therefore, at present,

efforts are focussed on developing methods for recovering those valuable compounds from raw biomass. In conventional extraction methods such as maceration, percolation, and Soxhlet extraction, the starting material is heated at high temperatures in the presence of organic solvents [9–11]. This allows higher extraction yields but, at the same time, leads to partial degradation of the thermolabile compounds, which reduces both the quality of the extract and the solid residue obtained at the end of the extraction process. In addition, they require a longer exposure time and higher energy consumption. For these reasons, novel "green technologies" have been developed that enable the extraction of bioactive compounds without the use of toxic chemicals, thus protecting the environment and human health, e.g., ultrasound, microwave-assisted, high-pressure homogenisers, enzymatic, and solid-liquid extraction using conventional and non-conventional solvents (biosolvents, supercritical fluids, and combinations of them) [12–16]. Extraction with compressed fluids is an example of a sustainable and efficient technique for the recovery of bioactives and secondary metabolites from raw biomass, since it prevents thermolabile compounds from degradation during the extractive step. In a recent study conducted by Stefanov et al. [13], supercritical carbon dioxide ($scCO_2$) and compressed propane were successfully tested in the extraction of fatty acids and phenolic compounds from burdock seeds and roots, and the possible use of ethanol and methanol-water as cosolvents was evaluated. The extracts obtained are characterised by high antioxidant activity and could find application in the chemical and pharmaceutical fields for the formulation of nutraceuticals and cosmetics. Thus, a sustainable pathway to the valorisation of this resource is identified. However, to fully exploit its potential, extraction residues obtained at the end of the recovery process should also be used in a sustainable way [12]. The latter represents a valuable source of hemicellulose, cellulose, and lignin that could be used to produce either high-value-added compounds or as supports for the design of new bioderived catalysts to be applied in industrial processes.

In this study, for the first time, three alternative routes to convert *A. lappa* biomass have been successfully tested, namely (i) enzymatic hydrolysis to glucose, (ii) direct ethanolysis of structural carbohydrates to produce ethyl levulinate, and (iii) pyrolysis to produce biochar as a support for the synthesis of sulfonated magnetic-iron-carbon catalysts (Fe-SMCC) to be applied in the dehydration of fructose for the synthesis of 5-hydroxymethylfurfural (5-HMF).

Considering the high content of hemicellulose and cellulose in the analysed samples (about 30–50% of the total composition), *A. lappa* biomass can be used as a potential source of glucose to be applied as a substrate to produce bioethanol as well as for the synthesis of new chemical compounds [17–21]. Generally, this conversion is carried out by the hydrolysis of complex carbohydrates (hemicellulose and cellulose), through mechanical and/or chemical treatments with concentrated acids (HCl, H_2SO_4) [22–24], alkaline solutions [25,26] and ionic liquids [27,28]. However, these applications require drastic conditions (100–140 °C) that result in partial loss of the product and increased overall energy consumption. In this context, enzymatic hydrolysis is becoming one of the most widely used methods as it requires less energy consumption and mild environmental conditions [29–31]. In addition, the absence of strong acids or bases does not adversely affect the subsequent conversion processes, such as fermentation, for the formation of valuable products. For these reasons, the biomass analysed was directly tested in enzymatic hydrolysis for glucose production.

In addition, with the gradual depletion of fossil fuels, the use of structural carbohydrates in biomass for the sustainable production of biofuels and chemicals has gained increasing interest in recent years. Several high-value monomers, including 5-HMF, levulinic acid, and γ-valerolactone, can be obtained by the catalytic conversion of cellulose and hemicellulose [32–34]. Among these, ethyl levulinate is a versatile chemical used in industry as an eco-friendly solvent, plasticiser, perfume, liquid hydrocarbon fuel, and petroleum additive [35–37]. Furthermore, it is a precursor for the preparation of more complex chemicals by means of hydrolysis reactions, condensation, or other chemical reactions [38,39]. The synthesis of ethyl levulinate involves the direct acid-catalysed ethanolysis of raw biomass in the presence of homogeneous and heterogeneous catalysts. Mineral acids such as HCl and H_2SO_4 were widely used to produce levulinate esters in high yields [40,41]. However, these catalysts also had obvious shortcomings, such as corrosion, product separation, and catalyst recycling. Furthermore, they also promote the formation of humins as a result of secondary side reactions of hydrolysis and/or condensation [42]. Heterogeneous acid catalysts can be easily separated and reused at the end of the reaction cycle. Several catalysts, such as heteropolyacids, metal oxides, zeolites, and ionic liquids, were used to produce ethyl levulinate [43–45]. However, the synthesis of these catalysts turns out to be very expensive and requires long reaction times, adversely affecting the cost of the final product. In recent studies by di Bitonto et al. [46], the use of $AlCl_3 \cdot 6H_2O$ and H_2SO_4 at low concentrations (1 wt%) was found to be particularly effective in the direct conversion of municipal food waste by obtaining an ethyl levulinate yield of 60 mol% under very mild conditions (180 °C, 4 h). Experimental studies have clearly shown that the presence of H_2SO_4 (Brønsted acid) enables the depolymerisation of hemicellulose and cellulose into monomeric units with the formation of ethyl glucoside, while $AlCl_3 \cdot 6H_2O$ promotes its subsequent isomerisation into ethyl fructoside, which is easily converted to ethyl levulinate [47]. 5-Ethoxymethylfurfural (5-EMF) is a useful intermediate considered a promising alternative to conventional fossil fuels for its high stability and energy density [48,49]. The combined use of Brønsted and Lewis acids in the esterification process, therefore, not only allows for high yields but also represents a useful and economical route for the production of ethyl levulinate from lignocellulosic biomass.

Sulfonated carbon catalysts are considered promising alternatives to conventional homogeneous acid catalysts because they are inexpensive, less corrosive, and have a low environmental impact [50,51]. They can be obtained by the sulfonation of a variety of carbon-based materials, including carbon nanotubes [52], graphene [53], mesoporous carbons [54], and polyaromatic molecules [55]. To date, the use of biomass for the production of catalytic materials is becoming increasingly attractive due to its valuable advantages [56,57]. The presence of functionalised groups $-SO_3H$ on the surface, obtained by simple treatment with concentrated sulfuric acid, makes them particularly versatile catalysts in several industrial contexts. In addition, the porous carbon-based structure significantly increases the surface area of the catalysts, thus improving their catalytic activity. Sulfonated carbon catalysts can be easily recovered by filtration or centrifugation; however, these processes have partial separation efficiency and high energy consumption. Magnetic separation is widely used to improve catalyst recovery [58]. The introduction of magnetic materials such as Fe_3O_4 and Fe onto the carbon-sulfonate surfaces and the evaluation of the relevant catalytic properties in converting fructose into 5-Hydroxymethylfurural (5-HMF) were investigated. 5-HMF is a platform molecule that finds several applications in the synthesis of fine chemicals, polymers, biofuels, and pharmaceuticals [59,60]. Components of cellulose and other polysaccharides such as glucose and fructose can be converted into 5-HMF by dehydration reactions in the presence of homogeneous acid catalysts such as organic acids [61], ionic liquids [62], and mineral acids [63]. However, these processes have several drawbacks in terms of sustainability, including corrosion of the equipment, high toxicity, and recovery of the catalyst. Heterogeneous acid catalysts like supported heteropolyacids [64], zeolites [65], metal oxides [66], and ion-exchange resins [67] are gradually becoming more competitive as they are less corrosive and easily recoverable at the end of the process.

The development and consequent implementation of those routes will allow the full exploitation of this abundant resource through eco-sustainable processes with low environmental impact. Thus, affordable and truly sustainable valorisation alternatives within the concepts of the circular economy approach will be realised as a single process or in an optimised network of processes.

2. Results

2.1. Chemical Characterisation of Raw Biomass and Extraction Residues

Preliminary characterisation of raw biomass (burdock seeds and roots) and residual solids obtained after the extraction of bioactive compounds with compressed fluids was carried out to determine the starting chemical composition and identify the most valuable components to be exploited. The results are reported in Table 1.

Table 1. Chemical characterisation of raw biomass (burdock seeds and roots) and extraction residues obtained after the preliminary extraction of bioactive compounds by using compressed fluids ($scCO_2$ and propane).

Samples	Seeds					Roots		
	Raw Biomass	Extraction Residues				Raw Biomass	Extraction Residues	
		№ 1	№ 2	№ 3	№ 4		№ 5	№ 6
Solvent	-	$scCO_2$ + EtOH	Propane	Propane	Propane	-	$scCO_2$ + MeOH/H_2O	$scCO_2$ + MeOH/H_2O
Extraction Conditions	-	40 °C 200 bar	40 °C 60 bar	80 °C 200 bar	Mix of T and P	-	40 °C, 200 bar	80 °C, 200 bar
Ref. in Article *J CO_2 Util.* [13]	-	Run 2	Run 4	Run 8	-	-	Runs 15–17	Runs 18–20
Moisture content, %	10.2 ± 0.4	4.6 ± 0.2	4.5 ± 0.2	3.4 ± 0.1	3.9 ± 0.1	9.7 ± 0.3	6.2 ± 0.1	6.0 ± 0.1
Total Solids composition (wt%)								
Total Lipids	17.4 ± 0.6	6.7 ± 0.3	8.3 ± 0.3	6.2 ± 0.2	9.5 ± 0.4	0.3	-	-
Proteins	29.7 ± 1.1	17 ± 0.6	18.6 ± 0.7	14.7 ± 0.5	12.7 ± 0.5	21.6 ± 0.8	15.6 ± 0.6	14.4 ± 0.4
EHS *	18.4 ± 0.8	19.1 ± 0.7	21 ± 0.8	20.4 ± 0.7	21.9 ± 0.8	18.1 ± 0.6	17.9 ± 0.6	18.8 ± 0.7
Cellulose	13.5 ± 0.6	24.7 ± 1.0	26.7 ± 1.1	26.5 ± 0.6	24.9 ± 0.8	14.1 ± 0.6	19.2 ± 0.7	21.9 ± 1.0
Lignin	11.6 ± 0.5	19.5 ± 0.8	16.5 ± 0.6	21.2 ± 0.6	19.6 ± 0.8	23.1 ± 0.9	28.7 ± 1.2	26.5 ± 1.1
Ashes	4.7 ± 0.1	5 ± 0.1	4.9 ± 0.1	4.3 ± 0.1	4.2 ± 0.1	16.5 ± 0.6	12.6 ± 0.4	13.1 ± 0.4

* EHS: easily hydrolysable sugars (inuline, oligofructose, and hemicellulose).

Burdock seeds are characterised by a high lipid and protein content of 17.4 ± 0.6 and 29.7 ± 0.6 wt%, respectively. At the same time, the lipid component is almost absent in burdock roots (0.3 wt%), with proteins and lignin being the main constituents with values of 21.6 ± 0.8 and 23.1 ± 0.9 wt%, respectively.

After the extraction process, a partial enrichment of the easily hydrolysable sugars (EHS), cellulose, and lignin content in the residues produced at the end of the process was observed in all extraction residues analysed (samples 1–6). Treatment with compressed fluids to extract bioactive components such as unsaturated fatty acids and polyphenols leads to a partial deconstruction of the original structure and, thus, to a partial breaking of the intermolecular bonds that link these molecules to the raw biomass. As a result, the solids remaining at the end of the extraction process consist mainly of EHS (17.9–21.9 wt%), cellulose (19.2–26.7 wt%), and lignin (16.5–28.7 wt%). In detail, as for roots in particular, a significant content of oligo-fructose chains and inulins was determined (3–5%). These biomass residues can be used as a potential resource to be utilised further since they could easily be converted into fine chemicals and value-added compounds.

2.2. Study of Enzymatic Hydrolysis for Glucose Production

In this study, the commercial cellulase Novozymes Cellic® CTec2 was used for its well-known ability to easily break the β 1–4 glycosidic bond of complex carbohydrates (hemicellulose, cellulose) with the formation of monomeric units [29]. The effect of enzyme loading (5 and 30 FPU/$g_{substrate}$) and reaction time (24, 48, and 72 h) on the degradation yield was evaluated and compared with the results of enzymatic hydrolysis of commercial cellulose. The results obtained are shown in Table 2.

Table 2. Enzymatic hydrolysis of raw biomass and extraction residues for glucose production. Reaction conditions: 1 g of sample, 10 mL of citrate buffer (50 mmol, pH 4.8), concentration enzyme (Novozymes Cellic® CTec2), 5–30 FPU/$g_{substrate}$, 50 °C, 72 h, 300 rpm.

Substrate	Total Carbohydrate Content (wt%)	Enzymatic Degradation Yield (wt%)					
	Time (h)	24		48		72	
	FPU/$g_{substrate}$	5	30	5	30	5	30
Cellulose	100	71.6 ± 2.8	84.5 ± 0.9	80.8 ± 0.8	100	88.2 ± 1.9	100
Burdock seeds	31.9 ± 1.4	40.1 ± 1.3	88.2 ± 2.2	46.0 ± 1.8	93.8 ± 1.7	49.6 ± 1.0	100
№ 1	43.8 ± 1.7	43.5 ± 0.9	81.2 ± 1.6	45.2 ± 1.6	94.1 ± 1.4	46.5 ± 1.4	96.7 ± 1.8
№ 2	47.7 ± 1.9	40.9 ± 1.1	81.7 ± 1.4	44.0 ± 1.1	83.9 ± 1.5	47.1 ± 1.3	89.3 ± 1.2
№ 3	46.9 ± 1.3	25.7 ± 0.7	48.4 ± 0.6	28.3 ± 0.8	51.7 ± 0.8	29.5 ± 0.7	53.7 ± 0.8
№ 4	46.8 ± 1.6	28.5 ± 0.9	51.9 ± 0.8	27.8 ± 0.9	54.8 ± 0.7	30.2 ± 0.8	56.0 ± 1.4
Burdock roots	32.2 ± 1.2	58.5 ± 1.4	88.3 ± 1.2	68.4 ± 1.4	95.6 ± 2.1	68.5 ± 1.9	100
№ 5	37.1 ± 1.3	73.2 ± 1.8	82.3 ± 1.4	87.7 ± 1.6	87.3 ± 0.9	94.5 ± 2.1	100
№ 6	40.7 ± 1.7	75.5 ± 1.6	98.4 ± 2.5	79.5 ± 1.1	100	86.6 ± 1.6	100

Using 5 FPU/$g_{substrate}$ of the loaded enzyme, partial hydrolysis of all substrates was observed, achieving a degradation yield of about 25–40% at 50 °C after 24 h of reaction. When the reaction time was extended to 72 h, a further increase in yield was attained, reaching values of 45–70%. A comparison with experimental data on the enzymatic hydrolysis of cellulose (88.2%) reveals a partial slowdown in the catalytic activity of the enzyme used. This inactivation could be due either to the formation of enzyme-substrate complexes with other molecules present or to the adsorption of the enzyme on non-cellular components such as lignin. In any case, the increase of the loaded enzyme to 30 FPU/$g_{substrate}$ allows for easy conversion of the hemicellulose and cellulose present in glucose by obtaining degradation yields of up to 80–100%. However, it is worth noting that burdock residues from extraction with $scCO_2$ + EtOH (or MeOH in the case of roots) at 40 °C and 200 bar were not only unaffected but were even beneficial regarding hydrolysis kinetics when compared with data collected using the original biomass samples. This result is of utmost importance because the perfect synergy between an extraction with compressed fluids and the enzymatic hydrolysis of residues is demonstrated to its fullest.

2.3. Direct Ethanolysis for the Synthesis of Ethyl Levulinate

The residual biomass samples recovered after the corresponding extraction processes described in Table 2 were tested in direct ethanolysis for the synthesis of ethyl levulinate by using $AlCl_3 \cdot 6H_2O$ and H_2SO_4 as catalysts.

The mechanism and the results of the experimental tests conducted at 190 °C for 2.5 h in the presence of CO_2 (3.5 g) are shown in Figures 1 and 2, respectively.

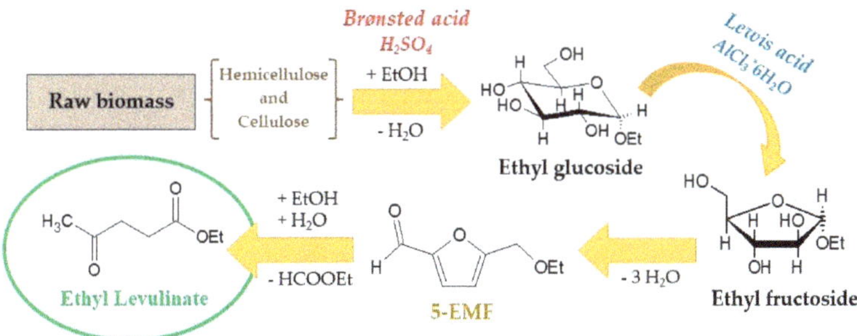

Figure 1. Reaction mechanism proposed for the conversion of raw biomass to ethyl levulinate by using H_2SO_4 and $AlCl_3 \cdot 6H_2O$ as catalysts [47].

From the treatment of residual solids, a yield of 40–50%mol of ethyl levulinate was obtained with a content of 5-EMF of 18–20%mol. As a result, about 60–70% of carbohydrates present in the starting biomass are converted into useful products. Consequently, the application of this process for the utilisation of residual biomass thus represents a rapid advance towards the development of an eco-sustainable economy in which the extraction residues resulting from the extraction of bioactive compounds operated through supercritical fluids can be converted into biofuels and green solvents to be applied in different industrial contexts.

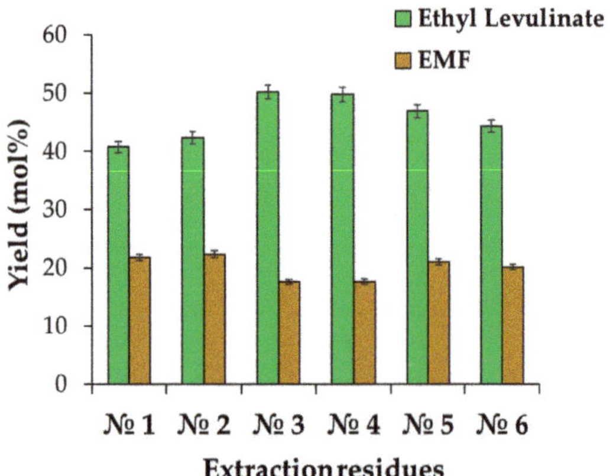

Figure 2. Experimental results obtained from the direct ethanolysis of the extraction residues by using $AlCl_3 \cdot 6H_2O$ and H_2SO_4 as catalysts. Reaction conditions: 0.85 g of sample, molar ratio starting carbohydrates:acid ethanolic solution: $AlCl_3 \cdot 6H_2O$ = 1:230:0.4, 3.5 g of CO_2, 190 °C, 2.5 h, 300 rpm.

2.4. Synthesis and Characterisation of Iron Sulfonated Magnetic Carbon Catalysts

The extraction residues recovered from *A. lappa* seeds and roots were used for the synthesis of iron sulfonated magnetic carbon catalysts (Fe-SMCC). A schematic illustration of the synthetic route for Fe-SMCC is shown in Figure 3.

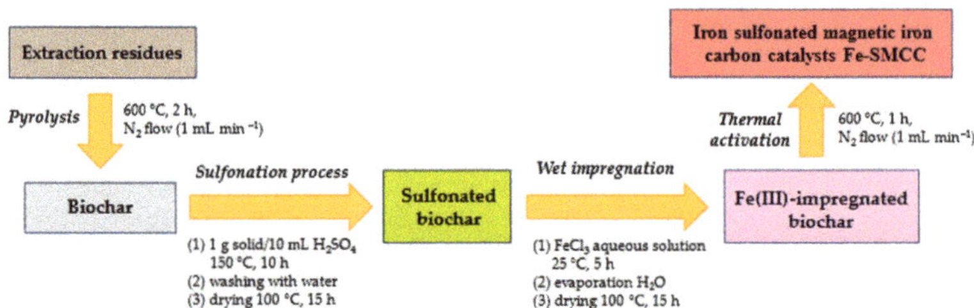

Figure 3. Synthetic route for the synthesis of iron sulfonated magnetic carbon catalysts (Fe-SMCC) from the extraction residues of Burdock seeds and roots.

The carbonaceous material (biochar) was obtained from the pyrolysis of the extraction residues at high temperatures (600 °C, 2 h) under nitrogen (N_2) flow. Subsequently, the biochar obtained was treated with concentrated sulfuric acid (150 °C, 10 h), resulting in the formation of strongly acidic sulfonic groups on the catalyst surface. Figure 4a shows the FTIR spectra of biochar and sulfonated biochar. Both samples exhibit a typical signal at 1568 cm^{-1}, attributable to the stretching signal of C=C bonds of the aromatic rings present in the carbonised materials deriving from lignocellulosic biomass. After the sulfonation process, the presence of new signals at 1123 and 1100 cm^{-1} evidences the incorporation efficiency of -SO_3H groups on the biochar surface [68,69].

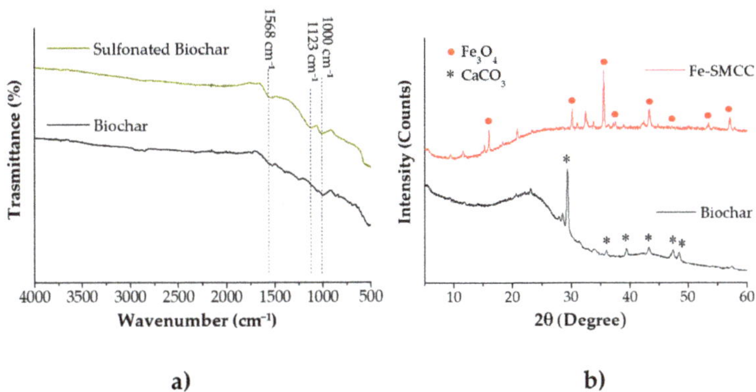

Figure 4. (**a**) Comparison of FTIR spectra of biochar and biochar obtained after the sulfonation process. (**b**) XRD spectra of biochar and iron-supported magnetic carbon catalyst (Fe-SMCC).

Then, the sulfonated biochar was impregnated with an aqueous solution of iron (III) chloride (25 °C, 5 h) and converted into Fe-SMCC by further thermal activation (600 °C, 1 h, N_2). Figure 4b reports the XRD patterns of the starting biochar and Fe-SMCC. The presence of a weak broad diffraction peak at 2θ of ~23–28° in both spectra can be attributed to amorphous carbon structures containing aromatic carbon sheets [70]. In addition, signals located at 2θ of 29.3°, 35.9°, 43.0°, 47.4°, and 48.5° have identified the presence of calcium carbonate ($CaCO_3$) in the starting biochar. The diffraction peaks in Fe-SMCC at 2θ of 16.1°, 30.2°, 35.5°, 37.7°, 43.2°, 47.5°, 53.5°, and 57.1° have highlighted the formation of magnetite Fe_3O_4 on the catalyst, confirming the effectiveness of the synthesis process. The SEM images of biochar, biochar sulfonate, and Fe-SMCC are shown in Figure 5.

Biochar (Figure 5a) exhibits an irregular and heterogeneous morphology with a well-developed alveolar structure resulting from the loss of volatile compounds during heat treatment [71]. After the sulfonation process (Figure 5b), a partial reduction in porosity was observed on the surface of the carbonaceous material. This effect is related to the partial removal of some parts of the amorphous carbon present in the starting biochar after the chemical treatment with sulfuric acid [72,73]. At the same time, the binding of -SO_3H groups on the surface of biochar involves a partial reduction in the porosity of the support. Nanoparticles of Fe_3O_4 were observed on the surface of Fe-SMCC at the end of the process (Figure 5c'), obtained as a result of wet impregnation of sulfonated biochar with $FeCl_3$ and the subsequent activation process. The effectiveness of the synthesis process was also confirmed by elementary analyses. Table 3 shows the elementary compositions obtained by EDS analysis of biochar, biochar sulfonate, and Fe-SMCC for both types of extraction residues (burdock seeds and roots).

Table 3. Elemental analysis of biochar, sulfonated biochar, and Fe-SMCC obtained from the extraction residues of Burdock seeds and roots.

Samples	Extraction Residues									
	Seeds					Roots				
Elemental Composition (wt%)	C	O	Ca	S	Fe	C	O	Ca	S	Fe
Biochar	83.0	14.2	0.96	-	-	69.3	24.8	1.6	0.32	-
Sulfonated biochar	76.8	19.7	-	2.34	-	68.2	26.8	-	2.64	-
Fe-SMCC	65.6	21.4	-	1.27	5.51	70.3	16.3	0.86	1.03	4.18

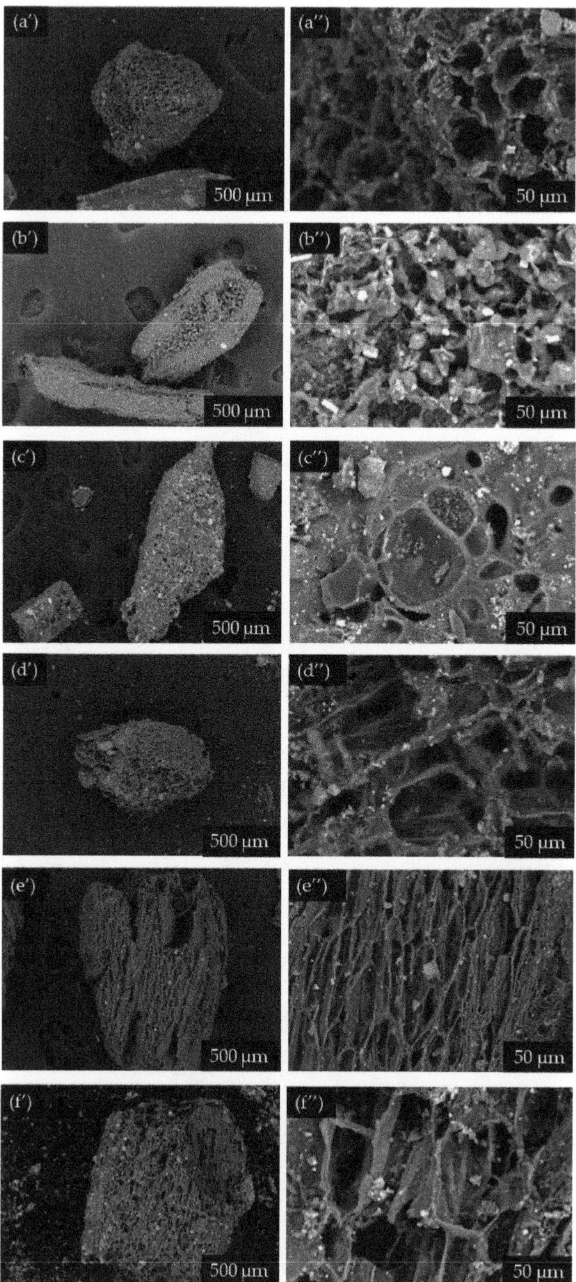

Figure 5. SEM images of biochar ((**a′**,**a″**) of seeds and (**d′**,**d″**) of roots, respectively), sulfonated biochar ((**b′**,**b″**) for seeds and (**e′**,**e″**) for roots), and Fe-SMCC ((**c′**,**c″**) for seeds and (**f′**,**f″**) for roots).

The chemical composition of biochar is clearly different before and after the sulfonation process. The increase in sulphur (S) and oxygen (O) content by 2.34–2.64 wt% and 19.7–26.8 wt%, respectively, indicates the correct incorporation of sulfonic groups on the biochar surface. The presence of iron (Fe) in sulfonated materials after the impregnation, washing, and thermal activation of 4.18–5.51 wt%, confirmed the formation of magnetite in Fe-SMCC. Finally, the total amount of active sites present on the surface of the synthesised catalysts was determined by the Boehm titration method [74] (Table 4), resulting in an overall acid density between 0.74 and 1.36 $mmol_{SO_3H}/g$.

Table 4. Determination of acid properties for Fe-SMCC.

Catalysts	Total Acid Density $(mmol_{SO_3H}/g)$
Fe-SMCC № 1	0.86 ± 0.02
Fe-SMCC № 2	0.98 ± 0.03
Fe-SMCC № 3	0.95 ± 0.02
Fe-SMCC № 4	1.36 ± 0.04
Fe-SMCC № 5	1.32 ± 0.04
Fe-SMCC № 6	0.78 ± 0.02

Dehydratation of Fructose for the Synthesis of 5-HMF Catalysed by Fe-SMCC

Fe-SMCC was tested in the dehydration of fructose for the synthesis of 5-HMF (Figure 6).

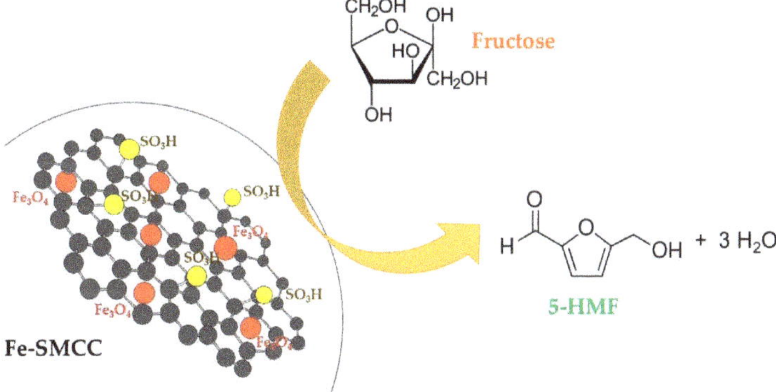

Figure 6. Use of Fe-SMCC in the dehydration of fructose for the synthesis of 5-HMF.

Although water is the most environmentally friendly solvent, fructose dehydration in an aqueous medium is generally not selective, generating hydrolysis and/or condensation by-products (i.e., humins) [59,60]. In order to improve the yield of 5-HMF, the dehydration reaction was carried out by using methyl isobutyl ketone (MIBK) and γ-valerolactone (GVL) as cosolvents. In addition, the catalytic activity of commercial resin (Amberlyst-15) in the dehydration process was also investigated. The results obtained are reported in Table 5.

Under the reaction conditions adopted (0.2 mmol fructose, organic phase = MIBK, volume ratio MIBK to H_2O = 2 to 1, 130 °C, 6 h), in the absence of a catalyst, a fructose conversion of 28.8 ± 1.2% was obtained with a 5-HMF yield of 6.6 ± 0.3% and a selectivity of 22.9 ± 2.0% (Entry 1). When Amberlyst-15 was used as a catalyst, the fructose conversion was 91.8 ± 2.3% with a 5-HMF yield of 14.5 ± 0.5%mol. However, the selectivity towards the production of 5-HMF was found to be only 15.8 ± 0.8 mol%, lower than the non-catalysed process. The acidic active sites present on the catalyst surface [67] induced further hydrolysis of the 5-HMF by generating formic and levulinic acid as reaction by-products (32.3 ± 1.6 and 28.6 ± 1.4 mol%, respectively). On the other hand, the results obtained using Fe-SCMM (Entries 3–8) showed that, despite the lower conversion of fructose of 52.5–62.5 mol%, the systems tested were more active and selective towards the production of 5-HMF, with values in the range of 16.2–23.1 mol% and 30.1–42.0 mol%, respectively, confirming the fundamental role of the catalysts in the selective dehydration process.

Compared to Amberlyst-15, which contains only sulfonic groups (4.3 $mmol_{SO_3H}/g$, as experimentally determined), the simultaneous presence of Brønsted (sulfonic groups) and Lewis (magnetite) acid sites allows to effectively catalyse the dehydration reaction, obtaining 5-HMF as the main product. These values can be further improved by using GVL as a cosolvent reaction (entries 9–14), with yields in 5-HMF of 28.9–41.9 mol% and selectivity of 40.1–55.5 mol%. However, in the case of MIBK, a convenient separation of 5-HMF was obtained at the end of the dehydration process (Figure 7). More than 70% of 5-HMF was present in the higher organic phase (R = 2.4–2.9), allowing easy recovery and isolation by distillation [66].

Table 5. Reactivity tests for the dehydration of fructose for the synthesis of 5-HMF. Reaction conditions: 1 mL of aqueous solution of fructose (0.2 mmol), 2 mL of organic solvent (MIBK or GVL), 130 °C, 6 h, 300 rpm.

E	Solvent	Catalyst	Fructose Conversion (mol%)	5-HMF			Formic Acid	Levulinic Acid
				Yield (mol%)	Selectivity (%)	R	Yield (mol%)	Yield (mol%)
1		In the absence of a catalyst	28.8 ± 1.2	6.6 ± 0.3	22.9 ± 2.0	2.2	-	-
2		Amberlyst-15	91.8 ± 2.3	14.5 ± 0.5	15.8 ± 0.8	2.4	32.3 ± 1.6	28.6 ± 1.4
3		Fe-SMCC № 1	53.6 ± 1.7	18.7 ± 0.3	34.9 ± 1.7	2.4	-	-
4	MIBK	Fe-SMCC № 2	57.0 ± 0.5	21.1 ± 0.8	37.0 ± 1.7	2.8	-	-
5		Fe-SMCC № 3	55.0 ± 0.7	23.1 ± 0.8	42.0 ± 2.0	2.8	-	-
6		Fe-SMCC № 4	58.3 ± 0.5	21.9 ± 1.6	37.6 ± 3.1	2.9	-	-
7		Fe-SMCC № 5	52.5 ± 1.9	16.2 ± 0.6	30.9 ± 2.3	2.6	-	-
8		Fe-SMCC № 6	62.5 ± 1.4	18.8 ± 0.4	30.1 ± 1.3	2.5	-	-
9		Fe-SMCC № 1	69.0 ± 1.4	28.9 ± 1.7	41.9 ± 3.3	-	-	-
10		Fe-SMCC № 2	75.3 ± 1.6	31.4 ± 1.3	41.7 ± 2.6	-	-	-
11	GVL	Fe-SMCC № 3	55.5 ± 1.2	41.9 ± 0.8	55.5 ± 1.9	-	-	-
12		Fe-SMCC № 4	74.2 ± 1.6	35.5 ± 1.4	47.8 ± 2.9	-	-	-
13		Fe-SMCC № 5	76.0 ± 1.9	30.5 ± 0.9	40.1 ± 2.2	-	-	-
14		Fe-SMCC № 6	77.1 ± 1.8	31.5 ± 1.1	40.9 ± 2.4	-	-	-

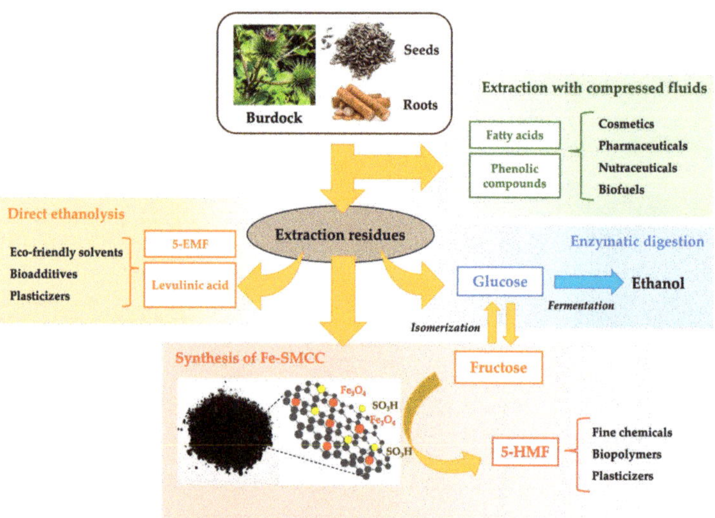

Figure 7. Integrated biorefinery approach for the complete valorisation of burdock seeds and roots.

3. Discussion

Biorefinery Approach for the Integral Valorisation of Burdock Seeds and Roots

A. lappa seeds and roots have always attracted attention for their therapeutic and pharmaceutical properties. Over the years, as an alternative to their direct consumption, several studies have mainly focused on the extraction of bioactive components (e.g., fatty acids, polyphenolic compounds) to be used in the manufacture of various products, such as nutraceuticals and pharmaceuticals. However, there is still a gap regarding the fate of extraction residues, which are supposed to be disposed of, since till now, it has not been demonstrated how they can be utilized on an industrial scale. Inevitably, that would lead to an increase in economic costs and a significant impact on the environment. In detail, considering that extractable components represent only about 15–20% of the total biomass, the possible conversion of the residual biomass to value added compounds will have a crucial impact on the overall feasibility evaluation. Moreover, hemicellulose, cellulose, inulin-type fructans in the case of roots [4,75] and lignin, which are the main constituents, can be exploited for the production of fine chemicals to be used in several other fields, namely new materials and liquid biofuel. The development of an integrated biorefinery approach for the full exploitation of burdock biomasses would not only reduce the costs of disposal but convert them from an environmental burden to a valuable resource, in accordance with the principles of the circular economy. The conceptual process diagram suggested for the valorisation of burdock seeds and roots is shown in Figure 7. As shown, the first step in an integrated biorefinery is the recovery of bioactive components using compressed liquids (supercritical CO_2 and propane). The extraction residues obtained are mainly composed of hemicellulose (17.9–21.9 wt%), cellulose (19.2–26.7 wt%), inulin-type fructans in the case of roots (2–5%), and lignin (16.5–28.7 wt%), as reported in Table 1.

Several sustainable methods are studied to enable the full exploitation of these residual biomasses in a multi-product biorefinery scheme, in which bio-ethanol, 5-HMF, 5-EMF, levulinic acid, and ethyl levulinate could eventually be obtained.

The first process investigated was the enzymatic digestion of the extraction residues, which allowed a complete conversion of complex carbohydrates (hemicellulose and cellulose, representing about 30–50% of the total biomass) into simple sugars by obtaining a glucose yield of 90–100%. In detail, the extraction process operated through compressed fluids can even be seen as an efficient pre-treatment of biomasses since the hydrolysis' kinetics improved by 15–25% with respect to the direct hydrolysis operated on the original burdock biomasses. The obtainment of glucose (and fructose from roots) can be functional to ethanol production through fermentation, according to well-known routes [76]. Ethanol can either be considered a final product (biofuel) or used in other processes involved in biorefining burdock residues. The best way to extract from burdock roots oligofructose species, known as prebiotic fibres, uses ethanol as a cosolvent [77]. Also, ethanol could be used in the direct alcoholysis of *A. lappa* seeds and root residues to obtain new platform molecules. In fact, operating at relatively mild conditions, namely 190 °C and 2.5 h, in the presence of very cheap catalysts, namely $AlCl_3 \cdot 6H_2O$ and H_2SO_4, about 70–80 mol% of the starting polymeric carbohydrates can be successfully converted into ethyl levulinate and 5-EMF, which will find application as bio-additives in bioderived materials. On the other hand, as a result of the enzymatic digestion of burdock seeds and roots, the glucose (and fructose in the specific case of roots) aqueous solutions could be directly used for the production of levulinic acid using an appropriate combination of catalysts ($AlCl_3 \cdot 6H_2O$ and H_2SO_4) [78], ethyl levulinate [79], or 5-HMF using the more selective heterogeneous ferromagnetic catalyst obtained from burdock seeds and root residues.

All of these options may efficiently compete and complete the flexible valorisation of burdock seeds and roots by offering a number of alternative products that can be appropriately modulated and selected following specific market requirements. With the simple inclusion of cheap and largely available reagents, namely *Cellulases*, $AlCl_3 \cdot 6H_2O$, and H_2SO_4 and well-known processes (fermentation of sugars, alcoholysis, and thermochemical valorisation), an *A. lappa* biorefinery burdock could represent an opportunity to improve the present economic scenario, especially in regions and countries where the plant is already cultivated widely. For example, in Fengxian's district, known as the "Hometown of Burdock" in China, the burdock cultivation area is over 3000 hm^2, producing 150,000 t of roots annually [8].

Thus, the novel ideas advocated in this work about the sustainable industrial application of the processes investigated and proposed could allow a potential production of 1000–2000 t extractives, 12,000–13,000 t ethanol, 10,000 t levulinate (as acid and/or ethyl esters), and 5000 t hydroxymethylfurfural, turning a "traditional medicinal plant" into a modern competitive provider of new pharmaceuticals and superfoods, as well as fine chemicals for the production of new bioderived materials and biofuels.

4. Materials and Methods

4.1. Reagents and Instruments

All reagents were of analytical grade ($\geq$99%) and were used directly without further purification or treatment. Fourier-transform infrared spectroscopy (FTIR) spectra were recorded on a Nicolet Summit Thermo Fisher Scientific FTIR spectrophotometer (Thermo Fisher Scientific, Waltham, MA, USA) equipped with an Everest Diamond ATR module. Scanning electron microscopy (SEM) and Energy Dispersive X-ray (EDX) spectra were performed by using a tabletop microscope, the Hitachi TM4000Plus (Hitachi, Tokyo, Japan). Carbohydrates, organic acids, and 5-HMF were determined by high-performance liquid chromatography (HPLC) with a JASCO (Easton, MD, USA) instrument equipped with a Hi-Plex H column (300 mm, 4 mm; Agilent, Santa Clara, CA, USA) and two different types of detectors: the RI-150 refractive index detector and the UV-150 detector (detection wavelengths of 235 and 260 nm). The samples were injected automatically (100 μL) with an autosampler AS 2055. The column was thermostatically controlled at 55 °C, and a 0.6 mL/min flow rate was applied, using a 0.01 M sulfuric acid (H_2SO_4) solution as the mobile phase. Gas-chromatographic (GC) analysis for the determination of ethyl levulinate were done with a Shimadzu GC 2010 Plus gas chromatograph equipped with an Agilent SH-Rtx-Wax capillary column (Shimadzu (Kyoto, Japan), 30 m $\times$ 0.32 mm $\times$ 0.25 μm) and a flame ionisation detector (FID). The injector and detector temperatures were set at 250 and 280 °C, respectively. The oven temperature was programmed with the following temperature ramp: 30 °C for 3 min, then increased to 240 °C at 10 °C min^{-1} and held at 240 °C for 15 min. A split ratio of 1:10 was used with Helium (He) as carrier gas (flow rate 1.0 mL/min). X-ray diffraction (XRD) analysis was carried out by using an Empyrean (Malvern169 Panalytical, Chipping Norton, Australia) diffractometer equipped with a PIXcel1D-Medipix3 detector operating with CuKα radiation (λ = 1.5406 Å, 45 kV, 40 mA) and a 2θ scanning range of 5 to 60°. The results obtained were analysed with HighScore Plus 4.8 software and the PDF2 database.

4.2. Biomass Characterisation

Burdock seeds were provided by a local farm in the town of Ivaipora (Parana, Brazil), while the roots were provided by a certified herbal pharmacy in Sofia (Bulgaria). Both biomasses were initially air-dried in an oven at 60 °C for 48 h, ground in a mixer (Philips Walita RI1774 (Philips, Amsterdam, The Netherlands)), and sieved to obtain a final particle size of 0.42–0.71 mm. The extraction residues (seeds and roots) were recovered after the preliminary extraction of bioactive components with compressed fluids (scCO$_2$ and propane), according to the experimental procedures described by Stefanov et al. [13]. Finally, all samples were analysed in terms of residual lipids, proteins, hemicellulose, lignin, and ash content, as described by di Bitonto et al. [47]. The average chemical composition of starting biomass and extraction residues is reported in Table 1. Each analysis was carried out in triplicate, allowing the average value and the relative standard deviation to be determined.

4.3. Enzymatic Tests

Enzymatic hydrolysis of raw biomass and extraction residues was carried out at 50 °C under the following conditions: 250 μL of commercial cellulase (Cellic® CTec2, >1000 Biomass Hydrolysis Units, density of 1.209 g/L; Novozymes, Lyngby, Denmark) were initially dissolved in citrate buffer solution (50 mmol, pH 4.8, up to 5 mL) and left for 1 h at 50 °C to activate the enzyme. Part of this solution was opportunely up-taken and diluted up to 5 mL of citrate buffer before being added to a mixture containing 0.1 g of solid substrate and 5 mL of citrate buffer to achieve the appropriate enzymatic load (5 and 30 FPU/g cellulose). The filter paper activity of the enzyme was previously experimentally determined [80]. Then, 200 μL of sodium azide solution (20 mg/mL) were added as an antimicrobial agent. The reaction was carried out at 50 °C for 72 h under constant stirring speed (300 rpm). To study the enzymatic kinetics, samples (1 mL) were withdrawn at 24, 48, and 72 h, filtered using 0.2 μm nylon filters, and analysed for the determination of hydrolysed glucose by HPLC analysis. The reaction yield was calculated according to Equation (1):

$$\text{Enzymatic degradation yield(wt\%)} = \frac{m_{\text{hydrolysed glucose}}}{m_{\text{biomass}}\%_{\text{cellulose}}} \frac{m_{\text{hydrolysed glucose}}}{m_{\text{biomass}}\%_{\text{cellulose}}} \times 100 \qquad (1)$$

where $m_{\text{hydrolyseds glucose}}$ is the mass of glucose experimentally detected after the enzymatic hydrolysis, m_{biomass} is the weighted amount of sample used, and $\%_{\text{cellulose}}$ is the cellulose content determined in the starting samples as reported in Table 1.

4.4. Ethanolysis Reaction

The direct ethanolysis of extraction residues was performed in a 50 mL stainless stirrer reactor (Parr series 4590, model 4597 (Parr Instrument Company, Moline, IL, USA)), equipped with a stirring speed controller, temperature control, and pressure indicator. In a typical reaction, 0.6 g of sample (total carbohydrate content = 43.8 wt%) were mixed with 20 mL of an ethanolic solution of 1 wt% H_2SO_4 and 0.14 g of $AlCl_3 \cdot 6H_2O$, resulting in a final molar ratio of carbohydrate:acidic ethanolic solution:catalyst of 1:230:0.4 [46,47]. Then, the reactor was loaded with 3.5 g of CO_2 ($\geq$99%, Air Liquid) by using a syringe pump (Teledyne Isco model 260D (Teledyne ISCO, Lincoln, NE, USA)) held at a constant temperature (15 °C) and pressure (150 bar). The reaction was carried out at 190 °C for 2.5 h with stirring at 300 rpm. At the end of the process, the system was cooled to room temperature and depressurised by opening a needle valve (pressure release rate of about 5 bar/min). The reaction mixture was collected, and the organic solution was separated from the residual solid by centrifugation (3000 rpm, 10 min; Rotofix 32 Hettich centrifuge (Hettich Lab, Tuttlingen, Germany)), and 5-EMF content was determined spectroscopically by using a UV–Vis SP 8001 spectrophotometer (Metertech, Taipei, Taiwan) [47]. Finally, ethyl levulinate was quantified by gas chromatographic analysis of the extract (1 μL) obtained after five successive extractions of the organic mixture with heptane (5 mL). Determinations were performed by external calibration with ethyl levulinate standard solutions in heptane ranging from 10 to 1000 ppm.

4.5. Synthesis of Iron Sulfonated Magnetic Carbon Catalysts

Iron sulfonated magnetic carbon catalysts (Fe-SMCC) were synthesised according to the synthetic route shown in Figure 5. Firstly, biochar was obtained by pyrolysis of extraction residues at 600 °C for 2 h with a heating rate of 10 °C/min and an N_2 flow of 1 mL/min. Subsequently, the porous carbon material was sulfonated with concentrated sulfuric acid at 150 °C for 10 h (1 g solid/10 mL H_2SO_4) to generate strongly acidic sulfone groups on the biochar surface. After the process completion, the sulfonated biochar was washed with deionised water until no more sulphate ions were detected in the wash water, and it was dried in an oven at 100 °C for 15 h. This activated carbon was used as a support material for the synthesis of Fe-SMCC by wet impregnation. 1 g of the sample was suspended into 50 mL of an aqueous solution of iron (III) chloride hexahydrate

(1.2 g of $FeCl_3 \cdot 6H_2O$, weight ratio Fe to biochar of 25%) at room temperature for 5 h. The suspension was then evaporated at 110 °C and the resultant solid dried at 100 °C for 15 h. Fe-SMCC were obtained after the thermal activation of Fe (III)-impregnated biochar at 600 °C for 1 h under N_2 flow (1 mL/min). The Boehm titration method was used for the determination of acid properties of the catalysts synthetised [65].

Dehydratation Process

The synthesised catalysts were tested in the reaction of dehydration of fructose for the synthesis of 5-HMF. In a Pyrex glass reactor of 10 mL, 0.1 g of supported catalyst was mixed with 1 mL of aqueous fructose solution (0.2 mmol) and 2 mL of organic solvent (methyl isobutyl ketone and γ-valerolactone), by obtaining a final volume ratio of organic to aqueous phase of 2 to 1. The reaction was carried out at 130 °C for 6 h with stirring at 300 rpm. At the end of the process, the system was cooled to room temperature in an ice-water bath, and the resultant organic mixture was analysed by HPLC for the determination of residual fructose (RI detection), organic acids (UV detection 235 nm) and 5-HMF (UV detection 260 nm). The conversion of fructose, the yield of the reaction products and the selectivity towards the production of 5-HMF were calculated by using Equations (2)–(4):

$$\text{Fructose conversion (mol\%)} = \frac{\text{mmol}_{\text{starting fructose}} - \text{mmol}_{\text{residual fructose}}}{\text{mmol}_{\text{starting fructose}}} \times 100 \tag{2}$$

$$\text{Yield product(mol\%)} = \frac{\text{mmol}_{\text{product}}}{\text{mmol}_{\text{starting fructose}}} \times 100 \tag{3}$$

$$\text{Selectivity 5-HMF(mol\%)} = \frac{\text{mmol}_{\text{5-HMF}}}{\text{mmol}_{\text{starting fructose}}} \times 100 \tag{4}$$

In addition, in the case of MIBK, the partition coefficient (R) of 5-HMF between the two phases (aqueous and organic) obtained at the end of the dehydration process was determined according to Equation (5).

$$R = \frac{\text{mmol}_{\text{5-HMF organic phase}}}{\text{mmol}_{\text{5-HMF aqueous phase}}} \tag{5}$$

5. Conclusions

An integrated strategy for biorefining of lignocellulosic biomasses of *Arctium lappa* burdock seeds and roots was developed and presented. A cascade of technologies was experimentally studied using as possible feedstocks the residual biomasses obtained after recovery of fatty acids and antioxidants applying compressed fluids. Valorisation routes of the residual cakes to attain glucose, ethyl levulinate and new magnetic catalysts were specifically investigated. It was demonstrated that cellulose can be almost totally hydrolysed in less than 48 h at 50 °C, by commercial enzymes into simple sugars (e.g., glucose and xylose) using a load of enzyme of 30 $FPU/g_{\text{substrate}}$ on the residues obtained from extractions with sc-CO_2/alcohols. Alternatively, lignocellulosic residues can be directly reacted through ethanolysis operated at 190 °C for 2.5 h in the presence of $AlCl_3 \cdot 6H_2O$ and H_2SO_4, achieving high yields of ethyl levulinate (40–50%) and 5-EMF (18–22%). Finally, novel magnetic Iron (III) based sulfonated lignin-based biochars were also synthesized, characterised and found to efficiently catalyse the fructose conversion into 5-HMF in biphasic water-MIBK or GVL systems up to obtain reasonably yields (28.9–41.9 mol%) and selectivity (40.1–55.5 mol%). All these fundamental steps can concur in defining a network of processes for a multiproduct biorefinery of *Arctium lappa* seeds and roots to obtain platform chemicals and biofuels.

Author Contributions: Conceptualization, C.P. and R.P.S.; methodology, M.E., L.d.B., C.P. and E.S.; validation, C.P., L.d.B. and E.S.; formal analysis, L.d.B., C.P., E.S., H.E.R.-Á., D.I.M.-C., M.H. and M.E.; investigation, L.d.B., C.P., E.S. and M.H.; resources, M.E. and C.P.; data curation, L.d.B. and C.P.; writing—original draft preparation, L.d.B., C.P. and E.S.; writing—review and editing, M.E., A.B.-P., R.P.S., M.L.C., L.R.S.K. and C.P.; supervision, C.P.; project administration, C.P.; funding acquisition, C.P. All authors have read and agreed to the published version of the manuscript.

Funding: This work has received funding from the European Union's Horizon 2020 Research and Innovation Programme under the Marie Sklodowska-Curie Grant Agreement No. 778168.

Institutional Review Board Statement: Not applicable.

Informed Consent Statement: Not applicable.

Data Availability Statement: Data are contained within the article.

Conflicts of Interest: All authors declare that the research was conducted in the absence of any commercial or financial relationships that could be construed as a potential conflict of interest.

References

1. De Souza, A.R.C.; de Oliveira, T.L.; Fontana, P.D.; Carneiro, M.C.; Corazza, M.L.; de Messias Reason, I.J.; Bavia, L. Phytochemicals and biological activities of burdock (*Arctium lappa* L.) extracts: A review. *Chem. Biodivers.* **2022**, *19*, e202200615. [CrossRef]
2. Petkova, N.; Hambarlyiska, I.; Tumbarski, Y.; Vrancheva, R.; Raeva, M.; Ivanov, I. Phytochemical composition and antimicrobial properties of burdock (*Arctium lappa* L.) roots extracts. *Biointerface Res. Appl. Chem.* **2022**, *12*, 2826–2842. [CrossRef]
3. El Khatib, N.; Morel, S.; Hugon, G.; Rapior, S.; Carnac, G.; Saint, N. Identification of a sesquiterpene lactone from *Arctium lappa* leaves with antioxidant activity in primary human muscle cells. *Molecules* **2021**, *26*, 1328. [CrossRef] [PubMed]
4. Moro, T.M.A.; Clerici, M.T.P.S. Burdock (*Arctium lappa* L.) roots as a source of inulin-type fructans and other bioactive compounds: Current knowledge and future perspectives for food and non-food applications. *Food Res. Int.* **2021**, *141*, 109889. [CrossRef]
5. Jiang, Y.Y.; Yu, J.; Li, Y.B.; Wang, L.; Hu, L.; Zhang, L.; Zhou, Y.H. Extraction and antioxidant activities of polysaccharides from roots of *Arctium lappa* L. *Int. J. Biol. Macromol.* **2019**, *123*, 531–538. [CrossRef]
6. Ma, K.; Sheng, W.; Song, X.; Song, J.; Li, Y.; Huang, W.; Liu, Y. Chlorogenic Acid from Burdock Roots Ameliorates Oleic Acid-Induced Steatosis in HepG2 Cells through AMPK/ACC/CPT-1 Pathway. *Molecules* **2023**, *28*, 7257. [CrossRef] [PubMed]
7. Mir, S.A.; Dar, L.A.; Ali, T.; Kareem, O.; Rashid, R.; Khan, N.A.; Chashoo, I.A.; Bader, G.N. Arctium lappa: A review on its phytochemistry and pharmacology. In *Edible Plants in Health and Diseases: Volume II: Phytochemical and Pharmacological Properties*; Springer: Singapore, 2022; pp. 327–348. [CrossRef]
8. Zhang, X.; Herrera-Balandrano, D.D.; Huang, W.; Chai, Z.; Beta, T.; Wang, J.; Feng, J.; Li, Y. Comparison of nutritional and nutraceutical properties of burdock roots cultivated in Fengxian and Peixian of China. *Foods* **2021**, *10*, 2095. [CrossRef]
9. Antony, A.; Farid, M. Effect of temperatures on polyphenols during extraction. *Appl. Sci.* **2022**, *12*, 2107. [CrossRef]
10. Daud, N.M.; Putra, N.R.; Jamaludin, R.; Norodin, N.S.M.; Sarkawi, N.S.; Hamzah, M.H.S.; Nasir, H.M.; Zaidel, D.N.A.; Yunus, M.A.C.; Salleh, M.L. Valorisation of plant seed as natural bioactive compounds by various extraction methods: A review. *Trends J. Food Sci. Technol.* **2022**, *119*, 201–214. [CrossRef]
11. Xia, J.; Guo, Z.; Fang, S.; Gu, J.; Liang, X. Effect of drying methods on volatile compounds of burdock (*Arctium lappa* L.) root tea as revealed by gas chromatography mass spectrometry-based metabolomics. *Foods* **2021**, *10*, 868. [CrossRef]
12. Errico, M.; Coelho, J.A.; Stateva, R.P.; Christensen, K.V.; Bahij, R.; Tronci, S. Brewer's Spent Grain, Coffee Grounds, Burdock, and Willow—Four Examples of Biowaste and Biomass Valorization through Advanced Green Extraction Technologies. *Foods* **2023**, *12*, 1295. [CrossRef]
13. Stefanov, S.M.; Fetzer, D.E.; de Souza, A.R.C.; Corazza, M.L.; Hamerski, F.; Yankov, D.S.; Stateva, R.P. Valorization by compressed fluids of *Arctium lappa* seeds and roots as a sustainable source of valuable compounds. *J. CO2 Util.* **2022**, *56*, 101821. [CrossRef]
14. Chemat, F.; Abert-Vian, M.; Fabiano-Tixier, A.S.; Strube, J.; Uhlenbrock, L.; Gunjevic, V.; Cravotto, G. Green extraction of natural products. Origins, current status, and future challenges. *TrAC Trends Anal. Chem.* **2019**, *118*, 248–263. [CrossRef]
15. Zheng, Z.; Wang, X.; Liu, P.; Li, M.; Dong, H.; Qiao, X. Semi-preparative separation of 10 caffeoylquinic acid derivatives using high speed counter-current chromatogaphy combined with semi-preparative HPLC from the roots of burdock (*Arctium lappa* L.). *Molecules* **2018**, *23*, 429. [CrossRef] [PubMed]
16. Tiwari, B.K. Ultrasound: A clean, green extraction technology. *Trends Anal. Chem.* **2015**, *71*, 100–109. [CrossRef]
17. Rathour, R.K.; Devi, M.; Dahiya, P.; Sharma, N.; Kaushik, N.; Kumari, D.; Kumar, P.; Baadhe, R.R.; Walia, A.; Bhatt, A.K.; et al. Recent trends, opportunities, and challenges in sustainable management of rice straw waste biomass for green biorefinery. *Energies* **2023**, *16*, 1429. [CrossRef]
18. Broda, M.; Yelle, D.J.; Serwańska, K. Bioethanol production from lignocellulosic biomass—Challenges and solutions. *Molecules* **2022**, *27*, 8717. [CrossRef] [PubMed]

19. Mumtaz, M.; Baqar, Z.; Hussain, N.; Bilal, M.; Azam, H.M.H.; Iqbal, H.M. Application of nanomaterials for enhanced production of biodiesel, biooil, biogas, bioethanol, and biohydrogen via lignocellulosic biomass transformation. *Fuel* **2022**, *315*, 122840. [CrossRef]

20. Panakkal, E.J.; Cheenkachorn, K.; Chuetor, S.; Tantayotai, P.; Raina, N.; Cheng, Y.S.; Sriariyanun, M. Optimization of deep eutectic solvent pre-treatment for bioethanol production from Napier grass. *Sustain. Energy Technol. Assess.* **2022**, *54*, 102856. [CrossRef]

21. Gong, Z.; Lv, X.; Yang, J.; Luo, X.; Shuai, L. Chemocatalytic Conversion of Lignocellulosic Biomass to Ethanol: A Mini-Review. *Catalysts* **2022**, *12*, 922. [CrossRef]

22. Świątek, K.; Gaag, S.; Klier, A.; Kruse, A.; Sauer, J.; Steinbach, D. Acid hydrolysis of lignocellulosic biomass: Sugars and furfurals formation. *Catalysts* **2022**, *10*, 437. [CrossRef]

23. Becker, M.; Ahn, K.; Bacher, M.; Xu, C.; Sundberg, A.; Willför, S.; Rosenau, T.; Potthast, A. Comparative hydrolysis analysis of cellulose samples and aspects of its application in conservation science. *Cellulose* **2021**, *28*, 8719–8734. [CrossRef]

24. Yoon, S.Y.; Han, S.H.; Shin, S.J. The effect of hemicelluloses and lignin on acid hydrolysis of cellulose. *Energy* **2014**, *77*, 19–24. [CrossRef]

25. Liu, Q.; He, W.Q.; Aguedo, M.; Xia, X.; Bai, W.B.; Dong, Y.Y.; Song, J.Q.; Richel, A.; Goffin, D. Microwave-assisted alkali hydrolysis for cellulose isolation from wheat straw: Influence of reaction conditions and non-thermal effects of microwave. *Carbohydr. Polym.* **2021**, *253*, 117170. [CrossRef] [PubMed]

26. Kumar, P.; Barrett, D.M.; Delwiche, M.J.; Stroeve, P. Methods for pre-treatment of lignocellulosic biomass for efficient hydrolysis and biofuel production. *Ind. Eng. Chem. Res.* **2009**, *48*, 3713–3729. [CrossRef]

27. Chen, T.; Xiong, C.; Tao, Y. Enhanced hydrolysis of cellulose in ionic liquid using mesoporous ZSM-5. *Molecules* **2018**, *23*, 529. [CrossRef] [PubMed]

28. Cao, Y.; Zhang, R.; Cheng, T.; Guo, J.; Xian, M.; Liu, H. Imidazolium-based ionic liquids for cellulose pre-treatment: Recent progresses and future perspectives. *Appl. Microbiol.* **2017**, *101*, 521–532. [CrossRef]

29. Ilić, N.; Milić, M.; Beluhan, S.; Dimitrijević-Branković, S. Cellulases: From lignocellulosic biomass to improved production. *Energies* **2023**, *16*, 3598. [CrossRef]

30. Mustafa, A.; Faisal, S.; Ahmed, I.A.; Munir, M.; Cipolatti, E.P.; Manoel, E.A.; Pastore, C.; di Bitonto, L.; Hanelt, D.; Nitbani, O.N.; et al. Has the time finally come for green oleochemicals and biodiesel production using large-scale enzyme technologies? Current status and new developments. *Biotechnol. Adv.* **2023**, *69*, 108275. [CrossRef]

31. Reis, C.E.R.; Junior, N.L.; Bento, H.B.; de Carvalho, A.K.F.; de Souza Vandenberghe, L.P.; Soccol, C.R.; Aminabhavi, T.M.; Chandel, A.K. Process strategies to reduce cellulase enzyme loading for renewable sugar production in biorefineries. *J. Chem. Eng.* **2023**, *451*, 138690. [CrossRef]

32. Deng, W.; Feng, Y.; Fu, J.; Guo, H.; Guo, Y.; Han, B.; Jiang, Z.; Kong, L.; Li, C.; Liu, H.; et al. Catalytic conversion of lignocellulosic biomass into chemicals and fuels. *Green Energy Environ.* **2023**, *8*, 10–114. [CrossRef]

33. Khemthong, P.; Yimsukanan, C.; Narkkun, T.; Srifa, A.; Witoon, T.; Pongchaiphol, S.; Kiatphuengporn, S.; Faungnawakij, K. Advances in catalytic production of value-added biochemicals and biofuels via furfural platform derived lignocellulosic biomass. *Biomass Bioenergy* **2021**, *148*, 106033. [CrossRef]

34. Kohli, K.; Prajapati, R.; Sharma, B.K. Bio-based chemicals from renewable biomass for integrated biorefineries. *Energies* **2019**, *12*, 233. [CrossRef]

35. Ahmad, E.; Alam, M.I.; Pant, K.K.; Haider, M.A. Catalytic and mechanistic insights into the production of ethyl levulinate from biorenewable feedstocks. *Green Chem.* **2016**, *18*, 4804–4823. [CrossRef]

36. Badgujar, K.C.; Badgujar, V.C.; Bhanage, B.M. A review on catalytic synthesis of energy rich fuel additive levulinate compounds from biomass derived levulinic acid. *Fuel Process. Technol.* **2020**, *197*, 106213. [CrossRef]

37. Adeleye, A.T.; Louis, H.; Akakuru, O.U.; Joseph, I.; Enudi, O.C.; Michael, D.P. A Review on the conversion of levulinic acid and its esters to various useful chemicals. *Aims Energy* **2019**, *7*, 165–185. [CrossRef]

38. Sinisi, A.; Degli Esposti, M.; Braccini, S.; Chiellini, F.; Guzman-Puyol, S.; Heredia-Guerrero, J.A.; Morselli, D.; Fabbri, P. Levulinic acid-based bioplasticizers: A facile approach to enhance the thermal and mechanical properties of polyhydroxyalkanoates. *Mater. Adv.* **2021**, *2*, 7869–7880. [CrossRef]

39. Xuan, W.; Hakkarainen, M.; Odelius, K. Levulinic acid as a versatile building block for plasticizer design. *ACS Sustain. Chem. Eng.* **2019**, *7*, 12552–12562. [CrossRef]

40. Chang, C.; Xu, G.; Jiang, X. Production of ethyl levulinate by direct conversion of wheat straw in ethanol media. *Bioresour. Technol.* **2012**, *121*, 93–99. [CrossRef]

41. Le Van Mao, R.; Zhao, Q.; Dima, G.; Petraccone, D. New process for the acid-catalyzed conversion of cellulosic biomass (AC3B) into alkyl levulinates and other esters using a unique one-pot system of reaction and product extraction. *Catal. Lett.* **2011**, *141*, 271–276. [CrossRef]

42. Silva, J.F.L.; Mariano, A.P.; Maciel Filho, R. Less severe reaction conditions to produce levulinic acid with reduced humins formation at the expense of lower biomass conversion: Is it economically feasible? *Fuel Commun.* **2021**, *9*, 100029. [CrossRef]

43. Appaturi, J.N.; Andas, J.; Ma, Y.K.; Phoon, B.L.; Batagarawa, S.M.; Khoerunnisa, F.; Hazwan Hussin, M.; Ng, E.P. Recent advances in heterogeneous catalysts for the synthesis of alkyl levulinate biofuel additives from renewable levulinic acid: A comprehensive review. *Fuel* **2022**, *323*, 124362. [CrossRef]

44. Sajid, M.; Farooq, U.; Bary, G.; Azim, M.M.; Zhao, X. Sustainable production of levulinic acid and its derivatives for fuel additives and chemicals: Progress, challenges, and prospects. *Green Chem.* **2021**, *23*, 9198–9238. [CrossRef]

45. Yamanaka, N.; Shimazu, S. Conversion of Biomass-Derived Molecules into Alkyl Levulinates Using Heterogeneous Catalysts. *Reactions* **2023**, *4*, 667–678. [CrossRef]

46. Di Bitonto, L.; Antonopoulou, G.; Braguglia, C.; Campanale, C.; Gallipoli, A.; Lyberatos, G.; Taikou, N.; Pastore, C. Lewis-Brønsted acid catalysed ethanolysis of the organic fraction of municipal solid waste for efficient production of biofuels. *Bioresour. Technol.* **2018**, *266*, 297–305. [CrossRef] [PubMed]

47. Di Bitonto, L.; Locaputo, V.; D'Ambrosio, V.; Pastore, C. Direct Lewis-Brønsted acid ethanolysis of sewage sludge for production of liquid fuels. *Appl. Energy* **2020**, *259*, 114163. [CrossRef]

48. Zuo, M.; Lin, L.; Zeng, X. The synthesis of potential biofuel 5-ethoxymethylfurfural: A review. *Fuel* **2023**, *343*, 127863. [CrossRef]

49. Liu, X.; Wang, R. 5-Ethoxymethylfurfural—A remarkable biofuel candidate. *Biomass Biofuels Biochem.* **2020**, 355–375. [CrossRef]

50. Chong, C.C.; Cheng, Y.W.; Lam, M.K.; Setiabudi, H.D.; Vo, D.V.N. State-of-the-Art of the Synthesis and Applications of Sulfonated Carbon-Based Catalysts for Biodiesel Production: A Review. *Energy Technol.* **2021**, *9*, 2100303. [CrossRef]

51. Stepacheva, A.A.; Markova, M.E.; Lugovoy, Y.V.; Kosivtsov, Y.Y.; Matveeva, V.G.; Sulman, M.G. Plant-Biomass-Derived Carbon Materials as Catalyst Support, A Brief Review. *Catalysts* **2023**, *13*, 655. [CrossRef]

52. Lee, S.H.; Shin, H.H.; Kim, S.G.; Lee, S.J. Fabrication and characteristics of sulfonated carbon foam via high internal phase emulsions with polypyrrole-modified carbon nanotubes. *Polymer* **2023**, *281*, 126144. [CrossRef]

53. Nongbe, M.C.; Ekou, T.; Ekou, L.; Yao, K.B.; Le Grognec, E.; Felpin, F.X. Biodiesel production from palm oil using sulfonated graphene catalyst. *Renew. Energy* **2017**, *106*, 135–141. [CrossRef]

54. Militello, M.P.; Martínez, M.V.; Tamborini, L.; Acevedo, D.F.; Barbero, C.A. Towards Photothermal Acid Catalysts Using Eco-Sustainable Sulfonated Carbon Nanoparticles—Part I: Synthesis, Characterization and Catalytic Activity towards Fischer Esterification. *Catalysts* **2023**, *13*, 1341. [CrossRef]

55. Yadav, N.; Yadav, G.; Ahmaruzzaman, M. Biomass-derived sulfonated polycyclic aromatic carbon catalysts for biodiesel production by esterification reaction. *Biofuels Bioprod. Biorefining* **2023**, *17*, 1343–1367. [CrossRef]

56. Parida, S.; Singh, M.; Pradhan, S. Biomass wastes: A potential catalyst source for biodiesel production. *Bioresour. Technol. Rep.* **2022**, *18*, 101081. [CrossRef]

57. Zou, R.; Qian, M.; Wang, C.; Mateo, W.; Wang, Y.; Dai, L.; Li, X.; Zhao, Y.; Huo, E.; Wang, L.; et al. Biochar: From by-products of agro-industrial lignocellulosic waste to tailored carbon-based catalysts for biomass thermochemical conversions. *J. Chem. Eng.* **2022**, *441*, 135972. [CrossRef]

58. Liu, M.; Ye, Y.; Ye, J.; Gao, T.; Wang, D.; Chen, G.; Song, Z. Recent Advances of Magnetite (Fe_3O_4)-Based Magnetic Materials in Catalytic Applications. *Magnetochemistry* **2023**, *9*, 110. [CrossRef]

59. Slak, J.; Pomeroy, B.; Kostyniuk, A.; Grilc, M.; Likozar, B. A review of bio-refining process intensification in catalytic conversion reactions, separations and purifications of hydroxymethylfurfural (HMF) and furfural. *Chem. Eng. J.* **2022**, *429*, 132325. [CrossRef]

60. Kong, Q.S.; Li, X.L.; Xu, H.J.; Fu, Y. Conversion of 5-hydroxymethylfurfural to chemicals: A review of catalytic routes and product applications. *Fuel Process. Technol.* **2020**, *209*, 106528. [CrossRef]

61. Zhang, X.; Cui, H.; Li, Q.; Xia, H. High-Yield Synthesis of 5-Hydroxymethylfurfural from Untreated Wheat Straw Catalyzed by FePO4 and Organic Acid in a Biphasic System. *Energy Fuels* **2023**, *37*, 12953–12965. [CrossRef]

62. Chen, L.; Xiong, Y.; Qin, H.; Qi, Z. Advances of Ionic Liquids and Deep Eutectic Solvents in Green Processes of Biomass-Derived 5-Hydroxymethylfurfural. *ChemSusChem* **2022**, *15*, e202102635. [CrossRef]

63. Zhang, Y.; Zhu, H.; Ji, Z.; Cheng, Y.; Zheng, L.; Wang, L.; Li, X. Experiments and Kinetic Modeling of Fructose Dehydration to 5-Hydroxymethylfurfural with Hydrochloric Acid in Acetone—Water Solvent. *Ind. Eng. Chem. Res.* **2022**, *61*, 13877–13885. [CrossRef]

64. Dibenedetto, A.; Aresta, M.; Pastore, C.; di Bitonto, L.; Angelini, A.; Quaranta, E. Conversion of fructose into 5-HMF: A study on the behaviour of heterogeneous cerium-based catalysts and their stability in aqueous media under mild conditions. *RSC Adv.* **2015**, *5*, 26941–26948. [CrossRef]

65. Yan, P.; Wang, H.; Liao, Y.; Wang, C. Zeolite catalysts for the valorization of biomass into platform compounds and biochemicals/biofuels: A review. *Renew. Sust. Energy Rev.* **2023**, *178*, 113219. [CrossRef]

66. Takagaki, A. Production of 5-hydroxymethylfurfural from glucose in water by using transition metal-oxide nanosheet aggregates. *Catalysts* **2019**, *9*, 818. [CrossRef]

67. Sampath, G.; Kannan, S. Fructose dehydration to 5-hydroxymethylfurfural: Remarkable solvent influence on recyclability of Amberlyst-15 catalyst and regeneration studies. *Catal. Comm.* **2013**, *37*, 41–44. [CrossRef]

68. Licursi, D.; Galletti, A.M.R.; Bertini, B.; Ardemani, L.; Scotti, N.; Di Fidio, N.; Fulignati, S.; Antonetti, C. Design approach for the sustainable synthesis of sulfonated biomass-derived hydrochars and pyrochars for the production of 5-(hydroxymethyl) furfural. *Sustain. Chem. Pharm.* **2023**, *35*, 101216. [CrossRef]

69. Xiong, X.; Iris, K.M.; Chen, S.S.; Tsang, D.C.; Cao, L.; Song, H.; Kwon, E.E.; Ok, Y.S.; Zhang, S.; Poon, C.S. Sulfonated biochar as acid catalyst for sugar hydrolysis and dehydration. *Catal. Today* **2018**, *314*, 52–61. [CrossRef]

70. Gao, W.; Wan, Y.; Dou, Y.; Zhao, D. Synthesis of partially graphitic ordered mesoporous carbons with high surface areas. *Adv. Energy Mat.* **2011**, *1*, 115–123. [CrossRef]

71. di Bitonto, L.; Reynel-Ávila, H.E.; Mendoza-Castillo, D.I.; Bonilla-Petriciolet, A.; Durán-Valle, C.J.; Pastore, C. Synthesis and characterization of nanostructured calcium oxides supported onto biochar and their application as catalysts for biodiesel production. *Renew. Energy* **2020**, *160*, 52–66. [CrossRef]

72. da Luz Corrêa, A.P.; Bastos, R.R.C.; da Rocha Filho, G.N.; Zamian, J.R.; da Conceição, L.R.V. Preparation of sulfonated carbon-based catalysts from murumuru kernel shell and their performance in the esterification reaction. *RSC Adv.* **2020**, *10*, 20245–20256. [CrossRef] [PubMed]

73. Ngaosuwan, K.; Goodwin, J.G., Jr.; Prasertdham, P. A green sulfonated carbon-based catalyst derived from coffee residue for esterification. *Renew. Energy* **2016**, *86*, 262–269. [CrossRef]

74. Leyva-Ramos, R.; Landin-Rodriguez, L.E.; Leyva-Ramos, S.; Medellin-Castillo, N.A. Modification of corncob with citric acid to enhance its capacity for adsorbing cadmium (II) from water solution. *Chem. Eng. J.* **2012**, *180*, 113–120. [CrossRef]

75. Nabeshima, E.H.; Moro, T.M.A.; Campelo, P.H.; Sant'Ana, A.S.; Clerici, M.T.P.S. Tubers and roots as a source of prebiotic fibers. *Adv. Food Nutr. Res.* **2020**, *94*, 267–293. [CrossRef] [PubMed]

76. Errico, M.; Ramírez-Márquez, C.; Torres Ortega, C.E.; Rong, B.-G.; Segovia-Hernandez, J.G. Design and control of an alternative distillation sequence for bioethanol purification. *J. Chem. Technol. Biotechnol.* **2015**, *90*, 2180–2185. [CrossRef]

77. Ishiguro, Y.; Onodera, S.; Benkeblia, N.; Shiomi, N. Variation of total FOS, total IOS, inulin and their related-metabolizing enzymes in burdock roots (*Arctium lappa* L.) stored under different temperatures. *Postharvest Biol. Technol.* **2010**, *56*, 232–238. [CrossRef]

78. Angelini, A.; Scelsi, E.; Ancona, V.; Aimola, G.; Pastore, C. Performic acid pre-treatment of poplar biomasses grown on a contaminated area for enhanced enzymatic digestibility: A viable route to obtain fine-products and recovery of contaminants. *J. Clean. Prod.* **2022**, *369*, 133346. [CrossRef]

79. Pastore, C.; D'Ambrosio, V. Intensification of Processes for the Production of Ethyl Levulinate Using $AlCl_3 \cdot 6H_2O$. *Energies* **2021**, *14*, 1273. [CrossRef]

80. Wood, T.M.; Bhat, K.M. Methods for Measuring Cellulase Activities. In *Methods in Enzymology*; Elsevier: Amsterdam, The Netherlands, 1988; Volume 160, pp. 87–112. [CrossRef]

molecules

Article

Extraction of High Value Products from *Zingiber officinale* Roscoe (Ginger) and Utilization of Residual Biomass

Alexandra Spyrou [1], Marcelle G. F. Batista [2], Marcos L. Corazza [2], Maria Papadaki [3,*] and Maria Antonopoulou [1,*]

[1] Department of Sustainable Agriculture, University of Patras, Seferi 2, GR30131 Agrinio, Greece; spyrou.a@upatras.gr
[2] Department of Chemical Engineering, Federal University of Parana, Curitiba CEP 81531-990, PR, Brazil; marcelleguth@ufpr.br (M.G.F.B.); corazza@ufpr.br (M.L.C.)
[3] Department of Agriculture, Nea Ktiria, University of Patras, GR30200 Messolonghi, Greece
* Correspondence: marpapadaki@upatras.gr (M.P.); mantonop@upatras.gr (M.A.); Tel.: +30-26310-58428 (M.P.); +30-26410-74114 (M.A.)

Abstract: *Zingiber officinale* Roscoe (ginger) is a plant from the *Zingiberaceae* family, and its extracts have been found to contain several compounds with beneficial bioactivities. Nowadays, the use of environmentally friendly and sustainable extraction methods has attracted considerable interest. The main objective of this study was to evaluate subcritical propane (scPropane), supercritical CO_2 ($scCO_2$), and supercritical CO_2 with ethanol ($scCO_2$ + EtOH) as co-solvent methods for the extraction of high value products from ginger. In addition, the reuse/recycling of the secondary biomass in a second extraction as a part of the circular economy was evaluated. Both the primary and the secondary biomass led to high yield percentages, ranging from 1.23% to 6.42%. The highest yield was observed in the $scCO_2$ + EtOH, with biomass prior used to $scCO_2$ extraction. All extracts presented with high similarities as far as their total phenolic contents, antioxidant capacity, and chemical composition. The most abundant compounds, identified by the two different gas chromatography-mass spectrometry (GC-MS) systems present, were a-zingiberene, β- sesquiphellandrene, a-farnesene, β-bisabolene, zingerone, gingerol, a-curcumene, and γ-muurolene. Interestingly, the reuse/recycling of the secondary biomass was found to be promising, as the extracts showed high antioxidant capacity and consisted of significant amounts of compounds with beneficial properties.

Keywords: ginger valorization; supercritical extraction; subcritical extraction; antioxidant capacity; chemical composition; circular economy

check for updates

Citation: Spyrou, A.; Batista, M.G.F.; Corazza, M.L.; Papadaki, M.; Antonopoulou, M. Extraction of High Value Products from *Zingiber officinale* Roscoe (Ginger) and Utilization of Residual Biomass. *Molecules* **2024**, *29*, 871. https://doi.org/10.3390/molecules29040871

Academic Editor: Young-Min Kim

Received: 30 December 2023
Revised: 6 February 2024
Accepted: 13 February 2024
Published: 16 February 2024

1. Introduction

Zingiber officinale Roscoe (ginger) is a plant that is widely distributed worldwide and belongs to the *Zingiberaceae* family. It contains compounds that present a wide range of benefits, such as antioxidant, antifungal, antibacterial, and anticancer effects [1–4]. Ginger, both fresh and dried, is used for consumption as well as medicinal applications [5]. Ginger's extracts and essential oils have diverse chemical compositions, and they mainly include compounds belonging to terpenoids and phenolics groups which possesses significant biological activity [6]. Moreover, polysaccharides have also been reported to be components of ginger extracts [1]. The bioactive compounds present in ginger extracts/oils lead to a vast number of applications in the medicinal and cosmetic sector, as well as the food industry and the agriculture sector [7–11].

As a plant with significant value, ginger has attracted intensive research interest. Thus, many studies have focused on the application of different extraction methods with the aim of maximizing the extraction efficiency [1]. In the past years, new technologies for the extraction of beneficial compounds have attracted a lot of attention. Specifically, environmentally friendly, low-cost, and sustainable extraction methods have been used to extract and isolate compounds of interest [6]. Among them, super- and sub- critical extraction methods have aroused remarkable interest from the scientific community, because only small changes in the pressure and temperature can increase the selectivity of the extraction process [12]. These types of extractions have numerus advantages, with the most interesting being the decrease of the thermal degradation of the targeted compounds [12], which is a major disadvantage of the conventional extraction procedures [13,14]. The lower temperatures also lead to the reduction of the extraction cost [15].

In the literature, it is well-documented that compounds extracted from ginger possess various biological activities and can have diverse applications. However, the extraction of bioactive compounds with high recovery remains a challenge [5]. Ginger solvent extraction has been the most studied. More recently, advanced methods such as supercritical and subcritical techniques, as well as pressurized fluid and ultrasound-assisted techniques, have shown promising results for the extraction of bioactive substances [5,16]. The evaluation of these techniques for the recovery of such compounds has led to significant interest [5]. It is worth noticing that these types of extractions lead to better quality products, and that the solvent can be recycled and reused [17,18].

The aim of the present study was to explore different methods of ginger extraction, using more advanced and environmentally friendly procedures. Therefore, a variety of sub- and supercritical extractions were performed and evaluated in terms of the yield percentage, the total phenolic content, the antioxidant capacity, and the chemical composition. The gas chromatography-mass spectrometry (GC-MS) technique was also used to identify the main components. Additionally, the utilization of the residual (secondary) biomass was studied, with the purpose being the production of additional added value products. Large amounts of residual biomass, which is potentially not environmentally friendly, are produced yearly [19,20]. There is increasing interest around the utilization of this type of biomass to further isolate natural bioactive compounds [21]. To the authors' best knowledge, this is the first study focusing on the extraction of ginger through using and comparing a variety of sub- and supercritical extractions. The main focus of the present study was to evaluate the possibility of enhancing the high value bioactive compounds' recovery through the utilizion of the residual biomass, with a multistep extraction procedure of the secondary biomass. The recovery of those high value compounds is a key point for the valorization of the used biomass, with the possibility of their use in food, cosmetics, and other products. The main limitation regarding the reuse of biomass is the high cost. The proposed procedure is a low-cost, environmentally friendly extraction method that follows the framework of circular economy.

2. Results and Discussion

2.1. Soxhlet Extraction

In Table 1, the extraction conditions and extraction yields using Soxhlet, in combination with different solvents, are presented. The Soxhlet extraction was performed to select the most suitable co-solvent in the supercritical extraction. The highest yield was observed using water as solvent, followed by ethanol (17.93% and 17.70%, respectively) (Table 1, Figure 1). Lower extraction yields were noticed with ethyl acetate and hexane (8.28% and 4.82%, respectively). This difference could be due to the different polarities of the used solvents (10.2 for water, 5.2 for ethanol, 4.3 for ethyl acetate, and 0.0 for hexane) [22], and their ability to solubilize the oils from the biomass.

Table 1. Experimental conditions and extraction yields of *Zingiber officinale* Roscoe acquired under 6 h of Soxhlet extraction using different solvents.

Solvents	Polarity *	Boiling Point (°C) **	Type of Biomass	Yield (%) ± SD
Ethyl acetate	4.3	77.0	Primary	8.28 ± 0.48 [abc]
Ethanol	5.2	78.5	Primary	17.70 ± 1.78 [ad]
Hexane	0.0	63.9	Primary	4.82 ± 0.23 [bde]
Water	10.2	100.5	Primary	17.93 ± 6.67 [ce]

SD: standard deviation, [abcde]: values in each column that share the same letter are significantly different from each other (One-way Anova, $p < 0.05$),* [22], ** [23].

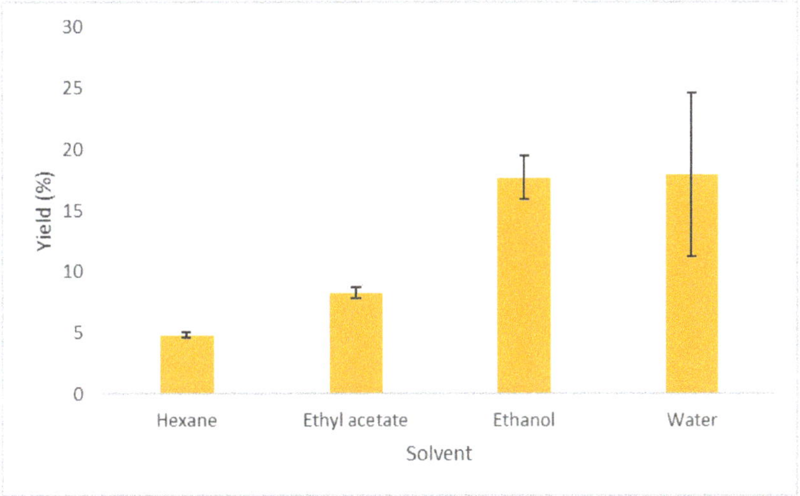

Figure 1. Extraction yield % of *Zingiber officinale* Roscoe using Soxhlet extraction with different solvents.

Similar to our results, Al-Areer et al. found that, [24] when performing Soxhlet extractions at 90 °C for 9 h, the highest extraction yield was observed when ethanol was used as solvent, followed by ethyl acetate and hexane. A different trend was observed by Lemma and Egza, [25] who noticed a higher yield in the hexane than the ethanolic extract when Soxhlet extractions were performed for about 4 h at the boiling point of each solvent.

Moreover, the antioxidant capacity, as well as the concentration of total phenols of the extracts, were evaluated. The results are shown in Table 2. The highest antioxidant activities were observed in the ethyl acetate and ethanolic extracts. Regarding the total phenolic content, the ethanolic extract presented the highest concentration.

Table 2. Antioxidant capacity and total phenolic content (mg of gallic acid equivalents (GAE) per 100 g of sample) of primary biomass of *Zingiber officinale* Roscoe extracts acquired under Soxhlet extraction using different solvents.

Solvents	IC$_{50}$ (µg mL^{-1}) ± SD	Total Phenolic Content (mg GAE 100 g^{-1}) ± SD
Ethyl Acetate	150.50 ± 6.06 [a]	1.96 ± 0.10 [ab]
Ethanol	191.16 ± 10.58 [b]	7.04 ± 0.79 [acd]
Hexane	201.05 ± 7.22 [c]	0.96 ± 0.16 [c]
Water	758.29 ± 29.59 [abc]	0.83 ± 0.42 [bd]

SD: standard deviation, [abcd]: values in each column that share the same letter are significantly different from each other (One-way Anova, $p < 0.05$).

The differences in the extraction yields, total phenolic contents, and antioxidant activity may occur because different solvents can extract different active compounds [26]. Water, even though it presented the higher extraction yield, showed the lowest total phenolic content and antioxidant activity. This can be due to the high temperature used in that case (100.5 °C), leading to the thermal decomposition of high bioactive compounds (Table 2). It is well-known that the high temperatures used in the Soxhlet extraction process (up to the boiling point of each solvent), along with the long extraction time, can lead to the thermal degradation of some compounds [27]. More specifically, the continuous evaporation and extraction of the target compounds caused by the solvents, the high extraction kinetics, and the prolonged extraction time can promote the decomposition of thermolabile target compounds [28]. Thermolabile target compounds can consist of polyphenols that present high antioxidant capacity [29].

Moreover, it is reported in the literature that many factors can affect antioxidant activities. Besides the amount and strength of the antioxidant compounds, the ability of the antioxidants to transfer a hydrogen to a free radical such as DPPH· can also be affected by the environment where the reaction takes place; for example, in different types of solvents [30].

Taking all of the above into consideration, along with the antioxidant capacity (Table 2, Figure S1) of each extract and their total phenolic content (Table 2), ethanol was selected as the co-solvent in the supercritical extraction.

2.2. Sub- and Supercritical Extraction

In Table 3, the extraction conditions for the sub- and supercritical extractions are presented. The primary biomass extraction resulted in high yield percentages (Table 4, Figure 2). The supercritical CO_2 extraction with ethanol as the co-solvent (scCO$_2$ + EtOH) gave an extraction yield of 6.06%, which was the highest of the three obtained, followed by subcritical propane (scPropane) and supercritical CO_2 (scCO$_2$), with yield percentages of 2.06% and 1.54%, respectively. It is worth noticing that scCO$_2$ + EtOH led to higher yields in less time (half the time in comparison with the other methods). Considering the increase in the yield percentage, and the decrease of the time needed, the addition of ethanol had a positive effect in the overall extraction.

Ethanol, when used as a co-solvent, solubilized the CO_2 and led to a decrease of the viscosity of the mixture of solvents (CO_2 + EtOH), and an increase of density [31,32]. The combination of the solvents accelerated the extraction process, led to less usage of CO, and, and enhanced the extraction yield. Additionally, the polar mixture of solvents led to an increase of the extracted amount of polar and soluble compounds [22]. In general, the usage of co-solvents may improve the extraction performance due to the enhanced transport of solute [33].

Table 3. Extraction yields of *Zingiber officinale* Roscoe acquired under subcritical and supercritical extraction using different extraction conditions.

Solvents	Temperature (°C)	Pressure (Bar)	Type of Biomass	Time of Extraction (Min)	Yield (%) ± SD
Propane	60	100	Primary	130	2.06 ± 0.01 [abcde]
CO$_2$	60	150	Primary	140	1.54 ± 0.01 [afghi]
CO$_2$ + EtOH	60	150	Primary	60	6.06 ± 0.01 [bfjkl]
CO$_2$ + EtOH	60	150	Secondary (Used once in scCO$_2$)	40	6.42 ± 0.01 [cgjmn]
CO$_2$ + EtOH	60	150	Secondary (Used once in scCO$_2$ + EtOH)	40	3.20 ± 0.01 [dhkmo]
CO$_2$ + EtOH	60	150	Secondary (Used twice in scCO$_2$ + EtOH)	40	1.23 ± 0.01 [eilno]

SD: standard deviation, [a–o]: values in each column that share the same letter are significantly different from each other (One-way Anova, $p < 0.05$).

Table 4. Total phenolic content of *Zingiber officinale* Roscoe extracts acquired under different extraction conditions (mg GAE 100 g^{-1}).

Solvents	Extraction Condition	Type of Biomass	Total Phenolic Content (mg GAE 100 g^{-1}) $\pm$ SD
Propane	100 bar/60 °C	Primary	1.76 $\pm$ 0.42
CO_2	150 bar/60 °C	Primary	1.65 $\pm$ 0.27
CO_2 + EtOH	150 bar/60 °C	Primary	1.87 $\pm$ 0.18
CO_2 + EtOH	150 bar/60 °C	Secondary (Used once in scCO_2)	1.47 $\pm$ 0.04 [a]
CO_2 + EtOH	150 bar/60 °C	Secondary (Used once in scCO_2 + EtOH)	1.94 $\pm$ 0.08 [a]
CO_2 + EtOH	150 bar/60 °C	Secondary (Used twice in scCO_2 + EtOH)	1.74 $\pm$ 0.05

SD: standard deviation, [a]: values in each column that share the same letter are significantly different from each other (One-way Anova, $p < 0.05$).

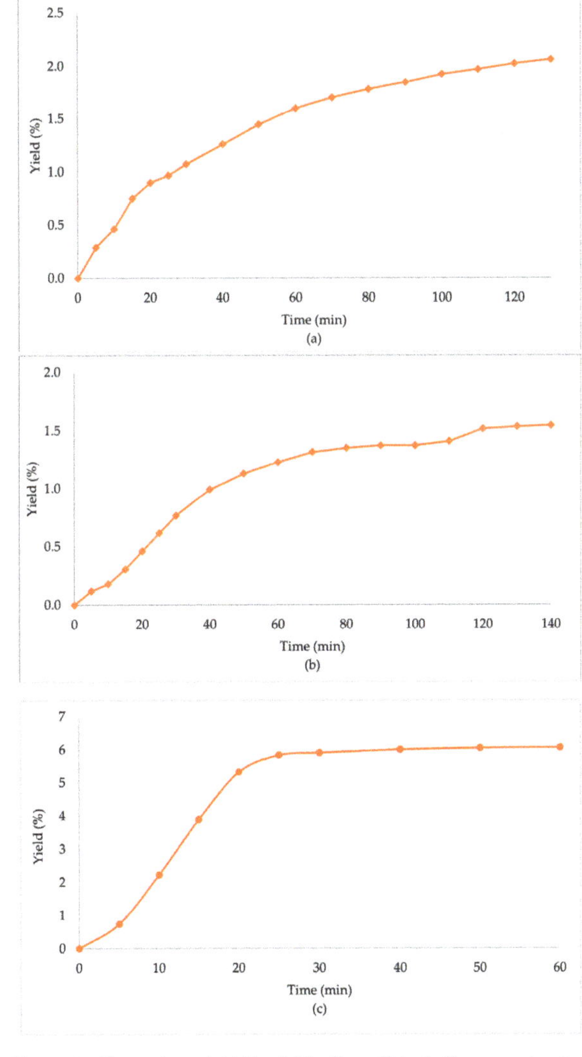

Figure 2. Extraction yield % of *Zingiber officinale* Roscoe extracts acquired under (**a**) scPropane extraction, (**b**) scCO_2 and (**c**) scCO_2 +EtOH.

Due to the enhanced performance of the $scCO_2$ + EtOH extraction, when a secondary biomass was used, high extraction yields were observed in all cases (Table 4, Figure 3). Specifically, in the case of the $scCO_2$ + EtOH extraction of the secondary biomass that was used once before in $scCO_2$ extraction, the extraction yield was significantly higher than that of the primary biomass. CO_2 behaves as a nonpolar solvent and presents with a low extraction performance for polar compounds [34]. The co-solvent (EtOH) that was used in the secondary biomass extraction had a positive influence on the extraction procedure, since it lowered the viscosity of the solvents, increased the density, and altered the overall polarity of the mixture of solvents, which led to the enhancement of the extraction yield.

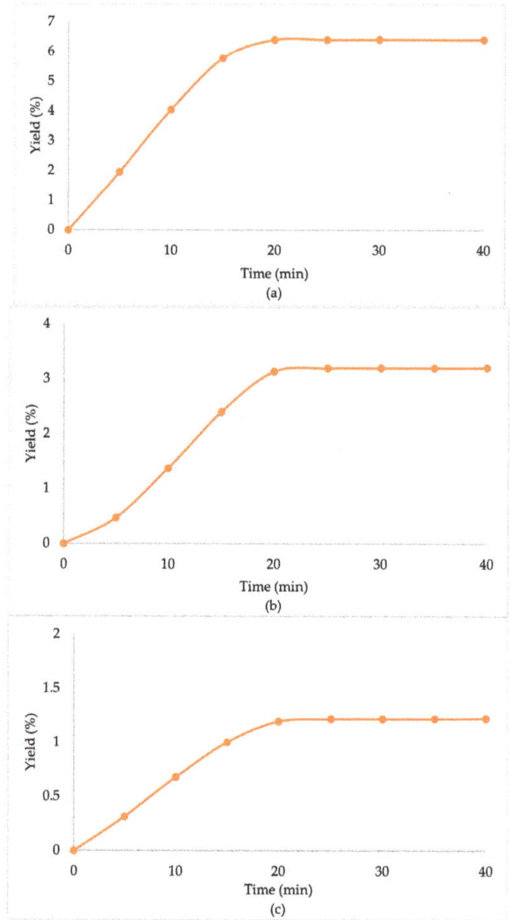

Figure 3. Extraction yield % of *Zingiber officinale* Roscoe extracts acquired under $scCO_2$ +EtOH, with secondary biomass, previously used at (**a**) $scCO_2$ once, (**b**) $scCO_2$ + EtOH once and (**c**) $scCO_2$ + EtOH twice.

When the biomass, prior used in $scCO_2$ + EtOH, was extracted in a multistep extraction twice with $scCO_2$ + EtOH, the extraction yield was lower than the primary biomass, but still significant. Mainly, this was observed due to the high amounts of polar compounds that the mixture of solvents obtained through the first extraction, and when only CO_2 was left in the chamber, the capability to extract polar compounds was minimized. In continuation, when the extraction procedure was repeated, and the mixture of solvents renewed, their capability to extract those compounds improved once again and the extraction yield remained significant.

High yield percentages for ginger have been reported by Mesomo et al., [15] and Mesomo et al., [11] using CO_2 and Propane as solvents, in various pressures and temperatures. For CO_2, the yield percentages ranged from 0.22% to 3.21%, and for Propane from 1.98% to 2.70%. For CO_2, the pressure led to a positive effect on the yield, and for propane, both pressure and temperature. Similarly, Salea et al., [35] using $scCO_2$ in various conditions, calculated a range of yields from 1.55% to 2.95%. That range was attributed to the variety of pressure and temperature conditions used in each extraction.

Zancan et al. [36] performed $scCO_2$ and $scCO_2$ + EtOH, and did not observe significant differences in the yield percentages between the two methods, but ethanol favored the extraction of gingerols and shogaols.

2.3. Total Phenolic Contents and Antioxidant Capacity

Table 4 presents the total phenolic contents acquired through using different extraction conditions and solvents, as well as both primary and secondary biomass. The total phenolic contents were expressed as mg of gallic acid equivalent per 100 g of sample. In both the primary and secondary biomass, similar total phenolic contents were observed, with the highest being in $scCO_2$ + EtOH using secondary biomass, which was formerly used in the same type of extraction.

Stoilova et al., [37] found a total phenolic content of 871 mg/g dry extract acquired through high pressure CO_2 extraction.

A similar trend can be observed for the antioxidant capacity of the extracts (Table 5, Figures S2 and S3). Overall, according to the IC_{50} values, the $scCO_2$ + EtOH extract presented the highest antioxidant activity, and the $scCO_2$ extract the lowest.

Table 5. Antioxidant capacity of *Zingiber officinale* Roscoe extracts acquired under different extraction conditions.

Solvents	Extraction Condition	Type of Biomass	IC_{50} (μg mL^{-1}) $\pm$ SD
Propane	100 bar/60 °C	Primary	153.14 $\pm$ 0.21 [abcde]
CO_2	150 bar/60 °C	Primary	169.74 $\pm$ 5.16 [afghi]
CO_2 + EtOH	150 bar/60 °C	Primary	125.12 $\pm$ 3.32 [bf]
CO_2 + EtOH	150 bar/60 °C	Secondary (Used once in $scCO_2$)	131.83 $\pm$ 6.95 [cg]
CO_2 + EtOH	150 bar/60 °C	Secondary (Used once in $scCO_2$ + EtOH)	132.39 $\pm$ 1.91 [dh]
CO_2 + EtOH	150 bar/60 °C	Secondary (Used twice in $scCO_2$ + EtOH)	127.54 $\pm$ 3.32 [ei]

SD: standard deviation, [a–i]: values in each column that share the same letter are significantly different from each other (One-way Anova, $p < 0.05$).

The antioxidant properties of ginger extracts obtained by different extraction methods have been widely studied. Its antioxidant properties are due to compounds such as gingerols, flavonoids, and phenolic acids [38].

Mesomo et al., [15] studying different conditions of supercritical extraction, observed the highest antioxidant capacity with $scCO_2$. Zancan et al. [36] observed that, when no gingerols and shogaols were yet obtained by the extraction the antioxidant, activity was much lower. Stoilova et al. [37] calculated the IC_{50} for inhibition of DPPH to be 0.64 μg mL^{-1}.

2.4. GC-MS Analysis

Ginger includes both volatiles such as geraniol, borneol, terpineol, curcumene, zingiberol, α-farnesene, α-sesquiphellandrene, α- β-bisabolene, β-elemene etc., as well as non-volatile compounds such as gingerols, shogaols, zingerone, and paradols [7]. The composition of ginger extracts and oils differ significantly. The bioactive compounds of ginger essential oils are mainly monoterpenes and sesquiterpene hydrocarbons, and their chemical composition depends on the nature (fresh/dry) and place of origin of the ginger rhizome, as well as the extraction method employed [10]. They are also composed of oxygenated hydrocarbon compounds including aldehydes, phenols, esters, oxide ethers, alcohols, and ketones [10].

The chemical composition of the extracts/oils was determined by GC-MS analysis, and all the compounds identified have been compiled and presented in Table 6 along with their area percentage. The extracts obtained with all the tested systems were found to have quite similar chemical compositions, and the main substances were a-zingiberene, β-sesquiphellandrene, a-farnesene, β-bisabolene, zingerone, gingerol, a-curcumene, and γ-muurolene. Similar compounds have also been identified in the literature [11,38–40]. In the case of scPropane, some more compounds, i.e., monoterpenes and sesquiterpenes, were identified but only in traces. When a primary biomass was used, Soxhlet with ethanol extracted less compounds in comparison with scPropane, $scCO_2$, and $scCO_2$ + EtOH, signifying the increased efficiency of such advanced extraction techniques.

Table 6. Chemical composition of ginger extracts (Type of biomass: Primary) using different extraction methods (+:presence/% of total area).

Compounds	Soxhlet/EtOH	scPropane	$scCO_2$	$scCO_2$ + EtOH
GC-MS (System 1)				
Hexanal	+(0.634%)	+(0.736%)	+(0.134%)	+(1.054%)
Decanal	+(1.830%)	+(1.225%)	+(0.934%)	+(1.512%)
α-Terpineol		+(0.550%)		+(0.789%)
cis-Geraniol		+(0.471%)	+(1.269%)	+(0.851%)
Geranial		+(1.358%)	+(0.678%)	
Geraniol acetate		+(0.390%)	+(1.569%)	
β-Bisabolene		+(1.988%)	+(2.002%)	+(5.3147%)
a-Farnesene	+(7.397%)	+(6.919%)	+(7.318%)	+(8.003%)
a-Zingiberene	+(45.072%)	+(57.398%)	+(52.105%)	+(46.355%)
β-Sesquiphellandrene	+(14.685%)	+(13.542%)	+(15.905%)	+(17.111%)
γ-Cadinene		+(2.486%)	+(2.838%)	+(3.094%)
Valencene		+(1.544%)	+(1.985%)	+(2.072%)
Zingerone	+(30.383%)	+(11.394%)	+(13.262%)	+(13.842%)
GC-MS (System 2)				
Cedrene	+(0.885%)	+(2.835%)		+(1.285%)
γ-Patchoulene	+(1.325%)		+(1.984%)	
a-Curcumene	+(26.152%)	+(28.745%)	+(25.244%)	+(22.418%)
Germacrene D		+(2.568%)		
γ-Muurolene	+(22.562%)		+(20.682%)	+(16.854%)
α-Bergamotene	+(1.862%)			
α-Eudesmol	+(1.859%)	+(5.624%)	+(2.856%)	+(3.145%)
Gingerol	+(40.152%)		+(30.658%)	+(29.745%)
Borneol		+(3.452%)		
(+)-Cycloisosativene		+(3.526%)		
Nerolidyl acetate		+(5.125%)	+(3.256%)	+(2.256%)
γ-Elemene		+(6.514%)		+(3.652%)
β-Farnesene	+(5.203%)	+(7.418%)		
(−)-allo-Aromadendrene		+(3.215%)	+(2.562%)	+(1.523%)
α-Himachalene		+(4.158%)		+(3.052%)
Elemol		+(3.103%)		
γ-Gurjunene		+(8.774%)		+(6.135%)
γ-Eudesmol		+(5.263%)	+(2.365%)	+(4.090%)
Cubenol		+(2.259%)		
Cedr-8-en-13-ol		+(4.156%)	+(4.526%)	
Cubenol		+(3.265%)	+(3.269%)	
Urs-12-en-28-ol			+(2.598%)	+(2.589%)
8-Isopropenyl-1,5-dimethyl-1,5-cyclodecadiene				+(3.256%)

Our results regarding the main components are in agreement with the bibliography [11]. Based on literature data, higher essential oil and β-zingiberene contents were obtained for the dried ginger rhizome than that of the fresh ones. Moreover, the drying method has been found to play a significant role in essential oil's yield and its chemical composition. Temperatures lower than 70 °C can increase the yield of ginger oil without having any effect on the transformation of 6-gingerol to 6-shogaol [41].

It is noteworthy that most of the above-mentioned compounds were identified after the reuse/recycling of the secondary biomass, highlighting the possibility to extract the maximum value from the used biomass. In Table 7, the identified compounds of the secondary biomass extractions are presented with their area percentage. After the first extraction of the secondary biomass, the compounds identified were less in comparison with the primary biomass's identified compounds, but their area percentage was higher.

Table 7. Chemical composition of ginger extracts (Type of biomass: Secondary) using different and sequential extraction methods (+: presence/% of total area).

Compounds	$scCO_2$ + EtOH after $scCO_2$	$scCO_2$ + EtOH after One $scCO_2$+ EtOH	$scCO_2$ + EtOH after Two $scCO_2$ + EtOH
GC-MS (System 1)			
Hexanal	+(0.720%)		
a-Farnesene	+(12.464%)	+ (9.598%)	
a-Zingiberene	+(48.558%)	+(37.036%)	
β- Sesquiphellandrene	+(8.409%)	+(5.637%)	
γ-Cadinene	+(16.834%)	+(14.008%)	
Zingerone	+(13.015%)	+(33.721%)	+(100%)
GC-MS (System 2)			
Cedrene	+(1.526%)		
a-Curcumene	+(25.032%)	+(24.195%)	
γ-Muurolene	+(14.012%)	+(13.589%)	
α-Bergamotene	+(5.141%)	+(6.524%)	
α-Eudesmol	+(2.154%)	+(3.528%)	
Gingerol	+(52.135%)	+(52.164%)	+(100%)

The main identified compounds (a-zingiberene, β-sesquiphellandrene, a-farnesene, β-bisabolene, zingerone, gingerol, and a-curcumene) that can be found in almost all of the studied extracts have been previously reported to possess significant antioxidant properties. This is in agreement with our results [42–47]. Badrunanto et al., [45] when studying the antioxidant components of Indonesian ginger essential oil, observed that, amongst others, a-zingiberene, β-sesquiphellandrene, a-farnesene, β-bisabolene, and a-curcumene, presented a high correlation with the antioxidant activity of the oil. Similarly, Misharina et al. [46] highlighted the antioxidant activity of zingiberene, β-sesquiphellandrene and β-bisabolene. Gingerol analogues have been associated with ginger extracts' antioxidant activity [42,43,47]. Specifically, Danwilai et al. [42] studied the antioxidant activity of ginger extract oral supplements in cancer patients who were receiving adjuvant chemotherapy. These supplements contained 20 mg day^{-1} 6-gingerol, and results found a significant increase regarding antioxidant activity, and a decrease of oxidative marker levels. Moreover, Wang et al. [43] observed that 10-gingerol presented with about 34.2% DPPH radical scavenging activity, and 6-gingerol about 16.3%. Furthermore, they noticed that the antioxidant activity of those ginger compounds contributed to the antimicrobial activity against *Acinobacter baummannii* infections. 6-gingerol, 8-gingerol, and 10-gingerol's antioxidant activity was studied by Dugasani et al. [47], who observed that the DPPH scavenging potential was in the order of 10-gingerol > 8-gingerol > 6-gingerol. Rajan et al. [44] studied zingerone's antioxidant activity using a DPPH free radical method, and observed a dose dependent increase of the compound's antioxidant activity.

In general, it is well reported in the literature that the components of ginger extracts/essential oils have significant bioactivities and health-promoting properties, and thus can have applications in various sectors [7–11]. Based on literature data, ginger extracts/oils containing most of the bioactive compounds found in the present have been reported to have significant pharmacological, medicinal, and cosmetic applications, as they have been found to possess antimicrobial and antiseptic activity, anti-carcinogenic potential, neuroprotective activity, anti-obese activity, anti-diabetic effect, and analgesic activity as well as provide cardiovascular protection [8,11]. Another significant application is in the food industry, as the bioactive compounds of ginger can provide oxidative and storage stability, sensorial properties, preservation, oxidative resistance, and anti-bacterial activity in consuming products [7]. As well, another notable factor is its application in agriculture for the control of plant diseases which minimizes simultaneously the possible negative effects on the environment, animals, and human health [9,10].

3. Materials and Methods

3.1. Materials

Ethyl acetate (99.5%) was obtained from NEON comercial (Suzano, SP, Brazil), ethanol (99.5%) from ACS CIENTIFICA (Sumare, SP, Brazil), hexane (98.5%) from exodo cientifica (Sumare, SP, Brazil), CO_2 (99.95%) and propane (99.5%) from White Martins S.A. (Curitiba, PR, Brazil), Folin-Ciocalteu's reagent from CARLOS ERBA REAGENTS (Val-de-Reuil, France), Sodium carbonate anhydrous (>99.5%) from Fluka (Buchs, Switzerland), 1,1-Diphenyl-2-picrylhydrazyl Free Radical (DPPH) from TCI EUROPE N.V. (Zwijndrecht, Belgium), and gallic acid from Sigma Aldrich (Darmstadt, Germany). The Soxhlet extractor was purchased from Qualividros (Passos, MG, Brazil).

3.2. Sample Preparation

Very fresh *Zingiber officinale* Roscoe rhizomes were purchased from the local market at Curitiba, Brazil. After being transferred to the laboratory, the rhizomes were washed carefully with water and cut into smaller sized pieces to dry evenly. After being dried at 30 °C with air circulation until the moisture content became less than 10% (Table 8), the samples were grounded in a blender and stored in plastic bags until use.

Table 8. Moisture content of *Zingiber officinale* Roscoe before and after drying.

Sample	Average Moisture Content (%) $\pm$ SD
Fresh	89.05 ± 1.20
Dried	8.96 ± 0.22

SD: standard deviation.

3.3. Soxhlet Extraction

Soxhlet extraction was performed according to the AOAC [48] method. Briefly, 5 g of biomass and 150 mL of solvent were used in each extraction. The total duration was 6 h, using different temperatures depending on the solvent (Table 1). Specifically, the extractions took place at each solvent's boiling point. The solvents used were ethyl acetate, ethanol, hexane, and water and, in each case, the procedure was repeated three times. A rotary vacuum evaporator (IKA RV 10 combined with IKA HB 10-IKA, Campinas, SP, Brazil) was used to concentrate the Soxhlet extracts, and they were then dried in an air circulation oven at 40 °C for about 24 h. The extraction yields were calculated with the Equation (1):

$$\text{Yield (\%)} = [\text{mass of dried extract (g)}/\text{initial mass of biomass (g)}] \times 100 \qquad (1)$$

3.4. Sub- and Supercritical Extraction

Sub- and supercritical extractions were performed at an extraction unit (inner volume 80 cm^3, length 16 cm and $\varnothing$ = 2.52 cm) which consisted of the extractor and a syringe pump. Additionally, the extraction system was equipped with pressure and temperature sensors in order to monitor the conditions. A thermostatic bath was attached to the system to better control the temperature. Furthermore, a needle valve controlled the flow inside the extractor, which was directly proportional to the pressure. The extractor was loaded with 10–15 g of biomass. In the cases a co-solvent was used, its mass was 20–25 g. The temperature was set at 60 °C. For the scPropane extraction, the pressure was set at 100 bar, and for the $scCO_2$ and $scCO_2$ +EtOH extraction the pressure was set at 150 bar. The CO_2 or the propane was loaded in the vessel until the desired pressure was achieved. In the beginning of each extraction process, a 30 min static extraction was performed, and after that a dynamic extraction was carried out with a flow rate of approximately 1–2 mL min^{-1}. The dynamic extraction ended when no more extract was collected in the sampling tubes. The extraction yields were calculated by the Equation (1).

Extraction of the used primary biomass was also performed to evaluate the possibility of obtaining compounds of interest even with secondary biomass. Specifically, biomass prior used in $scCO_2$ extraction was extracted a second time using $scCO_2$ + EtOH. Moreover, a multistep approach was performed for the primary biomass that was used in $scCO_2$ + EtOH, which then was extracted again twice with $scCO_2$ + EtOH.

3.5. Total Phenolic Contents and Antioxidant Capacity

The total phenolic contents were determined using the Folin-Ciocalteu method as described by Box, 1983 [49]. Briefly, 5 mL of 10 µg mL^{-1} extract diluted in ethanol was added in glass bottles with 0.25 mL of Folin-Ciocalteu Reagent. After gentle stirring and 2 min of waiting, 0.75 mL of Na_2CO_3 200 g L^{-1} was added and the solution was left in the dark for 1 h. The absorbance was measured at 765 nm in a Hitachi U2000 Spectrophotometer (Hitachi, Ltd., Santa Clara, CA, USA). The results were calculated using a calibration curve of gallic acid (0.5–10 mg L^{-1}), and expressed as mg of gallic acid equivalents per 100 g of sample.

The free radical quenching ability of the extracts was determined using the DPPH reagent, as described by Mensor et al. [50]. Briefly, 2.5 mL of extract diluted in ethanol in concentrations ranging from 5 to 250 µg mL^{-1} (apart from the extract from Soxhlet that water was used as solvent, where the concentrations were ranging between 5 and 1000 µg mL^{-1}, because of the low antioxidant capacity it presented) and DPPH in a final concentration of 0.3 mM were mixed. After 30 min in the dark, the absorbance was measured at 518 nm (Hitachi U2000 Spectrophotometer). The antioxidant capacity (AC) was calculated using the equation:

$$AC\ (\%) = 100 - \{[(abs_{sample} - abs_{blank}) \times 100]/abs_{control}\} \tag{2}$$

where abs_{sample} is the absorbance of the solution consisting of the extract, DPPH, and ethanol, abs_{blank} the absorbance of the extract and ethanol, and $abs_{control}$ the absorbance of DPPH and ethanol.

After the results were plotted, the concentration (µg mL^{-1}) of each extract needed to inhibit 50% (IC_{50}) of radicals' production was calculated by linear regression.

3.6. Statistics

Significant differences in the yield percentages, total phenolic contents and IC_{50} values were assessed by post-hoc multiple comparison tests (Bonferroni test, $p < 0.05$, ANOVA) using the IBM SPSS statistics Inc. (Version: 28.0.1.0 (142)) software package.

3.7. GC-MS Analysis

Analysis of the samples (injection volume 1 µL) were performed using two GC-MS systems. The first system was composed of an Agilent 6890 Series GC system and an Agilent 5973 Network mass selective detector (MSD) (Agilent Technologies, Santa Clara, CA, USA). The second system was an Agilent 5975B inert MSD integrated to an Agilent 6890N Network GC (Agilent Technologies, Santa Clara, CA, USA). More information about the analytical columns, the temperature programs, and the conditions adopted in the MSD are reported in Supplementary Materials. Enhanced Data Analysis software was used for the analysis of the chromatograms, and NIST MS Search 2.0 software was used for compound identification.

4. Conclusions

In this work, compressed solvents technology was investigated for the extraction of high value products from *Zingiber officinale* Roscoe. Soxhlet extraction was used in preliminary experiments for the selection of the most appropriate co-solvent for $scCO_2$. Using Soxhlet extraction, the highest yields were observed when water and ethanol were used, i.e., 17.93% and 17.7%, respectively. Ethanol was selected as the co-solvent for the supercritical extraction due to its high yield percentage, total phenolic content, and antioxidant capacity. When compressed solvents were used, the most efficient was $scCO_2$ + ethanol, reaching about 6% extraction yield. High antioxidant activity was also observed for the ginger extracts in all cases.

The proposed procedure for the extraction of the secondary biomass used CO_2 and ethanol as co-solvents, resulting in high extraction yields (reaching 6.42%) and an accelerated extraction time. Except the high extraction yields, another advantage of the reusage of the biomass is that, even after two extractions, high value bioactive compounds, such as zingerone and gingerol, were detected, and the extracts presented high antioxidant capacity with IC_{50} values up to 132.39 $\mu g\ mL^{-1}$.

The extracts obtained with all the tested methods presented similar chemical compositions, and the most abundant substances were a-zingiberene, β-sesquiphellandrene, a-farnesene, β-bisabolene, zingerone, gingerol, a-curcumene, and γ-muurolene. All these compounds present beneficial properties and can have real applications in various sectors, such as the food and pharmaceutical industries. After the first reuse, high value compounds were identified as well, similar to those of the primary biomass. Taking into consideration the high yield percentages and the antioxidant capacity, as well as the chemical composition, it is safe to conclude that the addition of ethanol as a co-solvent in the $scCO_2$ extraction had a positive effect both in the extraction of the primary biomass and in the secondary biomass extraction. Based on the results, the reuse of a secondary biomass (raw material) presents high significance. Indeed, the maximum utilization of biomass can contribute to the achievement of the goals of circular economy. Furthermore, although not tested in this work, the residual biomass after the secondary extraction could potentially be used as a raw fertilizing/pest repelling compound, enhancing the soils that are employed in agriculture.

Supplementary Materials: The following supporting information can be downloaded at: https://www.mdpi.com/article/10.3390/molecules29040871/s1, Supplementary Materials S1. GC-MS Analysis, Figure S1: Antioxidant capacity (%) of *Zingiber officinale* Roscoe extracts ($\mu g\ mL^{-1}$) acquired under Soxhlet extraction using (a) ethyl acetate, (b) ethanol, (c) hexane and (d) water as solvents. Figure S2: Antioxidant capacity (%) of *Zingiber officinale* Roscoe extracts ($\mu g\ mL^{-1}$) acquired under (a) scPropane extraction, (b) $scCO_2$ and (c) $scCO_2$ +EtOH. Figure S3: Antioxidant capacity (%) of *Zingiber officinale* Roscoe extracts ($\mu g\ mL^{-1}$) acquired under $scCO_2$ +EtOH, with biomass previously used at (a) $scCO_2$ once, (b) $scCO_2$ +EtOH once and (c) $scCO_2$ +EtOH twice.

Author Contributions: Conceptualization, M.A., M.P. and M.L.C.; methodology, A.S., M.G.F.B., M.A., M.P. and M.L.C.; software, A.S., M.G.F.B., M.A., M.P. and M.L.C.; validation, A.S., M.G.F.B., M.A., M.P. and M.L.C.; formal analysis, A.S., M.A., M.P. and M.L.C.; investigation, A.S., M.G.F.B., M.A., M.P. and M.L.C.; resources, M.A., M.P. and M.L.C.; data curation, A.S., M.A., M.P. and M.L.C.; writing—original draft preparation, A.S., M.A. and M.P.; writing—review and editing, A.S., M.A., M.P. and M.L.C.; visualization, A.S. and M.A.; supervision, M.A., M.P. and M.L.C.; project administration, M.A. and M.P. funding acquisition, M.A. and M.P. All authors have read and agreed to the published version of the manuscript.

Funding: This project has received funding from the European Union's Horizon 2020 research and innovation programme under the Marie Sklodowska-Curie grant agreement No. 778168.

Institutional Review Board Statement: Not applicable.

Informed Consent Statement: Not applicable.

Data Availability Statement: Data are contained within the article.

Acknowledgments: The authors would like to thank the Laboratory of Instrumental Analysis of the University of Patras as well as the Laboratory of Photo-Catalytic Processes and Environmental Chemistry of Institute of Nanoscience and Nanotechnology, NCSR Demokritos, for GC-MS analysis.

Conflicts of Interest: The authors declare no conflict of interest.

References

1. Hu, W.; Yu, A.; Wang, S.; Bai, Q.; Tang, H.; Yang, B.; Wang, M.; Kuang, H. Extraction, Purification, Structural Characteristics, Biological Activities, and Applications of the Polysaccharides from *Zingiber officinale* Roscoe. (Ginger): A Review. *Molecules* **2023**, *28*, 3855. [CrossRef]
2. Ballester, P.; Cerdá, B.; Arcusa, R.; García-Muñoz, A.M.; Marhuenda, J.; Zafrilla, P. Antioxidant Activity in Extracts from *Zingiberaceae* Family: Cardamom, Turmeric, and Ginger. *Molecules* **2023**, *28*, 4024. [CrossRef] [PubMed]
3. Sacchetti, G.; Maietti, S.; Muzzoli, M.; Scaglianti, M.; Manfredini, S.; Radice, M.; Bruni, R. Comparative evaluation of 11 essential oils of different origin as functional antioxidants, antiradicals and antimicrobials in foods. *Food Chem.* **2005**, *91*, 621–632. [CrossRef]
4. Ficker, C.; Smith, M.L.; Akpagana, K.; Gbeassor, M.; Zhang, J.; Durst, T.; Assabgui, R.; Arnason, J.T. Bioassay-guided isolation and identification of antifungal compounds from ginger. *Phytother. Res.* **2003**, *17*, 897–902. [CrossRef] [PubMed]
5. Garza-Cadena, C.; Ortega-Rivera, D.M.; Machorro-García, G.; Gonzalez-Zermeño, E.M.; Homma-Dueñas, D.; Plata-Gryl, M.; Castro-Muñoz, R. A comprehensive review on Ginger (*Zingiber officinale*) as a potential source of nutraceuticals for food formulations: Towards the polishing of gingerol and other present biomolecules. *Food Chem.* **2023**, *413*, 135629. [CrossRef] [PubMed]
6. Maghraby, Y.R.; Labib, R.M.; Sobeh, M.; Farag, M.A. Gingerols and shogaols: A multi-faceted review of their extraction, formulation, and analysis in drugs and biofluids to maximize their nutraceutical and pharmaceutical applications. *Food Chem. X* **2023**, *20*, 100947. [CrossRef] [PubMed]
7. Shaukat, M.N.; Nazir, A.; Fallico, B. Ginger Bioactives: A Comprehensive Review of Health Benefits and Potential Food Applications. *Antioxidants* **2023**, *12*, 2015. [CrossRef] [PubMed]
8. Rasool, N.; Saeed, Z.; Pervaiz, M.; Ali, F.; Younas, U.; Bashir, R.; Bukhari, S.M.; Khan, R.R.M.; Jelani, S.; Sikandar, R. Evaluation of essential oil extracted from ginger, cinnamon and lemon for therapeutic and biological activities. *Biocatal. Agric. Biotechnol.* **2022**, *44*, 102470. [CrossRef]
9. Gunasena, M.T.; Rafi, A.; Zobir, S.A.M.; Hussein, M.Z.; Ali, A.; Kutawa, A.B.; Wahab, M.A.A.; Sulaiman, M.R.; Adzmi, F.; Ahmad, K. Phytochemicals Profiling, Antimicrobial Activity and Mechanism of Action of Essential Oil Extracted from Ginger (*Zingiber officinale* Roscoe cv. Bentong) against *Burkholderia glumae* Causative Agent of Bacterial Panicle Blight Disease of Rice. *Plants* **2022**, *11*, 1466. [CrossRef]
10. Abdullahi, A.; Ahmad, K.; Ismail, I.S.; Asib, N.; Ahmed, O.H.; Abubakar, A.I.; Siddiqui, Y.; Ismail, M.R. Potential of Using Ginger Essential Oils-Based Nanotechnology to Control Tropical Plant Diseases. *Plant Pathol. J.* **2020**, *36*, 515–535. [CrossRef]
11. Mesomo, M.C.; Corazza, M.L.; Ndiaye, P.M.; Santa, O.R.D.; Cardozo, L.; Scheer, A.d.P. Supercritical CO_2 extracts and essential oil of ginger (*Zingiber officinale* R.): Chemical composition and antibacterial activity. *J. Supercrit. Fluids* **2013**, *80*, 44–49. [CrossRef]
12. Pronyk, C.; Mazza, G. Design and scale-up of pressurized fluid extractors for food and bioproducts. *J. Food Eng.* **2009**, *95*, 215–226. [CrossRef]
13. Ding, S.; An, K.; Zhao, C.; Li, Y.; Guo, Y.; Wang, Z. Effect of drying methods on volatiles of Chinese ginger (*Zingiber officinale* Roscoe). *Food Bioprod. Process.* **2012**, *90*, 515–524. [CrossRef]
14. Sikora, E.; Cieślik, E.; Leszczyńska, T.; Filipiak-Florkiewicz, A.; Pisulewski, P.M. The antioxidant activity of selected cruciferous vegetables subjected to aquathermal processing. *Food Chem.* **2008**, *107*, 55–59. [CrossRef]
15. Mesomo, M.C.; Scheer, A.d.P.; Perez, E.; Ndiaye, P.M.; Corazza, M.L. Ginger (*Zingiber officinale* R.) extracts obtained using supercritical CO_2 and compressed propane: Kinetics and antioxidant activity evaluation. *J. Supercrit. Fluids* **2012**, *71*, 102–109. [CrossRef]
16. Roselló-Soto, E.; Parniakov, O.; Deng, Q.; Patras, A.; Koubaa, M.; Grimi, N.; Boussetta, N.; Tiwari, B.K.; Vorobiev, E.; Lebovka, N.; et al. Application of Non-conventional Extraction Methods: Toward a Sustainable and Green Production of Valuable Compounds from Mushrooms. *Food Eng. Rev.* **2015**, *8*, 214–234. [CrossRef]
17. Riera, E.; Golás, Y.; Blanco, A.; Gallego, J.; Blasco, M.; Mulet, A. Mass transfer enhancement in supercritical fluids extraction by means of power ultrasound. *Ultrason. Sonochemistry* **2004**, *11*, 241–244. [CrossRef]
18. Kim, H.-J.; Lee, S.-B.; Park, K.-A.; Hong, I.-K. Characterization of extraction and separation of rice bran oil rich in EFA using SFE process. *Sep. Purif. Technol.* **1999**, *15*, 1–8. [CrossRef]
19. Santana-Méridas, O.; González-Coloma, A.; Sánchez-Vioque, R. Agricultural residues as a source of bioactive natural products. *Phytochem. Rev.* **2012**, *11*, 447–466. [CrossRef]
20. Liang, J.; Lu, Q.; Lerner, R.; Sun, X.; Zeng, H.; Liu, Y. Agricultural Wastes. *Water Environ. Res.* **2011**, *83*, 1439–1466. [CrossRef]
21. Saha, A.; Basak, B.B. Scope of value addition and utilization of residual biomass from medicinal and aromatic plants. *Ind. Crop. Prod.* **2020**, *145*, 111979. [CrossRef]
22. Guedes, A.R.; de Souza, A.R.C.; Barbi, R.C.T.; Escobar, E.L.N.; Zanoello, É.F.; Corazza, M.L. Extraction of *Synadenium grantii* Hook f. using conventional solvents and supercritical CO_2 + ethanol. *J. Supercrit. Fluids* **2020**, *160*, 104796. [CrossRef]

23. National Institute of Standards and Technology. *Security Requirements for Cryptographic Modules*; Federal Information Processing Standards Publications (FIPS PUBS) 140-2, Change Notice 2 December 03, 2002; Department of Commerce: Washington, DC, USA, 2001. [CrossRef]

24. Al-Areer, N.W.; Al Azzam, K.M.; Al Omari, R.H.; Al-Deeb, I.; Bekbayeva, L.; Negim, E.-S. Quantitative analysis of total phenolic and flavonoid compounds in different extracts from ginger plant (*Zingiber officinale*) and evaluation of their anticancer effect against colorectal cancer cell lines. *Pharmacia* **2023**, *70*, 905–919. [CrossRef]

25. Lemma, T.S.; Egza, T.F. Extraction and Characterization of Essential Oil from Ginger Rhizome Collected from Arba Minch Market. *Int. J. Eng. Trends Technol.* **2019**, *67*, 41–45. [CrossRef]

26. Fragkouli, R.; Antonopoulou, M.; Asimakis, E.; Spyrou, A.; Kosma, C.; Zotos, A.; Tsiamis, G.; Patakas, A.; Triantafyllidis, V. Mediterranean Plants as Potential Source of Biopesticides: An Overview of Current Research and Future Trends. *Metabolites* **2023**, *13*, 967. [CrossRef] [PubMed]

27. Zhang, Q.-W.; Lin, L.-G.; Ye, W.-C. Techniques for extraction and isolation of natural products: A comprehensive review. *Chin. Med.* **2018**, *13*, 20. [CrossRef]

28. Jha, A.K.; Sit, N. Extraction of bioactive compounds from plant materials using combination of various novel methods: A review. *Trends Food Sci. Technol.* **2021**, *119*, 579–591. [CrossRef]

29. Bahrin, N.; Muhammad, N.; Abdullah, N.; Talip, B.H.A.; Jusoh, S.; Theng, S.W. Effect of Processing Temperature on Antioxidant Activity of *Ficus carica* Leaves Extract. *J. Sci. Technol.* **2018**, *10*, 2. Available online: https://publisher.uthm.edu.my/ojs/index.php/JST/article/view/3005 (accessed on 6 February 2024). [CrossRef]

30. Thavamoney, N.; Sivanadian, L.; Tee, L.H.; Khoo, H.E.; Prasad, K.N.; Kong, K.W. Extraction and recovery of phytochemical components and antioxidative properties in fruit parts of *Dacryodes rostrata* influenced by different solvents. *J. Food Sci. Technol.* **2018**, *55*, 2523–2532. [CrossRef]

31. Araújo, M.N.; Azevedo, A.Q.P.L.; Hamerski, F.; Voll, F.A.P.; Corazza, M.L. Enhanced extraction of spent coffee grounds oil using high-pressure CO_2 plus ethanol solvents. *Ind. Crop. Prod.* **2019**, *141*, 111723. [CrossRef]

32. Juchen, P.T.; Araujo, M.N.; Hamerski, F.; Corazza, M.L.; Voll, F.A.P. Extraction of parboiled rice bran oil with supercritical CO_2 and ethanol as co-solvent: Kinetics and characterization. *Ind. Crop. Prod.* **2019**, *139*, 111506. [CrossRef]

33. Araus, K.A.; del Valle, J.M.; Robert, P.S.; de la Fuente, J.C. Effect of triolein addition on the solubility of capsanthin in supercritical carbon dioxide. *J. Chem. Thermodyn.* **2012**, *51*, 190–194. [CrossRef]

34. Mendes, R.L.; Nobre, B.P.; Cardoso, M.T.; Pereira, A.P.; Palavra, A.F. Supercritical carbon dioxide extraction of compounds with pharmaceutical importance from microalgae. *Inorg. Chim. Acta* **2003**, *356*, 328–334. [CrossRef]

35. Salea, R.; Veriansyah, B.; Tjandrawinata, R.R. Optimization and scale-up process for supercritical fluids extraction of ginger oil from *Zingiber officinale* var. Amarum. *J. Supercrit. Fluids* **2017**, *120*, 285–294. [CrossRef]

36. Zancan, K.C.; Marques, M.O.; Petenate, A.J.; Meireles, M.A. Extraction of ginger (*Zingiber officinale* Roscoe) oleoresin with CO_2 and co-solvents: A study of the antioxidant action of the extracts. *J. Supercrit. Fluids* **2002**, *24*, 57–76. [CrossRef]

37. Stoilova, I.; Krastanov, A.; Stoyanova, A.; Denev, P.; Gargova, S. Antioxidant activity of a ginger extract (*Zingiber officinale*). *Food Chem.* **2007**, *102*, 764–770. [CrossRef]

38. El-Ghorab, A.H.; Nauman, M.; Anjum, F.M.; Hussain, S.; Nadeem, M. A Comparative Study on Chemical Composition and Antioxidant Activity of Ginger (*Zingiber officinale*) and Cumin (*Cuminum cyminum*). *J. Agric. Food Chem.* **2010**, *58*, 8231–8237. [CrossRef]

39. Kim, J.S.; Lee, S.I.; Park, H.W.; Yang, J.H.; Shin, T.-Y.; Kim, Y.-C.; Baek, N.-I.; Kim, S.-H.; Choi, S.U.; Kwon, B.-M.; et al. Cytotoxic components from the dried rhizomes of *Zingiber officinale* Roscoe. *Arch. Pharmacal Res.* **2008**, *31*, 415–418. [CrossRef]

40. Agarwal, M.; Walia, S.; Dhingra, S.; Khambay, B.P.S. Insect growth inhibition, antifeedant and antifungal activity of compounds isolated/derived from *Zingiber officinale* Roscoe (ginger) rhizomes. *Pest Manag. Sci.* **2001**, *57*, 289–300. [CrossRef]

41. Huang, T.-C.; Chung, C.-C.; Wang, H.-Y.; Law, C.-L.; Chen, H.-H. Formation of 6-Shogaol of Ginger Oil Under Different Drying Conditions. *Dry. Technol.* **2011**, *29*, 1884–1889. [CrossRef]

42. Danwilai, K.; Konmun, J.; Sripanidkulchai, B.-O.; Subongkot, S. Antioxidant activity of ginger extract as a daily supplement in cancer patients receiving adjuvant chemotherapy: A pilot study. *Cancer Manag. Res.* **2017**, *9*, 11–18. [CrossRef]

43. Wang, H.-M.; Chen, C.-Y.; Chen, H.-A.; Huang, W.-C.; Lin, W.-R.; Chen, T.-C.; Lin, C.-Y.; Chien, H.-J.; Lu, P.-L.; Lin, C.-M.; et al. *Zingiber officinale* (ginger) compounds have tetracycline-resistance modifying effects against clinical extensively drug-resistant *Acinetobacter baumannii*. *Phytother. Res.* **2010**, *24*, 1825–1830. [CrossRef]

44. Rajan, I.; Narayanan, N.; Rabindran, R.; Jayasree, P.R.; Kumar, P.R.M. Zingerone Protects Against Stannous Chloride-Induced and Hydrogen Peroxide-Induced Oxidative DNA Damage In Vitro. *Biol. Trace Elem. Res.* **2013**, *155*, 455–459. [CrossRef]

45. Badrunanto; Wahyuni, W.T.; Farid, M.; Batubara, I.; Yamauchi, K. Antioxidant components of the three different varieties of Indonesian ginger essential oil: In vitro and computational studies. *Food Chem. Adv.* **2023**, *4*, 100558. [CrossRef]

46. Misharina, T.A.; Terenina, M.B.; Krikunova, N.I. Antioxidant properties of essential oils. *Appl. Biochem. Microbiol.* **2009**, *45*, 642–647. [CrossRef]

47. Dugasani, S.; Pichika, M.R.; Nadarajah, V.D.; Balijepalli, M.K.; Tandra, S.; Korlakunta, J.N. Comparative antioxidant and anti-inflammatory effects of [6]-gingerol, [8]-gingerol, [10]-gingerol and [6]-shogaol. *J. Ethnopharmacol.* **2010**, *127*, 515–520. [CrossRef] [PubMed]

48. Association of Official Analytical Chemists (AOAC). *Official Methods of Analysis of the Association of Official Analytical Chemists*, 17th ed.; Method 925.10; AOAC International: Gaithersburg, MD, USA, 2000.
49. Box, J. Investigation of the Folin-Ciocalteau phenol reagent for the determination of polyphenolic substances in natural waters. *Water Res.* **1983**, *17*, 511–525. [CrossRef]
50. Mensor, L.L.; Menezes, F.S.; Leitão, G.G.; Reis, A.S.; dos Santos, T.C.; Coube, C.S.; Leitão, S.G. Screening of Brazilian plant extracts for antioxidant activity by the use of DPPH free radical method. *Phytother. Res.* **2001**, *15*, 127–130. [CrossRef]

molecules

Article

Sustainable Transformation of Two Algal Species of Different Genera to High-Value Chemicals and Bioproducts

Flora V. Tsvetanova [1], Stanislava S. Boyadzhieva [1], Jose A. Paixão Coelho [2,3,*], Dragomir S. Yankov [1] and Roumiana P. Stateva [1,*]

[1] Institute of Chemical Engineering, Bulgarian Academy of Sciences, 1113 Sofia, Bulgaria; florablue@abv.bg (F.V.T.); maleic@abv.bg (S.S.B.); yanpe@bas.bg (D.S.Y.)

[2] Instituto Superior de Engenharia de Lisboa, Instituto Politécnico de Lisboa, Rua Conselheiro Emídio Navarro 1, 1959-007 Lisboa, Portugal

[3] Centro de Química Estrutural, Instituto Superior Técnico, Universidade de Lisboa, Av. RoviscoPais 1, 1049-001 Lisboa, Portugal

* Correspondence: jcoelho@deq.isel.ipl.pt (J.A.P.C.); thermod@bas.bg (R.P.S.)

Abstract: This study investigates the potential of two algae species from different genera, namely the recently isolated *Scenedesmus obliquus* BGP and *Porphyridium cruentum*, from the perspective of their integral sustainable transformation to valuable substances. Conventional Soxhlet and environmentally friendly supercritical fluid extraction were applied to recover oils from the species. The extracts were characterized through analytical techniques, such as GC-Fid and LC-MS/MS, which allowed their qualitative and quantitative differentiation. Thus, *P. cruentum* oils contained up to 43% C20:4 and C20:5 fatty acids, while those of *S. obliquus* BGP had only residual amounts. The LC-MS/MS analysis of phenolic compounds in the *S. obliquus* BGP and *P. cruentum* extracts showed higher content of 3-OH-4-methoxybenzoic acid and kaempferol 3-*O*-glycoside in the former and higher amounts of ferulic acid in the latter. Total phenolic content and antioxidant activity of the oils were also determined and compared. The compositional analysis of the oil extracts revealed significant differences and varying potentialities based on their genera and method of extraction. To the best of our knowledge our work is unique in providing such detailed information about the transformation prospects of the two algae species to high-value chemicals and bioproducts.

Keywords: microalgae; *Scenedesmus obliquus* BGP; *Porphyridium cruentum*; supercritical extraction; fatty acids; total phenolic content; antioxidant activity

check for updates

Citation: Tsvetanova, F.V.; Boyadzhieva, S.S.; Coelho, J.A.P.; Yankov, D.S.; Stateva, R.P. Sustainable Transformation of Two Algal Species of Different Genera to High-Value Chemicals and Bioproducts. *Molecules* **2024**, *29*, 156. https://doi.org/ 10.3390/molecules29010156

Academic Editor: Yu Yang

Received: 15 November 2023
Revised: 21 December 2023
Accepted: 23 December 2023
Published: 27 December 2023

1. Introduction

Microalgae, as a biological resource, have drawn considerable interest in the last years, particularly since they were granted the GRAS (Generally Recognized As Safe) status. Consequently, an avenue for using them as sustainable and appealing green factories for valuable compounds of nutritional and health benefits was widely opened. Moreover, unlike any other crop, they are known to have the lowest carbon, water, and arable footprint. Still, compared with other organisms, microalgae are not that well examined and studied, since just a few among the tens of thousands of microalgae species that exist worldwide have been described [1].

In compliance with the concept of circular economy, the development of effective and efficient biorefineries capable of producing ultra-low-footprint and high-value biochemicals from sustainable sources is essential. A biorefinery integrates various processes to transform biomass to high-added-value energy and non-energy-related compounds. Microalgal biomass is an underutilized resource with enormous commercial potential, as compared to terrestrial plants, due to their abundance.

The beneficial biomolecules accumulated by microalgae are widely applicable in different industrial fields such as nutritional, pharmaceutical, medicinal, cosmetic, etc. Among the most widely exploited in industry microalgae are the red ones, especially from the *Porphyridium* genus. They are known to produce several classes of essential compounds like long-chain polyunsaturated fatty acids (PUFA), exopolysaccharides, pigments, antioxidants, and various microelements. PUFAs contain low cholesterol levels and are important substances, especially for human well-being, as they contribute to cardiovascular, eye, and brain maintenance. For example, *Porphyridium* spp. are reported to synthesize omega 3 eicosapentaenoic (EPA, 20:5), docosahexaenoic (DHA, 22:6), and omega 6 arachidonic (ARA, 20:4) acids, which the human body is incapable of producing [1,2].

The recovery of valuable bioactives from microalgae by applying conventional and non-conventional methods has been a topic of extensive research in the past years. The methods examined include traditional extraction with an organic solvent (solid–liquid, Soxhlet, maceration), ultrasound, and microwave assisted extractions, as well as extractions applying compressed fluids, e.g., supercritical CO_2 ($scCO_2$) neat or with a co-solvent, as well as pressurized liquid extraction, etc. Details are published in a number of review articles, for example, [3,4], as well as research papers [2,5,6].

The main aim of our investigation was to compare two algal species of different genera from the perspective of their sustainable transformation to high-value substances with potential applications in the food, nutra-, pharmaceutical, energy, etc., industries.

As model targets, representatives of the genus *Scenedesmus* (green algae) and the genus *Porphyridium* (red algae) were chosen.

The selection of the two genera was motivated and substantiated since:

- *Scenedesmus* species are freshwater algae, which exhibit high growth rates, multiply quickly, adapt easily to changing conditions, and hence are suitable for semi-industrial and industrial cultivation. These algae are high in nutrition and synthesize biologically active substances such as carotenoids, chlorophylls, mycosporin-like amino acids, and polyphenols, with antioxidant and antiviral potential.
- The algae of the genus *Porphyridium* are representatives of a totally different group of marine deep-sea algae, which are characterized by a pigment composition different from that of green algae. Furthermore, they are rich sources of unique polyunsaturated fatty acids, carotenes, phycobiliproteins, amino acids, and minerals such as Ca, Mg, Zn, and K [7].
- These genera are interesting and challenging in their own right and provide examples of the generic issues to be faced when realizing any one-feed/multi-product biorefinery targeting the sustainable transformation of a biomass to valuables and bioactives with applications in various industries.

As representatives of the two genera, the green algae *Scenedesmus obliquus* BGP, a recently isolated Bulgarian strain, and the red algae *Porphyridium cruentum* were chosen.

Application of traditional techniques for recovering valuable compounds from *S. obliquus* BGP has been very limited, resulting in insufficient data. To gain insights into their performance potential for obtaining bioactives, secondary metabolites, and other important compounds sustainably, it is crucial to use advanced environmentally friendly techniques like supercritical fluid extraction, the efficiency and viability of which can be compared to traditional methods such as Soxhlet. Thus, new opportunities for obtaining high-quality extracts rich in valuable compounds from *S. obliquus* BGP will be uncovered.

Since the extraction method specificity (including types of solvents/co-solvents applied, etc.) is among the most important factors, the impact of the techniques employed on the yield and on the composition of selected oils was analyzed.

Thus, the comparison was based on essential indicators such as: (i) fatty acid composition from FAME GC-FID, (ii) identification and quantification of antioxidants by LC-MS/MS, (iii) total phenolic content (TPC) and antioxidant activity (AA), and (iv) several additional parameters and indices. The new data and information obtained allow us to reveal the two strains' importance and potential for sustainable utilization in the food, food supplements, pharmaceutical, cosmetics, energy, etc., industries.

To the best of our knowledge, no such detailed comparison of strains belonging to different genera, neither in terms of depth nor extent, as presented by us, has been published before.

2. Results and Discussion

This section starts with a brief comparison of the biochemical composition of the two species with data from the literature. Then, the efficiencies of the conventional Soxhlet and extractions with supercritical CO_2 (scCO_2), both neat and with a co-solvent, are compared based on yield. The influence of operating conditions (solvents, temperature, and pressure) is discussed as well. Subsequently, the chemical composition, total phenolic content (TPC), and antioxidant activity (AA) of certain extracts are presented.

Biochemical composition

Our previous results [8] demonstrated that the recently identified *S. obliquus* BGP synthesizes considerable amounts of proteins, carbohydrates, and lipids. In particular, with regard to the latter, the strain is one of the top lipid producers among the *Scenedesmus* genus species reported in the literature. Other authors have also analyzed the biochemical composition of *S. obliquus*. For example, in [9], the protein, lipids, and carbohydrates percent reported were 52.00, 9.70, and 7.90%, respectively, while the data of Silva et al. [10] showed that they were 37.15, 10.29, 21.93% and for the three biochemical groups. In another work [11], comparing the composition of *S. obliquus* with *Selenastrum bibraianum* the former was proved to have the highest lipid (42.6 $\pm$ 0.01%) and carbohydrate content (27.7 $\pm$ 0.02%) when cultured in Bristol media under control conditions.

Regarding *P. cruentum*, in their recent contribution, Ardiles et al. [12] compared the strain biochemical composition, determined by them, with the data of [13,14]. For example, the lipids content has been reported in wide ranges: 14.67 $\pm$ 0.24 (%, w/w) [12], 9–14 (% w/w) [13], and 5.3 $\pm$ 0.3 (% w/w) [14].

The qualitative deviations between the biochemical composition of *P. cruentum* and *S. obliquus* BGP determined by us and those shown in the literature are considerable in some cases. Still, it should be underlined that the amounts of the chemical species identified depend greatly not only on the cultivation conditions but also the strain. For example, it has been demonstrated that *Scenedesmus* spp. can accumulate lipids in the range of 15–40% of their dry matter [3]. However, when high stress levels are inflicted during cultivation, microalgae can accumulate lipids as high as 70–90% of their dry matter.

2.1. Extraction Yield

2.1.1. Soxhlet Extractions

The extraction yields achieved by one- and two-step Soxhlet extraction are given in Table 1. Considering the results of the compositional analysis of *P. cruentum*, which showed a low content of lipids, the low yield of the first step Soxhlet with *n*-hexane was expected (Table 1). On the other end is the yield achieved by the second-step Soxhlet with ethanol performed on the *P. cruentum* residual matrix, which was over 1.7-times higher than the respective one for the *S. obliquus* BGP.

Table 1. Influence of the type of Soxhlet method and solvent on the extraction yield of the two algal species.

Extraction Method	Extraction Conditions		*S. obliquus* BGP	*P. cruentum*
	Solvent	Temperature (°C)	Extraction Yield (wt %)	Extraction Yield (wt %)
Soxhlet	ethanol	78	23.6 $\pm$ 1.1	26.5 $\pm$ 1.15
Two-step Soxhlet	*n*-hexane (step 1) ethanol (step 2)	68 78	11.1 $\pm$ 0.5 16.9 $\pm$ 0.84 Cumulative yield: 28	2.57 $\pm$ 0.14 29.52 $\pm$ 1.47 Cumulative yield: 32

Extraction yield expressed in wt % (mean $\pm$ standard deviation).

Those results are not surprising since, as discussed in [15], the impact of solvent and extraction methods on the various lipidic classes recovery is considerable.

Considering that freshwater algae such as *S. obliquus* are generally rich in neutral and medium-chain PUFAs, like C16 and C18, while marine algae are rich in long-chain polar PUFAs (e.g., C20 and C22), the higher yield obtained from the second step Soxhlet of *P. cruentum* is due to the use of polar GRAS solvent ethanol, which enhances the recovery of polar PUFAs.

2.1.2. Supercritical Fluid Extraction (SFE)

SFE, as an advanced green technique, is widely applied to recover valuables from microalgae. For example, Tzima et al. [4], in their extensive and in-depth review, analyzed over 100 articles reporting SFE of carotenoids, chlorophylls, tocopherols, lipids, and fatty acids from microalgae, with a special section devoted to the genera *Scenedesmus* and certain *S. obliquus* species.

In another recent paper, the efficiencies of conventional, microwave-assisted, and SFE to recover bioactives from *S. obliquus* were compared [9], and it was shown that the results for the solid/liquid and microwave-assisted extractions were comparable, while in the SFE extraction, the lowest yield of bioactives was achieved. On the other hand, the SFE resulted in increased carotenoid content and enhanced antioxidant activity. Gilbert-Lopez et al. [16] examined the SFE by neat CO_2 of *S. obliquus*, while Guedes et al. [17] investigated the influence of pressure, temperature, CO_2 flow rate, and a polar co-solvent on the yields of carotenoids and chlorophylls in the SFE of a wild strain of *S. obliquus*. As reported in [18], the SFE extracts were rich in tricylglycerides but with low carotenoids and chlorophyll content in comparison with the gas-expanded liquid extraction [16].

With regard to *P. cruentum* recovery of polyunsaturated fatty acids (PUFAs), and total carotenoids from that strain by $scCO_2$ and subcritical *n*-butane was explored by Feller et al. [19]. Ardiles et al. [12] applied conventional methods like maceration and freeze/thaw, as well as microwave and ultrasound, for the recovery of phycoerythrin (PE) from *P. cruentum* and *P. purpureum*. Gallego et al. [5] advocated a green downstream approach to the valorization of *P. cruentum* biomass, applying pressurized liquids.

The yield of *S. obliquus* BGP achieved in our study applying neat $scCO_2$, as a function of temperature and pressure, is plotted vs. the extraction time and is displayed in Figure 1.

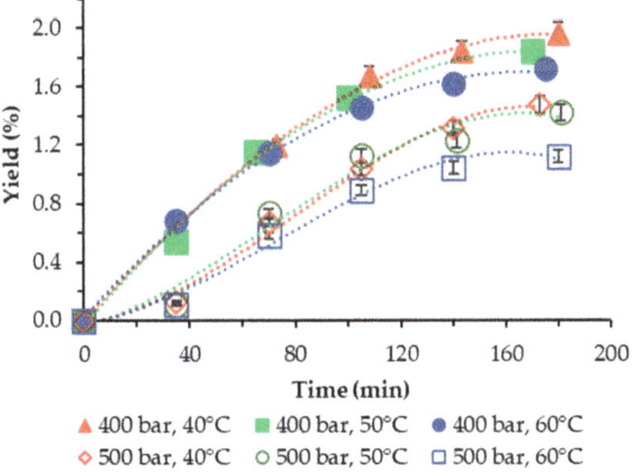

Figure 1. Cumulative extraction yield curves representing the influence of temperature and pressure on the S. *obliquus* BGP yield as a function of time.

In all cases studied, the yield was low and varied in the range of 1.12–1.96%. For both pressures applied, the negative influence of temperature was clearly demonstrated—lowest yields were registered at the highest temperature applied of 60 °C. The highest yield with neat scCO$_2$ was obtained at 40 °C and 400 bar.

Our results compare reasonably well with the results of Georgiopoulou et al. [9] who performed SFE of *S. obliquus* at T = (40, 50, 60) °C and p = (110; 250) bar and achieved a yield with neat scCO$_2$ in the range of 0.98–4.20%, as well with those of [16], where the temperatures applied are analogous to ours but the pressures were varied in the range p = (100; 250; 400) bar. The highest yield obtained was 1.15% at T = 40 °C and p = 400 bar.

SFE with neat scCO$_2$ of *P. cruentum* was not performed since the strain showed a low content of lipids, which was confirmed by the very low yield of the first-step Soxhlet with *n*-hexane achieved (Table 1).

Next, the influence of 10% ethanol as a co-solvent on the extraction yield was studied for both species. Based on the results obtained for *S. obliquus* BGP with neat scCO$_2$, the operating parameters chosen were p = 400 bar and T = (40, 60) °C, respectively.

In the case of *S. obliquus* BGP, the yield was increased considerably, and the influence of temperature exhibited a trend opposite to that displayed in the case of neat scCO$_2$. Thus, the highest yield (12.29%) was achieved at 60 °C. However, it was about two-times lower than one-step Soxhlet ethanol extraction. At 40 °C, the yield was lower but commensurable with that at 60 °C (10.33%). Georgiopoulou et al. also used 10% ethanol as a cosolvent at 60 °C and 250 bar and reported a yield of 9.75% [9]. The cumulative experimental kinetic extraction curves plotted vs. the extraction time for *S. obliquus* are displayed in Figure 2.

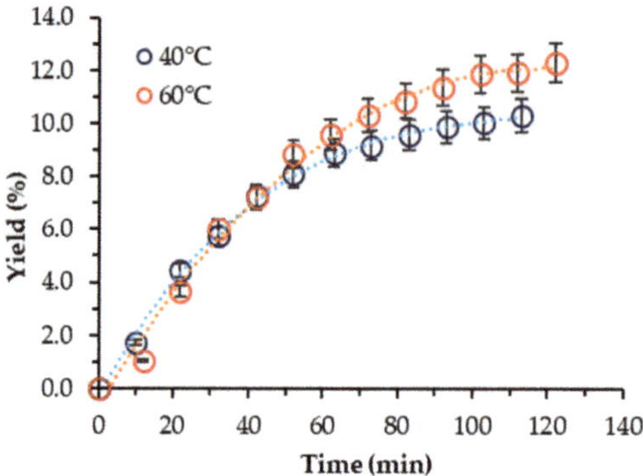

Figure 2. Cumulative extraction yield curves representing the influence of temperature at 400 bar and 10% cosolvent ethanol on the *S. obliquus* BGP yield, as a function of time.

For *P. cruentum*, the influence of pressure was first studied at 40 °C, with 10% co-solvent ethanol and a scCO$_2$ flow rate of 1 L/min. As shown (Figure 3), the highest yield registered was 1.93% at p = 400 bar, while the lowest was 1.04% at the highest pressure applied. Our initial expectations that the presence of the co-solvent would increase yields to values, if not commensurable then at least not that lower than the Soxhlet ethanol extraction, were not fulfilled. The reasons behind that assumption were that since the strain cellular walls were not as thick as those of *S. obliquus* BGP, the application of relatively high pressures and a co-solvent enhanced the mass transfer, which resulted in higher yield values. Still, the highest yield attained was almost 20-times lower than the one-step Soxhlet ethanol extraction. Gallego et al. [5] used pressurized ethanol (105 bars) at a wide range of temperatures (from 50 to 150 °C) and 20 min extraction time and reported increasing yields with increasing temperature—from 3.12 to 11.36%.

Though a direct comparison with our results cannot be made, the maximum yield registered by us at 10% ethanol is commensurable with the lowest value at 50 °C reported by Gallego et al. [5].

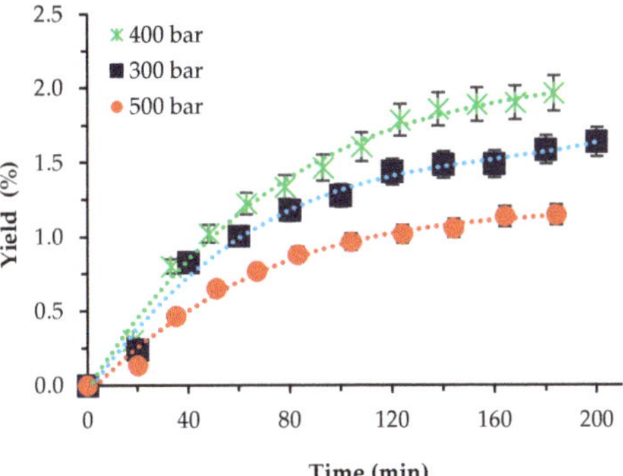

Figure 3. Cumulative extraction yield curves representing the influence of pressure at 40 °C, 10% cosolvent ethanol, and 1 L/min flow rate on the *P. cruentum* yield as a function of time.

Then, the influence of temperature at 400 bar was examined. Two temperatures were applied, T = 40 and 60 °C, respectively, at a 1 L/min scCO$_2$ flow rate. The trend was clear-cut—the higher temperature had a negative effect on the yield, as shown in Figure 4. Also, as demonstrated, the two-times-lower scCO$_2$ flow rate at 40 °C and 400 bar rendered a lower yield (1.33% vs. 1.96%) and required a longer time to achieve it.

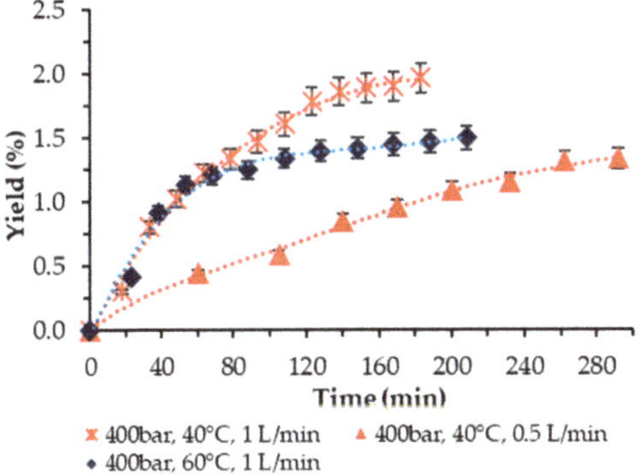

Figure 4. Cumulative extraction yield curves show the effect of temperature at 400 bar and 10% ethanol on *P. cruentum* yield over time, with flow rate variation.

2.2. Extracts Quali- and Quantification

2.2.1. GC-FID

Microalgae produce saturated and unsaturated fatty acids. Saturated fatty acids can be bio-transformed to biofuels, while unsaturated fatty acids are used in food, nutraceuticals, pharmaceuticals, cosmetics, etc. Since non-polar solvents (e.g., n-hexane) are best for the extraction of neutral saponifiable lipids, while polar solvents like ethanol perform better in the recovery of more polar lipids, two-step Soxhlet procedure was applied. In the first step, mainly neutral lipids extracted by n-hexane should be recovered and identified in the oils, while ethanol in the second step should extract the polar lipids from the residual biomass. Still, it should be underlined that in microalgal cells, the two types of lipids are not completely isolated, and in many cases, the application of polar solvents leads to the joint extraction of neutral and polar lipids.

The GC-FID results of the *S. obliquus* BGP and *P. cruentum* oils recovered by the two-step Soxhlet are displayed in Table 2. The results from the Soxhlet ethanol extractions showed a higher relative percentage of saturated (SFAs) and a lower percentage of unsaturated (mono-, di- and polyunsaturated—MUFAs, DUFAs, PUFAs) fatty acids in the red microalgae when compared to the green one (Table 2). Palmitic acid (C16:0) was dominant in the SFA percentage, both for the green and red strains.

Table 2. Fatty acid composition from FAME GC-FID analysis of *S. obliquus* BGP and *P. cruentum* Soxhlet oil extracts, expressed as relative percent of total fatty acids identified.

Fatty Acids	Soxhlet extraction					
	S. obliquus BGP			*P. cruentum*		
	96% EtOH	First Step *n*-Hexane	Second Step 96% EtOH	96% EtOH	First Step *n*-Hexane	Second Step 96% EtOH
C12:0	traces	1.0 ± 0.07	traces	nd	0.1 ± 0.01	nd
C14:0	0.2 ± 0.01	0.5 ± 0.03	0.2 ± 0.01	0.3 ± 0.02	0.3 ± 0.01	0.2 ± 0.01
C15:0	0.1 ± 0.01	0.1 ± 0.01	0.1 ± 0.01	0.3 ± 0.03	0.2 ± 0.01	0.2 ± 0.02
C16:0	22.3 ± 0.7	18.8 ± 0.6	24.1 ± 0.9	37.8 ± 0.9	23.6 ± 0.7	33.9 ± 0.8
C16:1-isom	5.0 ± 0.5	3.8 ± 0.1	5.7 ± 0.6	1.7 ± 0.1	1.2 ± 0.09	1.6 ± 0.1
C16:2	3.1 ± 0.1	2.6 ± 0.2	3.6 ± 0.5	nd	0.2 ± 0.01	0.2 ± 0.01
C16:3	3.0 ± 0.4	2.4 ± 0.2	3.5 ± 0.4	nd	nd	nd
C16:4	2.4 ± 0.2	1.7 ± 0.1	3.0 ± 0.3	nd	nd	nd
C17:0	0.2 ± 0.02	0.2 ± 0.01	0.3 ± 0.02	0.2 ± 0.01	0.2 ± 0.02	0.1 ± 0.01
C18:0	2.7 ± 0.3	3.2 ± 0.4	2.2 ± 0.3	1.0 ± 0.1	0.9 ± 0.03	0.6 ± 0.02
C18:1 (n-9)	32.8 ± 0.7	38.7 ± 0.9	27.6 ± 0.8	3.5 ± 0.4	4.5 ± 0.5	2.5 ± 0.3
C18:1 (n-7)	0.7 ± 0.02	0.6 ± 0.01	0.8 ± 0.03	1.3 ± 0.5	0.7 ± 0.03	0.7 ± 0.03
C18:2 (n-6)	14.3 ± 0.5	13.5 ± 0.4	14.8 ± 0.4	15.2 ± 0.5	21.2 ± 0.5	14.9 ± 0.6
C18:3 (n-6)	0.8 ± 0.04	0.6 ± 0.02	1.0 ± 0.04	0.2 ± 0.01	0.5 ± 0.03	0.3 ± 0.03
C18:3 (n-3)	9.5 ± 0.8	9.1 ± 0.7	9.9 ± 0.8	0.5 ± 0.02	0.3 ± 0.01	0.3 ± 0.01
C18:4	1.9 ± 0.1	1.9 ± 0.2	2.0 ± 0.2	nd	nd	nd
C20:0	0.1 ± 0.01	0.1 ± 0.01	0.1 ± 0.01	nd	nd	nd
C20:1	0.3 ± 0.01	0.4 ± 0.03	0.4 ± 0.01	0.2 ± 0.01	nd	nd
C20:2	nd	nd	nd	2.6 ± 0.3	1.0 ± 0.4	1.7 ± 0.5
C20:3-isom	nd	nd	nd	1.9 ± 0.3	2.2 ± 0.7	1.5 ± 0.4
C20:4	nd	nd	nd	23.6 ± 0.8	34.8 ± 0.9	31.4 ± 0.9
C20:5	nd	nd	nd	9.7 ± 0.6	8.1 ± 0.5	9.9 ± 0.9
C22:0	0.2 ± 0.01	0.6 ± 0.04	0.3 ± 0.01	nd	nd	nd
C22:1	0.4 ± 0.01	0.2 ± 0.01	0.4 ± 0.02	nd	nd	nd
SFA	25.8	24.5	27.3	39.6	25.3	35.0
MUFA	39.2	43.7	34.9	6.7	6.4	4.8
DUFA	17.4	16.1	18.4	17.8	22.4	16.8
PUFA	17.6	15.7	19.4	35.9	45.9	43.4
PUFA:SFA	1.03	1.02	1.04	1.36	2.69	1.71

nd—not detected.

For the three *S. obliquus* BGP oils analyzed, the percentages of SFAs, and $\sum$ MUFA + DUFA + PUFA did not differ substantially, while for *P. cruentum*, the lowest relative percentage of SFA (respectively, the highest $\sum$ MUFA + DUFA + PUFA) was detected in the oil recovered by the first-step Soxhlet n-hexane. Soxhlet with ethanol and second-step Soxhlet for that strain produced more or less commensurable percentages.

A more comprehensive examination of the results shows that the contribution of the MUFA, DUFA, and PUFA in the sum of unsaturated fatty acids was completely different for both strains. For example, the percentage of the nonpolar oleic acid (C18:1 ω9) was the highest contributor to the MUFAs % of S. obliquus. The highest relative percent was achieved by Soxhlet n-hexane, which was over nine-times higher than that of the corresponding *P. cruentum* oil recovered. Consequently, the MUFAs percentage in the *S. obliquus* oils was almost eight-times higher than that detected in the P. cruentum. On the other end is the PUFAs percentage —for P. cruentum, they were the dominant ones—almost three-times higher than the corresponding ones for S. obliquus.

Another interesting detail is that oils of both strains were relatively rich in linoleic acid (LA, C18:2 ω6) with commensurable relative percents in the range of 13.5–15.2%, with the Soxhlet n-hexane *P. cruentum* oil being the only exception, for which about 21% of LA was detected. The percentages of γ-linolenic acid (GLA C18:3 ω6) in all extracts of both strains were quite low, while for alpha linoleic acid (ALA C18:3 ω3), the trend was different—its relative percentages in all *S. obliquus* BGP extracts examined, though not very high, were still from 19 to 33 times higher than the corresponding ones in *P. cruentum* oils.

It is well known that the genus Porphyridium synthesizes the nutritionally important PUFAs eicosapentaenoic acid (EPA, 20:5 ω-3) and arachidonic acid (AA, 20:4 ω-6), the quantities of which, in certain strains, can comprise more than 40% of the total fatty acids.

In the extracts of the particular *P. cruentum* strain we studied, the highest percentages of AA were registered in the oils recovered by the first and second steps of Soxhlet (with n-hexane and ethanol, respectively). It should be noted that the AA percentage identified in the first-step Soxhlet was the second highest in all oils examined, being lower only than the percentage of the C16:0 obtained by Soxhlet ethanol. The relative percentages of EPA were over two-times lower than those of AA. Neither PUFAs were detected in the *S. obliquus* BGP oils.

As known, LA is an essential fatty acid that is not synthesized in the human body and should be provided by food. Its importance is further increased by the fact that humans can synthesize AA from LA. In this sense, both *S. obliquus* BGP and *P. cruentum* can be used as a sustainable source of LA. In addition, AA is important for human cell functioning since its metabolic breakdown leads to an enhanced production of prostaglandin E2, a hormone-like substance which takes part in a variety of bodily functions [17].

Hence, the ability of *P. cruentum* to synthesize AA and EPA (the first in high amounts) defines the strain as a promising alternative to oils from fish and land-based plant sources, since both acids are used to enrich functional foods.

The *S. obliquus* BGP and *P. cruentum* oils recovered by scCO$_2$ + 10% ethanol as a co-solvent performed at the operating conditions for which the maximum yield was achieved were also analyzed by GC-FID, and the results obtained are displayed in Table 3.

The *P. cruentum* oil obtained by scCO$_2$ + ethanol again showed higher relative percentage of SFAs when compared to the *S. oblicuus* one, with C16:0 being the major contributor.

In complete analogy to the Soxhlet extracted oils, the percentage of oleic acid was the highest contributor to the MUFAs percentage of *S. obliquus*. As should be expected, its percentage was commensurable only with that detected in the second-step Soxhlet + ethanol but considerably lower than the first-step Soxhlet, and again over nine-times higher than that of the corresponding *P. cruentum* oil recovered.

With regard to the important AA and EPA, the relative percentage of the first was lower than that registered by Soxhlet ethanol, while the second was higher. Again, in the *S. obliquus* BGP oils, those acids were not detected. Yet one additional interesting result was that while the PUFAs percentage of *P. cruentum* oils was within the percentages achieved by Soxhlet, for the *S. obliquus*, an almost twice higher percentage of PUFAs was achieved when compared to the Soxhlet (see Table 2 and Table 3, respectively). Concerning DUFA, their percentages were comparable for both species.

Tables 2 and 3 also show the PUFA: SFA ratios. It should be noted that the DUFAs C18:2 and C20:2 were taken into consideration when calculating those ratios since linolenic acid is an important essential acid, while eicosadienoic acid, though rare, is known to modulate the metabolism of other PUFAs.

It is hypothesized that all PUFAs in the diet can depress low-density lipoprotein cholesterol (LDL-C) and lower serum cholesterol levels, whereas all SFAs contribute to high levels of the latter. The results of a recent clinical study confirm the positive influence of PUFAs on high-density lipoprotein cholesterol (HDL-C) levels and total cholesterol—the levels of the former were increased, while those of the latter decreased [20].

The World Health Organization (WHO) reported the guidelines for a "balanced diet", in which the suggested ratio of PUFA: SFA is above 0.4 [21]. WHO also recommends that the proportions of SFA, MUFA, and PUFA in dietary fats should be 1:1.5:1 [22] in order to avoid the risk of developing cardiovascular and other chronic diseases.

Table 3. Fatty acid composition from FAME GC-FID analysis of *S. obliquus* BGP and *P. cruentum* scCO$_2$ + ethanol oil extracts, expressed as the relative percent of total fatty acids identified.

Fatty Acids	*S. obliquus* BGP	*P. cruentum*
	400 Bar, 40 °C, 10% EtOH	400 Bar, 40 °C, 10% EtOH
C12:0	nd	0.7 ± 0.03
C14:0	0.2 ± 0.01	0.4 ± 0.01
C15:0	0.1 ± 0.01	0.3 ± 0.02
C16:0	18.6 ± 0.4	29.0 ± 0.6
C16:1-isom	2.9 ± 0.2	1.2 ± 0.1
C16:2	2.5 ± 0.1	nd
C16:3	3.7 ± 0.3	nd
C16:4	5.9 ± 0.3	nd
C17:0	0.1 ± 0.01	0.2 ± 0.01
C18:0	1.6 ± 0.09	1.1 ± 0.09
C18:1 (n-9)	28.6 ± 0.7	3.3 ± 0.3
C18:1 (n-7)	0.9 ± 0.07	0.8 ± 0.05
C18:2 (n-6)	11.6 ± 0.3	17.5 ± 0.4
C18:3 (n-6)	0.4 ± 0.03	0.3 ± 0.02
C18:3 (n-3)	19.1 ± 0.8	0.2 ± 0.01
C18:4	2.2 ± 0.09	nd
C20:0	0.1 ± 0.01	nd
C20:1	0.3 ± 0.01	nd
C20:2	nd	2.1 ± 0.3
C20:3-isom	nd	2.2 ± 0.3
C20:4	nd	29.0 ± 0.8
C20:5	nd	11.7 ± 0.5
C22:0	0.7 ± 0.03	nd
C22:1	0.3 ± 0.02	nd
C24:0	0.2 ± 0.01	nd
SFA	21.6	31.7
MUFA	33.0	5.3
DUFA	14.1	19.6
PUFA	31.3	43.4
PUFA:SFA	1.54	1.99

nd—not detected.

The proportions calculated for *S. obliquus* BGP oil recovered by scCO$_2$ + ethanol were 21.6:33.0:33.3 (1:1.53:1.54), while for P. cruentum, they were 31.7:5.3:63.0 (1:0.17:1.99), resepctively.

Though none of the above comply with the WHO recommendations still the proportions of *S. obliquus* BGP are better..

Table 2 shows that the PUFA: SFA values of *S. obliquus* BGP oils were generally lower (up to about 2.5 times) than the corresponding ones of the red microalgae. For the latter, the highest PUFA: SFA ratio was registered in the oil recovered by the first step Soxhlet n-hexane—almost twice as that of Soxhlet ethanol. The influence of the extraction techniques on the PUFA: SFA ratios is clearly demonstrated when the data displayed in Tables 2 and 3 are compared. Thus, the PUFA: SFA of the *S. obliquus* oil (Table 3) was over 1.5-times higher than those of the Soxhlet extractions (1.54 vs. 1.03; 1.02; 1.04). Obviously, the latter enhanced the recovery of SFAs, while the SFE enhanced that of PUFAs. The trend observed for *P. cruentum* was similar when the PUFA: SFA ratio of the SFE oil was compared to that of the Soxhlet ethanol and second-step ethanol (1.99 vs. 1.36; 1.71) but was lower than the Soxhlet n-hexane. Other important indices for the characterization of algal oils were calculated and are presented in Tables 4 and 5, respectively.

For example, the oxidizability, allylic position equivalent, and bis-allylic position equivalent indices (OX, APE, and BAPE) allow the calculation of the oxidation stability index (OSI). The rate of oxidation of fatty acids of oils/biodiesel depends on the number of double bonds per mole and their relative positions.

Table 4. Parameters and indices of *S. obliquus* BGP and *P. cruentum* Soxhlet oil extracts.

Oil Parameters and Indices	Soxhlet Extraction					
	S. obliquus BGP			*P. cruentum*		
	96% EtOH	First Step *n*-Hexane	Second Step 96% EtOH	96% EtOH	First Step *n*-Hexane	Second Step 96% EtOH
OX	0.26	0.34	0.37	0.17	0.23	0.16
APE	1.16	1.25	1.08	0.41	0.54	0.37
BAPE	0.35	0.33	0.37	0.17	0.23	0.16
OSI	3.89	3.90	3.89	3.90	3.90	3.90
UI	131.1	126.6	134.9	193	239.9	219.8
h/H	2.58	3.24	2.23	1.47	3.03	1.8
IA	0.31	0.29	0.34	0.65	0.33	0.53

Table 5. Parameters and indices of S. *obliquus* BGP and *P. cruentum* oils obtained by SCE at 400 bar, 40 °C, and 10% ethanol.

Oil Parameters	*S. obliquus* BGP	*P. cruentum*
	400 Bar, 40 °C, 10% EtOH	400 Bar, 40 °C, 10% EtOH
OX	0.51	0.19
APE	1.21	0.45
BAPE	0.51	0.19
OSI	3.89	3.9
UI	163.2	227.1
h/H	3.22	2.21
IA	0.25	0.46

The corresponding indices for the oils recoverd from both strains were calculated according to [23–25] and are shown in Tables 4 and 5.

$$OX = (0.02 \times C18:1 + C18:2 + 2 \times C18:3)/100 \tag{1}$$

$$APE = (2 \times (C18:1 + C18:2 + C18:3))/100 \tag{2}$$

$$BAPE = (C18:2 + 2 \times C18:3)/100 \tag{3}$$

$$OSI = 3.91 - 0.045 \times BAPE \tag{4}$$

A high OSI value of an oil indicates that it is stable and can be used for the production of biodiesel without the addition of any antioxidants to enhance the stability during the OSI period, up to which the oil/biodiesel quality would remain unchanged, and the biodiesel must be entirely utilized for engine operation. Following the ASTM standard, oils can be classified as best, moderate, and poor. The best oils are characterized with an OSI $\geq$ 3 h since they are considered relatively more stable and do not require the addition of antioxidants for stabilization, i.e., their biodiesel is expected to be entirely utilized before OSI expires and degradation begins.

According to Kumar and Sharma who compared the above indices for different microalgal oils, the best-performing oil was the *S. oblicuus* oil (OSI = 3.87), while the OSI indices of other two non-identified Scenedesmus species varied from 1.93 to 2.13 [26].

As demonstrated in Tables 4 and 5, *S. obliquus* BGP and *P. cruentum* oils belonged to the group of best oils since their OSI values were in the range of 3.89–3.9, regardless of the technique employed to recover them.

Tables 4 and 5 also show the indices of unsaturation (UI), which reflect the proportion of FAs with different degrees of unsaturation in the total FA composition of a species.

Consequently, unlike the PUFA: SFA ratio, which reflects the impact of highly unsaturated FAs, UI highlights the influence of acids with a low degree of unsaturation like MUFAs and DUFAs. In view of this, UI is usually applied to characterize the composition of macroalgal FAs and used as a reference whether the respective macroalgae may be used as alternative sources of high-quality PUFA instead of fish or fish oil.

UI was calculated according to [25]:

$$UI = 1 \times (\% \text{ monoenoics}) + 2 \times (\% \text{ dienoics}) + 3 \times (\% \text{ trienoics})$$
$$+ 4 \times (\% \text{ tetraenoics}) + 5 \times (\% \text{ pentaenoics}) + 6 \times (\% \text{ hexaenoics}) \tag{5}$$

As discussed by Chen and Liu [27], the UI value of seaweeds varies widely—from 45 to 368.68. Colombo et al. [28] used the UI to compare macroalgae caught in the cold waters of Canada with species from the temperate waters of South China. They showed that the UI values of the warm water aglae were in the range from 54 to 151, while for the Canadian algae, the UI varied in the range from 174 to 245, respectively [28]. Our results corroborate those findings as the oils of the deep sea species *P. cruentum* recovered, regardless of the techniques applied, had much higher UI indices when compared to those of *S. obliquus* BGP.

Another important index is the Σ hypercholesterolemic fatty acids/Σ hypercholesterolemic fatty acids ratio (h/H), which is related to cholesterol metabolism. Nutritionally, higher h/H values are considered more beneficial for human health. Chen and Liu [27] mentioned in their recent review that initially the h/H index was introduced by Santos-Silva et al. [29] to assess the effect of fatty acids composition of lamb meat on cholesterol, and they pointed out that compared with the PUFA:SFA, the h/H ratio might better reflect that effect on cardiovascular disease [27].

Later, since there was no C12:0 detected in the lamb meat, in order to characterize the relationship between hypocholesterolemic fatty acid (cis-C18:1 and PUFA) and hypercholesterolemic fatty acid [27], the originally proposed formula was extended as follows:

$$h/H = (cis\text{-}C18{:}1 + \Sigma PUFA)/(C12{:}0 + C14{:}0 + C16{:}0) \tag{6}$$

In the present study, we applied the formula advocated by Fernandez et al. [30]:

$$h/H = [(\Sigma\,(C18{:}1\,n\text{-}9, C18{:}1\,n\text{-}7, C18{:}2\,n\text{-}6, C18{:}3\,n\text{-}6, C18{:}3\,n\text{-}3, C20{:}3\,n\text{-}6,$$
$$C20{:}4\,n\text{-}6, C20{:}5\,n\text{-}3, C22{:}4\,n\text{-}6, C22{:}5\,n\text{-}3 \text{ and } C22{:}6\,n\text{-}3)/\Sigma\,(C14{:}0 \text{ and } C16{:}0) \tag{7}$$

The results obtained are presented in Tables 4 and 5, respectively.

The highest h/H was calculated for the *S. obliquus* BGP oil recovered by *n*-hexane in the first step of the two-step Soxhlet. The value was higher but still commensurable with the analogous oil extract of *P. cruentum* (3.24 vs. 3.03). Obviously, the over nine-times-higher quantity of C18:1 (n-9) and over 30-times-higher quantity of C18:3 (n-3) in the *S. obliquus* BGP outperformed the C20:4 present in a substantial amount in the *P. cruentum* but was not detected in the *S. obliquus* BGP at all.

The h/H values of the oil extracts of *S. obliquus* BGP and *P. cruentum* recovered by scCO$_2$ + 10% ethanol were 3.22 and 2.21, respectively. The *S. obliquus* BGP h/H index was slightly lower than that of the *n*-hexane Soxhlet extraction, while for the *P. cruentum*, the value was decreased more substantially, owing to the higher content of SFA (C16:0) and the lower level of MUFA (C18:1 (n-9)) and DUFA (C18:2 (n-6)) in the SFE extracts.

It should be noted that, as reported by Chen and Liu [27], one of the highest h/H indices was calculated for *Camelina sativa* oil (11.2–15.0). For red seaweed, h/H was 4.22; for shellfish, it ranged between 1.9 and 4.75, except for *Loxechinus albus*, for which h/H = 0.21, a value lower than those of the other species. That could be a result of the fact that the main food source of *Loxechinus albus* is algae. For fish, the h/H is in the range from 0.87 to about 4.83, etc.

Hence, in order to perform a correct comparison of the h/H values, the method advocated in [27] was applied, and the new values calculated for *S. obliquus* BGP and

P. cruentum oils are displayed in Table 6. In all cases examined, the h/H indices of both microalgal species oils were higher than or in the worst case commensurable with the range of h/H indices calculated for the various fish species. The influence of the extraction technique applied was clearly demonstrated for both species—the highest h/H indices were calculated for oils recovered by scCO$_2$ + ethanol, with the *S. obliquus* BGP oil being the best performer.

Table 6. h/H indices of *S. obliquus* BGP and *P. cruentum* oils calculated.

Species	Extraction Method	h/H = (*cis*-C18:1 + ΣPUFA)/(C12:0 + C14:0+ C16:0)
S. obliquus BGP	Soxhlet 96% ethanol	2.27
	Soxhlet *n*-hexane (first-step)	2.71
	Soxhlet 96% ethanol second step after *n*-hexane	1.96
	400 bar, 60 °C, 10% ethanol	3.23
P. cruentum	Soxhlet 96% ethanol	1.07
	Soxhlet *n*-hexane (first-step)	2.13
	Soxhlet 96% ethanol second step after *n*-hexane	1.37
	400 bar, 40 °C, 10% ethanol	1.58

As noted above, a reliable comparison with analogous species was difficult as different authors have used different variants of the h/H formula. Moreover, it is not always clearly stated how the oil was recovered (method, solvents, etc.). Thus, in order to compare our data with the results presented in the seminal paper by Matos et al. [14], the h/H of the *P. cruentum* extracts obtained by the different techniques were yet once again recalculated applying the method used in the reference.

The following values for h/H were obtained: 1.38, 2.88, 1.73, and 2.1, respectively, where the order of techniques followed that of Table 4, with the scCO$_2$ + ethanol extraction being the last in the row. Matos et al. [14] used the Soxhlet extraction method with petroleum ether applied after acid digestion with 4.0 N HCl for 6 h, and the h/H index for *P. cruentum* was calculated to be 1.9. Our h/H values were of the same order of magnitude, e.g., the oil recovered by the first step Soxhlet *n*-hexane had the best nutritional quality index equal to 2.88, followed closely by that obtained by scCO$_2$ + 10% ethanol at 40 °C and 400 bar with h/H = 2.1.

Finally, Tables 4 and 5 also display the index of atherogenicity (IA), which characterizes the atherogenic potential of fatty acids. As discussed in [27], IA is a more adequate indicator when compared to the PUFA/SFA ratio, which is too general and unsuitable for assessing the atherogenicity of food products. The lowest IA = 0.25 was calculated for the *S. obliquus* oil recovered by scCO$_2$ + 10% ethanol, which was lower than the IA indices of the majority of red and brown seaweeds examined. At the same time, it was either commensurable or slightly higher than those of the majority of the green seaweeds reported in the review.

The only exception being Ulva sp., for which an IA = 0.08 was calculated. With regard to P. cruentum, the IA lowest value = 0.33 was calculated for the oil recovered by Soxhlet n-hexane, while for the oil obtained by scCO$_2$ + 10% ethanol, IA = 0.46, which, though lower, is still commensurable with the IA value reported for *P. cruentum* oil in [14].

2.2.2. Analysis and Quantification of Antioxidants

A knowledge of the oils' composition regarding the presence of phenolics was obtained from the LC-MS/MS analyses of the extracts. The results are shown in Table 7.

The presence of constituents of some important groups of antioxidants was tested, namely hydroxycinnamic, caffeoylquinic, and hydroxybenzoic acids, as well as representatives of four subgroups of flavonoids. The quali- and quantification of the *S. obliquus* BGP and *P. cruentum* extracts showed that the oils of both species were not very rich in antioxidants. Furthermore, the quantities of the bioactives identified varied sometimes by orders of magnitude, which demonstrates not only the influence of the specific genus but also the impact of the recovery methods, operating conditions, and solvents applied. For example, the quantities of hydroxycinnamic and caffeoylquinic acid derivatives in *S. obliquus* BGP and in *P. cruentum* were commensurable, with the exception of ferulic and cinnamic acids, which were more abundant in P. cruentum. With regard to hydroxybenzoic acid derivatives, the picture was different. The highest amount in that group was detected for vanillic acid in the *S. obliquus* oil recovered by second step Soxhlet, followed by 3-OH-4-methoxybenzoic acid. Actually, the quantity of the former was the highest found in all oil tests. The respective amounts of those acids in *P. cruentum* were considerably much lower.

Table 7. LC–MS/MS analysis of phenolic compounds in *S. obliquus* BGP and *P. cruentum* extracts obtained by Soxhlet ethanol and SFE with co-solvent.

Compound Identified	S. obliquus BGP			P. cruentum		
	Soxhlet 96% Ethanol	Soxhlet 96% Ethanol Second Step after *n*-Hexane	400 Bar, 60 °C, 10% Ethanol	Soxhlet 96% Ethanol	Soxhlet 96% Ethanol Second Step after *n*-Hexane	400 Bar, 40 °C, 10% Ethanol
ng/mg						
Phenolic acids						
Hydroxycinnamic and caffeoylquinic acid derivatives						
o-coumaric acid	4.171	4.143	3.931	3.470	4.242	4.340
p-coumaric acid	0.242	0.148	0.040	0.550	0.061	0.262
m-coumaric acid	4.439	4.748	4.571	5.216	2.269	4.613
ferulic acid	0.221	0.458	0.412	17.030	2.164	1.246
cinnamic acid	2.934	1.633	3.275	0.559	9.769	2.026
3-*O*-caffeoylquinic (chlorogenic) acid	0.085	0.415	0.393	2.245	0.764	0.546
Hydroxybenzoic acid derivatives						
gallic acid	0.014	0.072	0.046	0.026	0.041	0.012
vanillic acid	2.880	167.174	14.911	n.d.	9.114	7.287
ellagic acid	3.206	0.471	0.839	0.281	0.394	0.530
gentisic acid	0.226	1.778	0.578	2.519	0.091	0.093
protocatechinic acid	0.602	0.453	0.118	2.265	0.111	0.192
o-hydroxybenzoic acid	4.964	5.433	4.459	10.913	3.254	5.702
m-hydroxybenzoic acid	0.060	1.279	1.786	0.792	0.463	0.841
syringic acid	0.246	0.063	1.130	2.696	1.122	0.210
3-OH-4-methoxybenzoic acid	2.773	94.075	12.126	2.011	8.290	2.744
Flavonoids						
Flavonols						
quercetin	0.603	0.180	0.724	0.084	0.093	0.160
myrecitrin	0.123	0.012	0.126	0.016	0.045	0.016
myrecitin	0.483	0.356	0.435	0.041	0.266	0.799
rutin	2.700	4.597	4.925	4.025	5.867	4.559
resveratrol	0.174	0.202	0.220	0.232	0.190	0.091
kaempferol	0.607	0.101	1.986	0.163	0.053	0.120
kaempferol-3-*O*-glycoside	31.063	13.782	99.507	22.810	0.974	21.391
fisetin	0.482	0.066	0.105	0.204	0.015	0.024
Flavones						
luteolin	0.648	0.103	0.199	0.100	0.051	0.105
apigenin	0.443	0.049	0.910	0.081	0.039	0.067
Flavan-3-ols						
catechin	n.d.	0.066	0.042	0.052	0.398	0.156
epicatechin	0.052	0.118	0.043	0.059	3.590	0.108
Flavanones						
hisperidin	0.123	0.012	0.003	0.003	0.001	0.004
naringenin	0.166	0.006	0.086	0.040	0.002	0.003

Relative standard deviation (RSD) = 2.3%.

With regard to flavonoids, the best repersented among all subgroups was kaempferol-3-O-glycoside. Its amounts in the oils of both species were of similar magnitude, the only exception being the *S. obliquus* BGP oil recovered by scCO$_2$ + 10% ethanol, the quantity of which was the second highest among all antioxidants detected in the oils of both species examined.

2.2.3. Total Phenolic Content (TPC) and Antioxidant Activity (AA)

The TPC and AA of the oils of *S. obliquus* BGP and *P. cruentum* are shown in Table 8. TPC is just a quantitative indicator of the total phenols in the oils and does not provide any information about their particular composition. Still, it showed some trends—for example, examining the results for the Soxhlet extractions of *S. obliquus* BGP, it is clear that the largest amount of polyphenols was extracted in the single-step extraction with ethanol (409.59 µg/mg), followed by scCO$_2$ + 10% ethanol, with the amount in the oil recovered by the second-step Soxhlet being the lowest. This trend was expected since, in general, phenolics are better recovered by polar solvents.

Table 8. Total phenolic content, antioxidant activity and IC50 of the extracts obtained from *S. obliquus* BGP and *P. cruentum* by different extraction techniques.

Species	Extraction Method	TPC	DPPH	
		Quercetin eq. [µg/mg]	Trolox eq. [mM]	IC50 mg Extract
S. obliquus BGP	Soxhlet 96% ethanol	409.59 ± 7.42	1.81	2.65
	Soxhlet 96% ethanol second step after *n*-hexane	155.85 ± 0.40	1.58	1.42
	400 bar, 60 °C, 10% ethanol	255.73 ± 6.33	1.98	3.52
P. cruentum	Soxhlet 96% ethanol	134.40 ± 0.80	1.38	0.31
	Soxhlet 96% ethanol second step after *n*-hexane	162.45 ± 3.40	1.09	-
	400 bar, 40 °C, 10% ethanol	182.09 ± 8.08	1.89	2.72

Relative standard deviation (RSD): RSD$_{DPPH}$ = ±3.01%; RSD$_{IC50}$ = ±1.6%.

The DPPH free radical scavenging activity for the species exhibited a somewhat different pattern; namely, the highest value was measured for the oil recovered by scCO$_2$ + ethanol. It was, however, commensurable with that for Soxhlet ethanol. The best (lowest) IC50 value = 1.42 was determined for the oil obtained by the second-step Soxhlet. That can be explained, to a certain extent, by the presence in relatively higher amounts of potent antioxidants like vanilic and 3-OH-4-methoxybenzoic acids.

TPC values calculated for the *P. cruentum* oils were not only lower but the trend of the influence of the extraction methods on the TPC was also different from that for S. obliquus. Thus, the oil obtained by scCO$_2$ with 10% ethanol had the highest TPC. Furthermore, in contrast to S. obliquus, the amount of polyphenols in the oil recovered by the second-step Soxhlet ethanol of the spent matrix was higher than the amount registered in the Soxhlet ethanol. On the other hand, the activity toward the DPPH radical for the species was analogous to the one exhibited by S. obliquus. With regard to IC50, the lowest value was calculated for the Soxhlet ethanol extract. The latter was the best among all IC50 values calculated.

Comparing the TPC of the two species, the data clearly show that oils of *S. obliquus* were richer in polyphenols. For both species, the SFE oil extracts had the highest scavenging ability against DPPH free radicals.

In [9], the TPC of *S. obliquus* oil obtained by neat scCO$_2$ and scCO$_2$ + 10% ethanol were reported. However, a direct comparison cannot be made since the operating conditions are not only different (e.g., lower pressures applied) but also the TPC is represented as mgGA/g$_{extr}$.

The in-depth comprehensive analyses performed on the two species' oils showed the impact of the techniques' specificity (operational parameters, nature of solvents, etc.) on their yield and composition. However, the picture of the latter is too complex and reflects not only the effect of the above but also the major influence of the specifics of the strain and the genera, intertwined with other factors like growing conditions, etc., which were not a topic of the present research.

As difficult as it is to generalize our findings, still the following observations and conclusions are valid:

i.   Both *S. obliquus* BGP and *P. cruentum* oils are "best oils" and can be used for biodiesel production without any antioxidants.

ii.   *P. cruentum* oils, regardless of the extraction technique used, can enrich functional foods since they have high levels of PUFAs, which are two- to three-times higher than those found in *S. obliquus* BGP oils. *P. cruentum* oil can also serve as a substitute for fish oil because of the high amounts of AA and EPA synthesized.

iii.   *S. obliquus* BGP and *P. cruentum* oils have commensurable high values of h/H that are in the upper range of the corresponding indices of shellfish and fish. These oils can be used as additives in human nutrition to prevent cardiovascular disease, particularly for people with high blood pressure.

iv.   *S. obliquus* BGP oil obtained through $scCO_2$ extraction with 10% ethanol exhibited the lowest IA of 0.25, which is lower than most red and brown seaweeds as reported in [27]. Therefore, it can be considered a suitable additive to foods or products that can help prevent plaque accumulation and reduce levels of total cholesterol and LDL-C or "bad" cholesterol.

v.   TPC and AA analysis of two strains' oils show the impact of genera and extraction methods. *S. obliquus* BGP ethanol oil has the highest TPC, over three-times higher than *P. cruentum*. The lowest IC50 is calculated for *P. cruentum* ethanol oil—over 4.5-times lower than *S. obliquus* BGP. The best AA performers are oils from both species obtained by SFE.

3. Materials and Methods

3.1. Microalgal Strains

Scenedesmus obliquus BGP is a previously not known strain of the genus *Scenedesmus*. It was newly isolated from a rainwater puddle in Sofia, Bulgaria, at an average temperature of 20 °C. The morphological analysis identified the new strain as *Scenedesmus obliquus* (Turpin) Kutzing [8], and the strain was subsequently named *S. obliquus* BGP. A description of the taxonomic and molecular analyses performed was outlined in great detail in a previous contribution by some of the present authors [8] and will not be presented here. Furthermore, it was demonstrated that the strain showed tolerance toward the influence of the most important environmental factors such as light intensity, temperature, and composition of the nutrient medium, and the optimum cultivation conditions were determined.

Lyophilized biomasses of both *S. obliquus* BGP and *P. cruentum* were generously donated to us by the Laboratory of Experimental algology, Institute of Plant Physiology and Genetics, Bulgarian Academy of Sciences.

The monoalgal, non-axenic cultures of red microalga *Porphyridium cruentum* (AG.) (Rhodophyta), strain VISCHER 1935/107, acquired by the Laboratory of Experimental algology, Institute of Plant Physiology and Genetics, Bulgarian Academy of Sciences from the culture collection of the Institute of Botany, Třeboň, The Czech Republic, was grown on the modified culture medium as reported in [31].

The lyophilization of *S. obliquus* BGP and *P. cruentum* biomass was performed in a LGA 05 lyophilizer (Janetzki, Leipzig, Germany), as explained in [8], and then stored in dry–dark conditions prior to use.

3.2. Chemicals and Reagents

The chemicals used for the GC–FID analyses were as follows: Supelco 37 Component FAME Mix (CRM47885), reference mixture of fatty acid methyl esters from Sigma-Aldrich (Darmstadt, Germany), toluene (pure, VWR International, Paris, France), sulfuric acid (98%, MerckKGaA, Darmstadt, Germany), sodium chloride (pure, MerkKGaA, Darmstadt, Germany), potassium bicarbonate (pure, VWR International, Paris, France), chloroform (99.8% VWR International, Paris, France), sodium sulfate (pure, Sigma-Aldrich Chemie GmbH, Taufkirchen, Germany), and helium (99.9999%, Air Liquide A/S).

LC MS/MS quantification of polyphenolic compounds used standards enumerated in detail in a previous article [32].

For TPC and antioxidant activity analysis, the following chemicals were used: Folin–Ciocalteu reagent 2 N, sodium carbonate (Merck, Darmstadt, Germany), DPPH (2,2-diphenyl-1-picrylhydrazyl), ATBS, ethanol HPLC grade (Panreac, Barcelona, Spain), gallic acid, Trolox (6-hydroxy-2,5,7,8-tetramethylchroman-2-carboxylic acid) (Sigma-Aldrich, St. Louis, MO, USA), potassium persulfate and absolute ethanol (Neon, Suzano, SP, Brazil), and quercetin dihydrate (Sigma-Aldrich Chemie GmbH, Steinheim, Germany) [32].

The rest of the reagents applied were of the highest purity: methanol $\geq$ 99.9%, ethanol $\geq$ 99.8%, and *n*-hexane $\geq$ 99% were purchased from Honeywell Riedel-de-Haen (Seelze, Germany); ethyl acetate $\geq$ 99.5% from JLS-Chemie Handel GmbH (Hannover, Germany); methyl tert-butyl ether $\geq$ 99.8%, and acetonitrile $\geq$ 99.9% from Sigma-Aldrich (Darmstadt, Germany); and bone dry grade CO_2 (99.99% pure; no water, Messer, Sofia, Bulgaria).

3.3. Biochemical Analyses

In our previous contribution [8], the biochemical composition of *S. obliquus* BGP was analyzed, and the values for proteins (24–45%), lipids (23–30%), and carbohydrates (25–28%) were determined. The biochemical composition of the *P. cruentum* strain VISCHER 1935/107 was reported in [31] to be: proteins (27–38%), lipids (9–12%), and carbohydrates (40–57%). The protocols of the analyses are presented in detail in [8] and will not be reproduced here.

3.4. Microalgal Extracts Recovery

3.4.1. Preliminary Preparation of the Material

The lyophilized microalgal samples were firstly crushed additionally to a mean particle diameter (dp) of 0.5 mm and then subjected to ultrasonication in an ultrasonic disintegrator UD 20 (Bandeline electronic, Berlin, Germany) with an ultrasonic field of 35 kHz. After each sonication, the sample treated was examined by a microscope to establish the level of cell disintegration. The procedure was explained in detail in our previous work [8].

3.4.2. Soxhlet Extraction

The protocol for performing the Soxhlet extractions was reported in detail in our previous work [32]. In brief, the Soxhlet apparatus used was ISOLAB NS29/32 + 34/35 (Merck KGaA, Darmstadt, Germany). In the one-step Soxhlet, ethanol was used, while in the two-step, extraction with *n*-hexane was first performed. Subsequently, the residual matrix was subjected to an extraction with ethanol.

In all experiments, the extraction cartridge was filled with 3.0 $\pm$ 0.1 g sonicated microalgal biomass (*S. obliquus* BGP or *P. cruentum*). The extraction was performed until a complete discoloration of the solvent was observed. The solvent from the liquid extract was evaporated under vacuum using Hei-VAP Rotary Evaporator (Heidolph Instruments GmbH&Co. KG, Schwabach, Germany). Next, the resulting dry extract was dried at 60 $\pm$ 2.0 °C to a constant weight in an air circulation oven, and the yield was calculated according to:

$$\text{Yield}(\%) = \frac{\text{mass of extract (g)}}{\text{mass of sample (g)}} \times 100 \tag{8}$$

The extracts were placed in glass vials and kept at 4 °C until analysis. Experiments were performed in triplicates and total extraction yield was expressed as the mean $\pm$ standard deviation. The results are presented in Table 1.

3.4.3. Supercritical Fluid Extraction

In our study, *S. obliquus* BGP was extracted with neat $scCO_2$ at T = (40, 50, and 60) °C and p = (400 and 500) bar. The $scCO_2$ flow rate was 1 L min^{-1}. SCE with a co-solvent ethanol was applied to both strains. The experiments were carried out in a flow apparatus (SFT-110-XW, Supercritical Fluid Technologies Inc., Newark, DE, USA), with two parallel 50 cm^3 internal volume extractors made from stainless steel tubing (7 cm long, internal diameter 3.02 cm) and temperature controllers for extraction vessels and restrictor valves, which could be adjusted up to 120 °C. The CO_2 pressure was guaranteed by a SFT Nex10 SCF pump actuate from a compressor model HYAC50-25, Hyundai, South Korea, while in experiments with a co-solvent an additional pump (LL-Class, USA) from SFT, Inc., Newark, DE, USA is used. A full description of the equipment is given in [32] and will not be elaborated further here. In all experiments, about 5 g dry sample of the respective species biomass was placed in the processing vessel.

3.5. Characterization and Quantification of the Extracts

The fatty acid composition of certain extracts of *S. obliquus* BGP and *P. cruentum* was determined by GC-FID, the methodology of which was described in detail in [33]. The identification and quantification of the different groups of phenolics by LC-MS/MS was presented exhaustively in its entirety in our earlier works [32].

In addition, the total phenolic content (TPC) of the extracts analyzed was determined using the Folin–Ciocalteu reagent, while the DPPH assay determined the free radical scavenging activity of the samples. The corresponding methodology of the three methods was presented fully in our previously published contribution [32].

4. Conclusions

Oils of two algae strains belonging to different genera—the recently isolated Bulgarian strain *S. obliquus* BGP and *P. cruentum*—were extracted using both a conventional method (Soxhlet with hexane and ethanol) and an advanced green technology (scCO$_2$ with and without a co-solvent ethanol). The use of scCO$_2$ results in lower yields compared to Soxhlet extraction, even with the addition of ethanol as a co-solvent, which is consistent with previous findings in the literature.

The quali- and quantitative analysis of the oils through GC-Fid and LC-MS/MS, as well as the determination of parameters and indices that made it possible to outline their viability, demonstrated the higher potential of *P. cruentum* algae as a sustainable source of bioactive compounds with possible applications in the food and/or pharmaceutical industry when compared to the *S. obliquus* BGP.

The fatty acid profiles of the oils of the two species differed significantly. In the case of *S. obliquus* BGP, the percentage of different C18 fatty acids (both saturated and unsaturated) ranged from 56.3% to 64.4%, a percentage much higher than that of *P. cruentum*, for which those values were between 19.3% and 28.3%. On the other hand, the latter contained up to 43% of C20:4 and C20:5 fatty acids, which were not detected in the *S. obliquus* BGP. The OSI values calculated for both *S. obliquus* BGP and *P. cruentum* oils position them among the best oils for the production of biodiesel.

Analysis of the fatty acids and polyphenols in *P. cruentum* oil indicates its superior potential for food and pharmaceutical applications compared to the green algae. However, the h/H indices calculated for the oils of both species show that they have the capacity to serve as additives to human nutrition. Hence, the algae of both species exhibit promising properties and could be exploited in the future in a one-feed, multi-product biorefinery with a wide variety of applications.

Author Contributions: Conceptualization, J.A.P.C., R.P.S., F.V.T., S.S.B. and D.S.Y.; methodology, R.P.S., F.V.T., S.S.B. and D.S.Y.; validation, R.P.S., J.A.P.C., F.V.T., S.S.B. and D.S.Y.; formal analysis, S.S.B., F.V.T., D.S.Y., J.A.P.C. and R.P.S.; investigation, F.V.T., S.S.B., R.P.S. and D.S.Y.; writing—original draft preparation, R.P.S., F.V.T., S.S.B., J.A.P.C. and D.S.Y.; writing—review and editing, R.P.S., J.A.P.C., F.V.T., S.S.B. and D.S.Y.; supervision, R.P.S., J.A.P.C., S.S.B., F.V.T. and D.S.Y.; project administration, D.S.Y., R.P.S., F.V.T., S.S.B. and J.A.P.C.; funding acquisition, D.S.Y., R.P.S., F.V.T., S.S.B. and J.A.P.C. All authors have read and agreed to the published version of the manuscript.

Funding: This project has received funding from The Bulgarian National Science Fund, Grant number KP-06-OPR 04/1, and the European Union's Horizon 2020 research and innovation programme under the Marie Sklodowska-Curie grant agreement No 778168.

Institutional Review Board Statement: Not applicable.

Informed Consent Statement: Not applicable.

Data Availability Statement: Data are contained within the article.

Acknowledgments: The authors would like to thank V. Lozanov and members of "Analysis and Synthesis of Biologically Active Substances" laboratory for their contribution to a part of the analyses.

Conflicts of Interest: The authors declare no conflicts of interest. The funders had no role in the design of the study; in the collection, analyses, or interpretation of the data; in the writing of the manuscript; or in the decision to publish the results.

References

1. Gaignard, C.; Gargouch, N.; Dubessay, P.; Delattre, C.; Pierre, G.; Laroche, C.; Fendri, I.; Abdelkafi, S.; Michaud, P. New horizons in culture and valorization of red microalgae. *Biotechnol. Adv.* **2019**, *37*, 193–222. [CrossRef]
2. Mobin, S.; Firoz, A. Some promising microalgal species for commercial applications: A review. *Energy Procedia* **2017**, *110*, 510–517. [CrossRef]
3. Koyande, A.; Show, P.; Guo, R.; Tang, B.; Ogino, C.; Chang, J. Bio-processing of algal bio-refinery: A review on current advances and future perspectives. *Bioengineered* **2019**, *10*, 574–592. [CrossRef]
4. Tzima, S.; Georgiopoulou, I.; Louli, V.; Magoulas, K. Recent advances in supercritical CO_2 extraction of pigments, lipids and bioactive compounds from microalgae. *Molecules* **2023**, *28*, 1410. [CrossRef]
5. Gallego, R.; Martínez, M.; Cifuentes, A.; Ibáñez, E.; Herrero, M. Development of a green downstream process for the valorization of Porphyridium cruentum biomass. *Molecules* **2019**, *24*, 1564. [CrossRef]
6. Li, T.; Xu, J.; Wang, W.; Chen, Z.; Li, C.; Wu, H.; Wu, H.; Xiang, W. A novel three-step extraction strategy for high-value products from red algae Porphyridium purpureum. *Foods* **2021**, *10*, 2164. [CrossRef]
7. Pignolet, O.; Jubeau, S.; Vaca-Garcia, C.; Michaud, P. Highly valuable microalgae: Biochemical and topological aspects. *J. Ind. Microbiol. Biotechnol.* **2013**, *40*, 781–796. [CrossRef]
8. Vasileva, I.; Boyadzhieva, S.; Kalotova, G.; Ivanova, J.; Kabaivanova, L.; Naydenova, G.; Jordanova, M.; Yankov, D.; Stateva, R. A new Bulgarian strain of *Scenedesmus* sp.—Identification, growth, biochemical composition, and oil recovery. *Bulg. Chem. Commun.* **2021**, *53*, 105–116. [CrossRef]
9. Georgiopoulou, I.; Louli, V.; Magoulas, K. Comparative study of conventional, microwave-assisted and supercritical fluid extraction of bioactive compounds from microalgae: The case of Scenedesmus obliquus. *Separations* **2023**, *10*, 290. [CrossRef]
10. Silva, M.; Martins, M.; Leite, M.; Miliao, G.; Coimbra, J. Microalga Scenedesmus obliquus: Extraction of bioactive compounds and antioxidant activity. *Rev. Cienc. Agronom.* **2021**, *52*. [CrossRef]
11. Khatoon, H.; Rahman, N.; Suleiman, S.; Banerjee, S.; Abol-Munafi, A. Growth and proximate composition of Scenedesmus obliquus and Selenastrum bibraianum cultured in different media and condition. *Proc. Natl. Acad. Sci. India Sect. B Biol. Sci.* **2019**, *89*, 251–257. [CrossRef]
12. Ardiles, P.; Cerezal-Mezquita, P.; Salinas-Fuentes, F.; Órdenes, D.; Renato, G.; Ruiz-Domínguez, M.C. Biochemical Composition and Phycoerythrin Extraction from Red Microalgae: A Comparative Study Using Green Extraction Technologies. *Processes* **2020**, *8*, 1628. [CrossRef]
13. Becker, E. Micro-algae as a source of protein. *Biotechnol. Adv.* **2007**, *25*, 207–210. [CrossRef]
14. Matos, Â.P.; Feller, R.; Moecke, E.H.S.; de Oliveira, J.V.; Junior, A.F.; Derner, R.B.; Sant'Anna, E.S. Chemical characterization of six microalgae with potential utility for food application. *J. Am. Oil Chem. Soc.* **2016**, *93*, 963–972. [CrossRef]
15. Callejón, M.; Medina, A.; Sánchez, M.; Moreno, P.; López, E.; Cerdán, L.; Molina-Grima, E. Supercritical fluid extraction and pressurized liquid extraction processes applied to eicosapentaenoic acid-rich polar lipid recovery from the microalga Nannochloropsis sp. *Algal Res.* **2022**, *61*, 102586. [CrossRef]
16. Gilbert-Lopez, B.; Mendiola, J.; Van Den Broek, L.; Houweling-Tan, B.; Sijtsma, L.; Cifuentes, A.; Herrero, M.; Ibáñez, E. Green compressed fluid technologies for downstream processing of Scenedesmus obliquus in a biorefinery approach. *Algal Res.* **2017**, *24*, 111–121. [CrossRef]
17. Guedes, A.; Giao, M.; Matias, A.; Nunes, A.; Pintado, M.; Duarte, C.; Malcata, F. Supercritical fluid extraction of carotenoids and chlorophylls a, b and c, from a wild strain of Scenedesmus obliquus for use in food processing. *J. Food Eng.* **2013**, *116*, 478–482. [CrossRef]
18. Raposo, M.; Morais, A.; Morais, R. Influence of sulphate on the composition and antibacterial and antiviral properties of the exopolysaccharide from Porphyridium cruentum. *Life Sci.* **2014**, *101*, 56–63. [CrossRef]
19. Feller, R.; Matos, A.; Mazzutti, S.; Moecke, E.; Tres, M.; Derner, R.; Oliveira, J.; Junior, A. Polyunsaturated ω-3 and ω-6 fatty acids, total carotenoids and antioxidant activity of three marine microalgae extracts obtained by supercritical CO2 and subcritical n-butane. *J. Supercrit. Fluids* **2018**, *133*, 437–443. [CrossRef]

20. Yang, Y.; Xia, Y.; Zhang, B.; Li, D.; Yan, J.; Yang, J.; Sun, J.; Cao, H.; Wang, Y.; Zhang, F. Effects of different n-6/n-3 polyunsaturated fatty acids ratios on lipid metabolism in patients with hyperlipidemia: A randomized controlled clinical trial. *Front. Nutr.* **2023**, *10*, 1166702. [CrossRef]

21. WHO. *Diet, Nutrition and the Prevention of Chronic Diseases: Report of a Joint WHO/FAO Expert Consultation*; WHO: Geneva, Switzerland, 2003; pp. 87–88.

22. Ander, B.; Dupasquier, C.; Prociuk, M.; Pierce, G.; Faha, F. Polyunsaturated fatty acids and their effects on cardiovascular disease. *Exp. Clin. Cardiol.* **2003**, *8*, 164–172.

23. Grover, S.; Kumari, P.; Kumar, A.; Soni, A.; Sehgal, S.; Sharma, V. Preparation and quality evaluation of different oil blends. *Lett. Appl. NanoBioScience* **2021**, *10*, 2126–2137.

24. Pinto, T.I.; Coelho, J.A.; Pires, B.I.; Neng, N.R.; Nogueira, J.M.; Bordado, J.C.; Sardinha, J.P. Supercritical carbon dioxide extraction, antioxidant activity, and fatty acid composition of bran oil from rice varieties cultivated in Portugal. *Separations* **2021**, *8*, 115. [CrossRef]

25. Knothe, G.; Dunn, R.O. Dependence of oil stability index of fatty compounds on their structure and concentration and presence of metals. *J. Am. Oil Chem. Soc.* **2003**, *80*, 1021–1026. [CrossRef]

26. Kumar, M.; Sharma, M. Assessment of potential of oils for biodiesel production. *Renew. Sustain. Energy Rev.* **2015**, *44*, 814–823. [CrossRef]

27. Chen, J.; Liu, H. Nutritional indices for assessing fatty acids: A mini-review. *Int. J. Mol. Sci.* **2020**, *21*, 5695. [CrossRef]

28. Colombo, M.L.; Giavarini, R.P.F.; De Angelis, L.; Galli, C.; Bolis, C. Marine macroalgae as sources of polyunsaturated fatty acids. *Plant Foods Hum. Nutr.* **2006**, *61*, 67–72. [CrossRef]

29. Santos-Silva, J.; Bessa, R.; Santos-Silva, F. Effect of genotype, feeding system and slaughter weight on the quality of light lambs: II. Fatty acid composition of meat. *Livest. Prod. Sci.* **2002**, *77*, 187–194. [CrossRef]

30. Fernández, M.; Ordóñez, J.; Cambero, I.; Santos, C.; Pin, C.; Hoz, L. Fatty acid compositions of selected varieties of Spanish dry ham related to their nutritional implications. *Food Chem.* **2007**, *101*, 107–112. [CrossRef]

31. Vasileva, I.; Ivanova, J. Biochemical profile of green and red algae–A key for understanding their potential application as food additives. *TJS* **2019**, *1*, 1–7. [CrossRef]

32. Boyadzhieva, S.; Coelho, J.A.P.; Errico, M.; Reynel-Avila, H.E.; Yankov, D.S.; Bonilla-Petriciolet, A.; Stateva, R.P. Assessment of Gnaphalium viscosum (Kunth) Valorization Prospects: Sustainable Recovery of Antioxidants by Different Techniques. *Antioxidants* **2022**, *11*, 2495. [CrossRef] [PubMed]

33. Momchilova, S.; Kazakova, A.; Taneva, S.; Aleksieva, K.; Mladenova, R.; Karakirova, Y.; Petkova, Z.; Kamenova-Nacheva, M.; Teneva, D.; Denev, P. Effect of Gamma Irradiation on Fat Content, Fatty Acids, Antioxidants and Oxidative Stability of Almonds, and Electron Paramagnetic Resonance (EPR) Study of Treated Nuts. *Molecules* **2023**, *28*, 1439. [CrossRef] [PubMed]

molecules

Article

Fractionation of Arctic Brown Algae (*Fucus vesiculosus*) Biomass Using 1-Butyl-3-methylimidazolium-Based Ionic Liquids

Artyom V. Belesov *, Daria A. Lvova, Danil I. Falev, Ilya I. Pikovskoi, Anna V. Faleva, Nikolay V. Ul'yanovskii, Anton V. Ladesov and Dmitry S. Kosyakov *

Laboratory of Natural Compound Chemistry and Bioanalytics, Core Facility Center 'Arktika', Northern (Arctic) Federal University, 163002 Arkhangelsk, Russia; darya.lvova2017@yandex.ru (D.A.L.); i.pikovskoj@narfu.ru (I.I.P.); a.bezumova@narfu.ru (A.V.F.); n.ulyanovsky@narfu.ru (N.V.U.)
* Correspondence: a.belesov@narfu.ru (A.V.B.); d.kosyakov@narfu.ru (D.S.K.)

Abstract: Arctic brown algae are considered a promising industrial-scale source of bioactive substances as polysaccharides, polyphenols, and low-molecular secondary metabolites. Conventional technologies for their processing are focused mainly on the isolation of polysaccharides and involve the use of hazardous solvents. In the present study a "green" approach to the fractionation of brown algae biomass based on the dissolution in ionic liquids (ILs) with 1-butil-3-methylimidazolium (bmim) cation with further sequential precipitation of polysaccharides and polyphenols with acetone and water, respectively, is proposed. The effects of IL cation nature, temperature, and treatment duration on the dissolution of bladderwrack (*Fucus vesiculosus*), yields of the fractions, and their chemical composition were studied involving FTIR and NMR spectroscopy, as well as size-exclusion chromatography and monosaccharide analysis. It was shown that the use of bmim acetate ensures almost complete dissolution of plant material after 24 h treatment at 150 °C and separate isolation of the polysaccharide mixture (alginates, cellulose, and fucoidan) and polyphenols (phlorotannins) with the yields of ~40 and ~10%, respectively. The near-quantitative extraction of polyphenolic fraction with the weight-average molecular mass of 10–20 kDa can be achieved even under mild conditions (80–100 °C). Efficient isolation of polysaccharides requires harsh conditions. Higher temperatures contribute to an increase in fucoidan content in the polysaccharide fraction.

Keywords: brown algae; *Fucus vesiculosus*; ionic liquids; 1-butil-3-methylimidazolium; biomass fractionation

Citation: Belesov, A.V.; Lvova, D.A.; Falev, D.I.; Pikovskoi, I.I.; Faleva, A.V.; Ul'yanovskii, N.V.; Ladesov, A.V.; Kosyakov, D.S. Fractionation of Arctic Brown Algae (*Fucus vesiculosus*) Biomass Using 1-Butyl-3-methylimidazolium-Based Ionic Liquids. *Molecules* **2023**, *28*, 7596. https://doi.org/10.3390/molecules28227596

Academic Editors: José A.P. Coelho and Roumiana P. Stateva

Received: 26 October 2023
Revised: 12 November 2023
Accepted: 13 November 2023
Published: 14 November 2023

1. Introduction

Marine micro- and macroalgae produce unique biogenic compounds possessing a wide range of biological activities, which makes them a promising source of pharmaceuticals, food supplements, cosmetics, antimicrobial agents, fertilizers, etc. This is especially true for brown algae, which are widespread in the seas of temperate and high latitudes and produce polyphenolic compounds, polysaccharides (fucoidan, alginic acids), and other valuable secondary metabolites in large quantities [1]. Due to harsh environmental conditions at high latitudes and long daylight hours during the growing season, arctic brown algae are capable of producing the largest amounts of bioactive substances and are of greatest interest as a feedstock for industrial-scale processing (biorefining).

Among the most important constituents of brown algae are polyphenols, which are present both as low molecular weight compounds and, predominantly, as phloroglucinol-based polymers—phlorotannins. The latter group of compounds possesses pronounced antioxidant properties, and has antidiabetic, antiproliferative, anti-HIV, radioprotective, and anti-allergic effects [1–3]. The content of polyphenols in the algal biomass varies depending on the brown algae species, age, and growth location, and can reach 20% of the dry weight (d.w.) [4]. The highest values were observed in the brown algae, namely *Fucus vesiculosus* (15–18%) and *Ascophyllum nodosum* (14–15%), growing in the White and Barents Seas [5]. Thus, isolation and subsequent valorization of polyphenols must be an integral component of brown algae biorefining technologies. Despite this, the technologies traditionally used for the industrial processing of macroalgae are aimed primarily at isolating the polysaccharide component and, in some cases, obtaining pigment (chlorophyll and carotenoids) extracts. Conventional polysaccharide extraction techniques typically involve treating the algal material with various solvents such as hot water or acidic or salt solutions at high temperatures for several hours [6] after preliminary removing of lipids with nonpolar extractants, including toxic halogenated hydrocarbons (chloroform, dichloromethane).

The development of novel approaches to algal biomass processing is hindered by the lack of effective methods for selective extraction of target components from plant raw material and their separation and purification. At the same time, due to growing environmental safety concerns, the developed technologies should rely on the use of "green" processes and solvents. The latter include, first of all, sub- and supercritical water and carbon dioxide, ionic liquids (ILs), and deep eutectic solvents [7–9]. Unusual physical and chemical properties of ILs (ionic structure, extremely low vapor pressure, incombustibility, thermostability) and high dissolving power toward various classes of biomolecules, as well as the possibility of efficient regeneration allow considering ILs as most promising media for algal biomass treatment. Moreover, the well-known ability of ILs to completely dissolve plant tissues [10] through breaking inter- and intramolecular bonds in biopolymers (mainly ether bonds between phenolic units) opens prospects for the efficient fractionation of algal biomass into polysaccharide and aromatic components [11,12] as the basis for biorefining technologies.

Dissolution of algal biomass in ionic liquids has been described previously for various algae species [13–16], including those belonging to brown algae, such as *Sargassum fulbellum*, *Laminaria japonica*, *Undaria pinnatifida*, *Saccharina japonica*. Dialkylimidazolium- and alkylpyridinium-based room temperature ILs (chlorides, tetrafluoroborates, and acetates) were used at temperatures of 100–150 °C and at a treatment duration of up to 6 h. However, the main objective of these studies was the isolation of polysaccharides for subsequent hydrolysis with further biofuel production, as well as lipid extraction with organic solvents without focusing on the obtaining and characterizing polyphenolic fractions.

In the present study, we propose an approach to the fractionation of brown algae biomass involving its dissolution in dialkylimidazolium ionic liquids with further selective antisolvent precipitation of polysaccharides and polyphenols. 1-Butyl-3-methylimidazolium (bmim) methyl sulfate ([bmim]MeSO$_4$), chloride ([bmim]Cl), and acetate ([bmim]OAc), which significantly differs in anion basicity and solvation properties, and have previously proven themselves well in solving problems of wood processing to produce cellulose and lignin [17], were chosen as studied ILs. Bladderwrack (*Fucus vesiculosus*) brown algae species, most widespread and harvested on an industrial scale in the White Sea, was used as an object of research aimed at optimizing conditions for algal biomass fractionation, isolating, and characterizing the resulting polyphenolic and polysaccharide fractions.

2. Results and Discussion

2.1. Solubility of Algal Biomass

The nature of IL anion and temperature of the reaction mixture are the most critical factors determining the completeness and dynamics of dissolving plant biomass. In our experiments, the powdered bladderwrack thallus samples were treated with the three ILs at 80, 100, 120, and 150 °C under constant stirring with the measurements of the solid residue during 24 h (the attained relative standard deviation was 10–15%). The obtained results (Table 1) demonstrate that the substantial dissolution of the plant material (>50%) can be achieved at temperatures ≥ 100 °C. Treatment at 150 °C for 24 h allowed dissolving up to 92% of the bladderwrack biomass when using [bmim]OAc as a solvent.

Table 1. Completeness of algal biomass dissolution in ILs under different treatment conditions.

IL	Amount of Solid Residue (%) at Different Temperatures and Treatment Durations															
	80 °C				100 °C				120 °C				150 °C			
	2 h	4 h	8 h	24 h	2 h	4 h	8 h	24 h	2 h	4 h	8 h	24 h	2 h	4 h	8 h	24 h
[bmim]OAc	78	76	76	72	67	64	52	49	61	54	48	47	36	30	19	8
[bmim]Cl	94	87	82	82	77	72	69	66	68	59	56	42	45	40	39	17
[bmim]MeSO$_4$	84	78	73	64	72	68	63	62	70	66	64	46	68	65	54	44

The comparison of the three ILs revealed two different patterns of their action depending on the treatment temperature. At higher temperatures (120–150 °C) the pronounced dependence on the IL's anion nature is observed, and the studied ILs, according to their effectiveness, can be arranged in the following series, corresponding to a decrease in the basicity of the anion [18,19]: [bmim]OAc > [bmim]Cl > [bmim]MeSO$_4$. The higher basicity of bmim acetate and chloride is expected to break hydrogen bonds in biopolymers and promotes dissolution of polysaccharides due to specific interactions with acetate and chloride anions. Chemical (covalent) interactions with the IL cation, occurring with the participation of a reactive carbene intermediate [20], also may contribute to the dissolution of biopolymers. The formation of carbene occurs due to the deprotonation of bmim upon interaction with highly basic anions and is promoted at elevated temperatures.

Treatment at 80 °C enables the dissolution of up to 30% of the algal biomass, and the solubility in different ILs varies as follows: [bmim]OAc ≈ [bmim]MeSO$_4$ > [bmim]Cl. This effect may be explained by the ability of the most basic (bmim acetate) and acidic (bmim methyl sulfate) ILs to cleave ether and ester bonds in biopolymers. At this temperature, it is likely that the polyphenolic component is preferentially dissolved in ILs, with the proportion of the polysaccharide component increasing as the duration of treatment increases (mainly for [bmim]OAc and [bmim]MeSO$_4$). At a temperature of 100 °C, a transitional system state is observed—the effectiveness of [bmim]Cl increases more significantly than for other ILs and thus the differences in their action are not so pronounced as at higher or lower temperatures. Considering the previous data on the thermostability of the ionic liquids used [20], prolonged heating of the reaction mixture at 120–150 °C cannot be recommended for [bmim]OAc, which is prone to partial decomposition under these conditions. However, samples isolated with [bmim]OAc at 120–150 °C were further used for comparison purposes.

According to the recorded FTIR spectra (Figure 1), the insoluble residue obtained after IL treatment consists of polysaccharides, predominantly alginates.

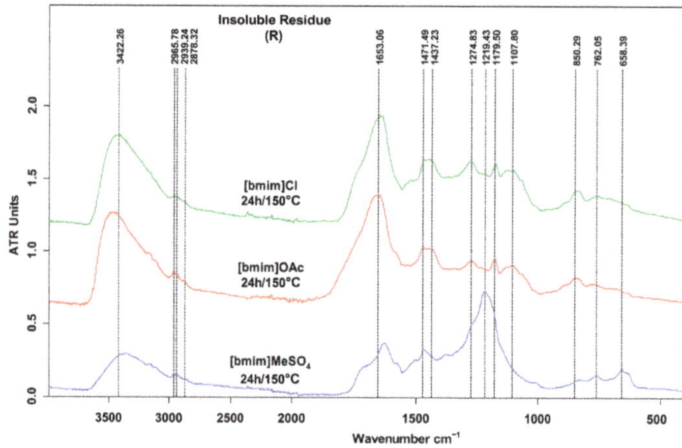

Figure 1. FTIR spectra of insoluble residue after IL treatment of algal biomass.

This is evidenced by intense absorption bands corresponding to carboxylate anions observed at ~1600 and ~1400 cm^{-1} (asymmetric and symmetric C-O-C stretching, respectively) along with a number of peaks in the area of C-O single-bond stretching (1200–1000 cm^{-1}). The absence of signals from aromatic compounds confirms the assumption about the preferential transition of polyphenols into the IL liquid phase.

2.2. Biomass Fractionation

For the selective isolation of polysaccharide and polyphenol components from the obtained solutions of plant material in ILs, a known approach involving sequential precipitation with different antisolvents was used. As regards the latter: Following the example of wood fractionation [17], acetone and water, which can effectively precipitate polysaccharides and phenolic compounds, respectively, were chosen. The advantages of acetone over some other organic solvents are its rather low environmental toxicity and availability as a bio-based solvent, which allows for considering this substance as a green solvent if properly recycled in the technological process [21,22]. Fractions F1 (insoluble in acetone) and F2 (insoluble in water) obtained after 24 h treatment in accordance with Scheme 1 were separated from the liquid phase of the reaction mixture with yields of up to 39% (F1) and 11% (F2) from the initial plant material (Table 2).

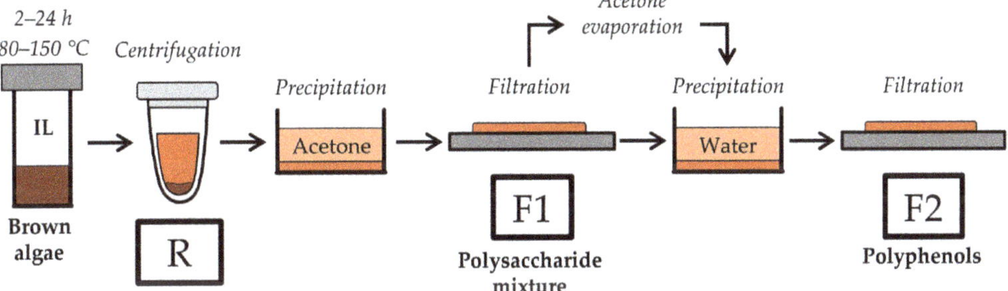

Scheme 1. Fractionation scheme of brown algae using 1-butyl-3-methylimidazolium ionic liquids.

Table 2. Yields of the algal biomass fractions obtained under different conditions.

IL	Fraction	Yield, %			
		80 °C	100 °C	120 °C	150 °C
[bmim]OAc	F1	19 ± 5	30 ± 5	32 ± 5	39 ± 5
	F2	8 ± 2	10 ± 2	11 ± 2	9 ± 2
[bmim]Cl	F1	2 ± 2	4 ± 2	4 ± 2	4 ± 2
	F2	9 ± 2	10 ± 2	10 ± 2	11 ± 2
[bmim]MeSO$_4$	F1	3 ± 2	4 ± 2	5 ± 2	8 ± 2
	F2	9 ± 2	10 ± 2	9 ± 2	9 ± 2

The presented data indicate that all the studied ILs are able to extract almost completely the polyphenolic component already at 80–100 °C. This is evidenced by the yield of the F2 fraction, which corresponds to the polyphenol content of the studied commercially available *Fucus vesiculosus* preparation. In this regard, it's not surprising that changing the processing temperature has no substantial effect on the yield of the F2 fraction. The insignificant decrease in the content of polyphenols isolated from the [bmim]OAc solution obtained at 150 °C may be caused by thermal degradation of phlorotannins or solvolysis of ether bonds in their structure.

Increasing the processing temperature had a more pronounced effect on the yield of polysaccharide fraction (F1). Thus, when going from 80 to 150 °C, at least a twofold gain in the content of isolated polysaccharides was attained, which, however does not correspond to the increase in the dissolved biomass percentage (for example, from 30 to 90% in the case of [bmim]OAc). This may be evidence that the polysaccharide component is partially hydrolyzed under the action of ILs in the presence of residual water. The greatest differences were observed for the samples treated with [bmim]MeSO$_4$ and [bmim]Cl, which is explained by the presence of strong acid anions and thus occurrence of the acidic medium promoting hydrolysis.

The chemical composition and properties of the isolated fractions are the most important factors that determine their suitability for practical applications. Since the highest degree of biomass dissolution was achieved after prolonged heating in ILs, the acetone (F1) and water (F2) insoluble fractions obtained after 24 h plant material treatment at various temperatures were used for further characterization by Fourier transform infrared (FTIR) and nuclear magnetic resonance (NMR) spectroscopy, size-exclusion chromatography, and ligand exchange chromatography for monosaccharide determination.

2.3. Chemical Composition of Fraction F1

FTIR spectroscopy measurements confirmed the predominance of the polysaccharides in fraction F1 (Figure 2).

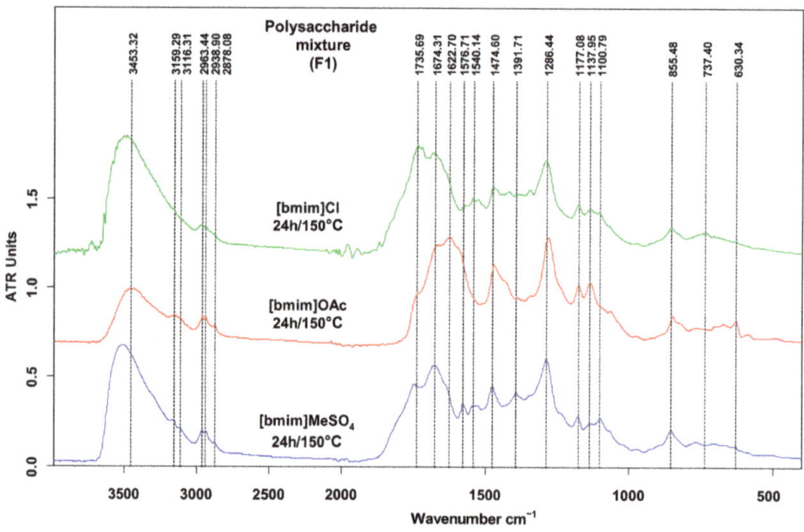

Figure 2. FTIR spectra of F1 fractions (polysaccharides mixture).

All the investigated samples showed broad absorption bands at 3500–3300 cm (hydrogen bonded O-H stretching) and typical signals of aliphatic C-H stretching of 3000–2800 cm^{-1}. The presence of a polysaccharide backbone is confirmed by an intense peak centered at 1025–1010 cm^{-1}, which is attributed to C-O-C in glycoside bridge stretching [23]. Due to overlap with other peaks related to the stretching of the carbon-oxygen single bond in C-O-C, C-O-H, and C-O-S (in fucoidan) structures, the intense broad absorption band is observed in this region (1010–1090 cm^{-1}) [24–26]. Identification of fucoidan in fraction F1 is also based on the combination of specific bands associated with sulfur–oxygen bonds at ~830 (C-O-S stretching) and 1250–1220 (S-O stretching in sulfate esters) [26]. The peak at ~1620 cm^{-1} arises from asymmetric O-C-O stretching in carboxylates (alginic acid). The noticeable peak at ~1750 cm^{-1} (C=O stretching vibrations) may indicate the presence of acetyl moieties [27]. At the same time, the presented FTIR spectra do not contain significant signals of aromatic compounds. This allows for concluding that the fraction F1 comprises polysaccharides (alginic acid, fucoidan), which completely coincides with our expectations.

For more detailed characterization, the polysaccharide fraction was subjected to acid hydrolysis with further monosaccharide determination. The obtained data (Table 3) indicate a decreased proportion of hydrolysable polysaccharides in fraction F1 isolated by using bmim chloride as a solvent. Another peculiarity is an increase in fucose (fucoidan) content with increasing temperature of the IL treatment, which may indicate the preferential isolation of alginic acid. The most selective IL for fucoidan extraction is [bmim]MeSO$_4$, which provides the highest (up to 279 mg g^{-1}) fucose content in the fraction F1 hydrolysates. The high glucose content is considered evidence of the presence of cellulose in noticeable amounts along with alginic acid and fucoidan.

Table 3. Monosaccharide content in the initial plant material and fraction F1 obtained after 24 h IL treatment at different temperatures.

IL	Temperature, °C	Glucose	Xylose	Galactose	Fucose	Mannose	Mannitol	Sum
No IL treatment (initial algae)	-	25	21	12	81	15	23	178
[bmim]OAc	80	38	20	10	31	0	8	107
	100	50	21	9	47	0	3	130
	120	49	20	7	70	0	1	147
	150	49	33	15	180	0	1	277
[bmim]Cl	80	31	20	8	97	0	4	160
	100	42	25	10	83	0	1	161
	120	62	29	9	64	0	1	164
	150	95	15	4	33	0	1	148
[bmim]MeSO$_4$	80	20	32	14	90	0	13	169
	100	18	48	20	191	0	10	286
	120	16	80	29	279	0	1	406
	150	9	71	27	216	0	1	324

The fraction F1 obtained after the [bmim]OAc treatment at the highest temperature (150 °C) exhibits a relatively high hydrolysable polysaccharide content (277 mg g^{-1}) along with the highest yield from the plant material (39%). It is worth noting that a sharp increase in the fucoidan content is observed when the IL treatment temperature increases from 120 to 150 °C, whereas the yield of the fraction increases insignificantly. This means that high treatment temperatures promote the transition of fucoidan into the solution, while alginic acid is extracted by bmim acetate under mild conditions.

The predominance of alginic acid (alginates) in the non-hydrolysable part of the fraction F1 was confirmed by FTIR spectra of the residue after acid hydrolysis (Figure 3).

A distinctive feature of alginates are intense absorption bands at ~1610 and ~1450 cm^{-1}, which are characteristic of the carboxylate anion and related to asymmetric and symmetric C-O stretching, respectively [27]. Weaker bands at 1700–1710 cm^{-1} may be attributed to the molecular form of carboxylic acids (C=O stretching) and considered additional evidence of the alginic acid presence in the studied sample. Other intense bands in the FTIR spectra presented in Figure 3 are related to vibrations of the O-H and C-O groups typical of polysaccharides and mentioned above in the discussion of the F1 fraction spectra (Figure 2).

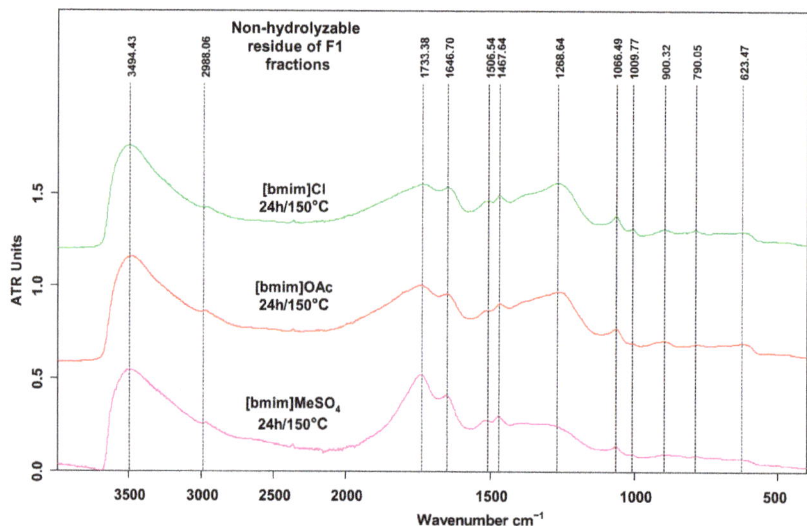

Figure 3. IR spectra of non-hydrolyzable residue of F1 fractions.

2.4. Chemical Composition of Fraction F2

Preparations containing high levels of polyphenolic compounds are of thegreatest interest. FTIR spectra of fraction F2 show the predominance of aromatic components in its composition (Figure 4) responsible for the intense absorption bands at ~1620 and ~1517 cm^{-1} corresponding to C-C stretches in the aromatic ring. Other most intense peaks at 1260–1100 cm^{-1} related to C-O-C stretching allow for attributing them to phlorotannins. Rather intense absorption at ~1450 and 1370 cm^{-1} (asymmetric and symmetric C-H bending in aliphatic structures, respectively) may indicate the presence of methyl or other hydrocarbon moieties as substituents or admixtures originating from other types of lipophilic extractives present in the algal biomass—carotenoids, steroids, lipids, etc. In general, the recorded FTIR spectra correspond well to those published in the literature for algal phlorotannins [28,29]. Considering that the yield of the acetone-soluble fraction is close to the content of polyphenols in the plant material, this gives reason to believe that fraction F2 is almost completely represented by phlorotannins and contains only minor admixture of other lipophilic compounds.

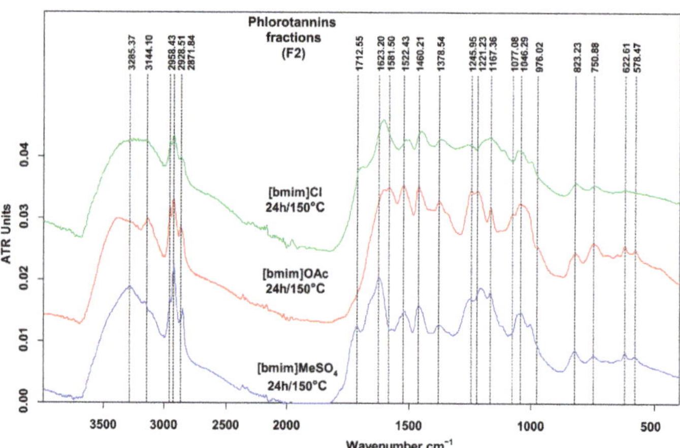

Figure 4. IR spectra of F2 fractions (phlorotannins).

This finding is confirmed by molecular weights and molecular weight distributions of the F2 fractions obtained by size-exclusion chromatography. As can be seen from Table 4, weight-average molecular weight (M_w) values of all the obtained fraction F2 preparations fall into a rather narrow range of 10–20 kDa.

Table 4. Number- (M_n) and weight-average (M_w) molecular weights, and polydispersity indices (PDI) of the fractions F2, obtained after IL treatment at different temperatures.

Ionic Liquid	Temperature, °C	M_n, kDa	M_w, kDa	PDI (M_w/M_n)
[bmim]OAc	80	6.3	10.1	1.6
	100	6.5	10.8	1.6
	120	6.3	12.8	1.7
	150	6.6	13.1	2.0
[bmim]Cl	80	4.1	10.3	2.5
	100	4.4	10.4	2.4
	120	4.5	10.7	2.4
	150	9.8	17.1	1.7
[bmim]MeSO$_4$	80	8.6	14.2	1.7
	100	9.1	16.9	1.9
	120	9.4	17.9	1.9
	150	11.1	20.3	1.8

The relatively low values of polydispersity indices (PDI) indicate the small proportion of low-molecular compounds and, as a consequence, the high homogeneity of the samples. It should be noted that an increase in the treatment temperature, especially in the range 120–150 °C, leads to an increase in the molecular weights of the isolated phlorotannins. This is explained by the side processes of polyphenol condensation at higher temperatures. The latter can be promoted by acidic medium; thus it is natural that the samples obtained with more acidic bmim chloride and methylsulfate at 150 °C demonstrate higher polymerization degrees when compared to [bmim]OAc.

For a more detailed characterization of the polyphenolic fractions, their two-dimensional HSQC (heteronuclear single quantum coherence) NMR spectra were recorded and analyzed for the samples obtained after 24 h treatment at 150 °C (Figure 5).

The residual presence of ILs in the sample results in the most intense signals on the spectrum. However, their chemical shift is located in the non-target region of the spectrum and does not interfere with the identification of target compounds. To determine the possibility of extracting polyphenolic compounds from algae using ILs, it is necessary to analyze the aromatic region of the spectrum at $\delta C/\delta H$ 90–140/5.0–7.5 ppm. It is known that the main polyphenolic compounds of algae are phlorotannins, which are characterized by clusters of signals with chemical shifts at 90–100/5.5–6.0 ppm, which is clearly observed in the experimental spectrum. In addition, the spectrum shows clear signals in the aromatic region, which can be correlated with the structures of amino acids such as phenylalanine and tyrosine based on their chemical shifts ($\delta C/\delta H$ 110–140/6.5–7.3 ppm). However, identification of all amino acids present in fraction F2 is currently impossible. Further research is underway. It is worth noting that the obtained spectra do not show any signals corresponding to polysaccharides.

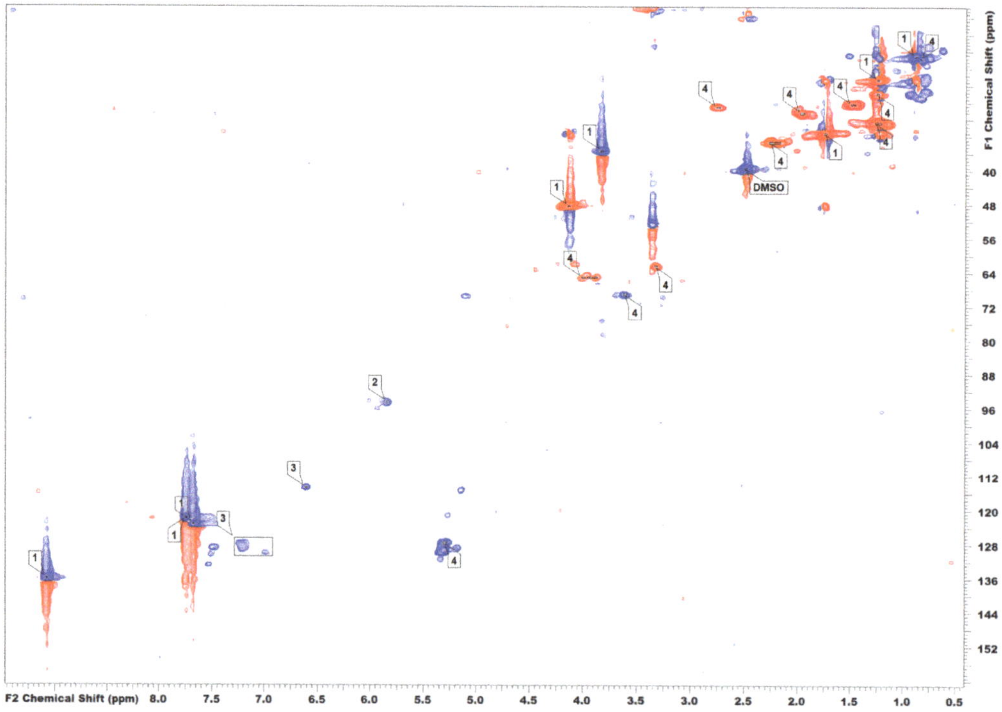

Figure 5. 2D HSQC spectrum of the F2 ([bmim]MeSO$_4$, 24 h/150 °C) sample, where: (1) ILs; (2) phlorotannin's; (3) amino acids; (4) fatty acids (unsaturated) and their glycerides.

3. Materials and Methods

3.1. Reagents and Materials

1-Butyl-3-methylimidazolium acetate, chloride, and methyl sulfate (BASF quality, >95%) were purchased from Sigma-Aldrich (Steinheim, Germany). Algae biomass fractionation was carried out with the use of "chem. pure" grade acetone (Komponent-Reaktiv, Moscow, Russia) and deionized water obtained with the Milli-Q system (Merk Millipore, Molsheim, France).

Brown alga bladderwrack (*Fucus vesiculosus*) thallus plant material was purchased from the Arkhangelsk Seaweed Plant (Arkhangelsk, Russia) as a dried and powdered commercial product with the following characteristics: alginates—35.4%, other carbohydrates—20.0%, ash—22.1%, polyphenols—10.0%, proteins—7.0%, water—4.0%, fats—1.5%.

3.2. Algal Biomass Fractionation

An accurately weighed 50 mg sample of algal biomass was placed in a 2.5 mL glass vial containing 1 mL of IL. The mixture was subjected to a thermal treatment at 80–150 °C for 2–24 h under constant agitation. After the treatment, the reaction mixture was separated by centrifugation into an insoluble residue fraction (if present) and a solution in IL. A 5-fold excess of acetone was added to the obtained solution, and the resulting precipitate (fraction F1) was separated from the liquid phase by filtration. Then the acetone was removed from the solution under a vacuum on a rotary evaporator, and a dark viscous liquid with a volume of ~1 mL was obtained. The fraction F2 was isolated through precipitation by adding 5-fold excess of water, cooling to 4 °C, and leaving it overnight in nitrogen atmosphere, followed by filtration. Both obtained fractions were dried in a vacuum oven to the constant weight. To enhance the reliability of the obtained results, at least three repetitions of the biomass dissolution and fractionation procedure were performed.

3.3. Size-Exclusion Chromatography

Determination of the molecular weights and molecular-weight distributions of the polyphenolic fractions (F2) was carried out by size-exclusion high-performance liquid chromatography using an LC-20 Prominence HPLC system (Shimadzu, Kyoto, Japan) consisting of an SIL-20A autosampler, an LC-20AD pump, a DGU A3 vacuum degasser, an STO-20A column thermostat, and an SPD-20A spectrophotometric detector. Separation was performed at 50 °C on a Polargel-M chromatographic column (Agilent, Santa Clara, CA, USA), 300 × 7.5 mm. Lithium bromide solution (0.0125 M) in DMF was used as a mobile phase with a flow rate of 1 mL min^{-1}. Detection was performed at a wavelength of 275 nm. The system was calibrated with monodisperse polystyrene standards (PSS, Mainz, Germany) in a molecular weight range of 0.35–187 kDa.

3.4. Monosaccharide Analysis

Six target monosaccharides (glucose, xylose, galactose, arabinose, mannose, fructose) were determined by high-performance ligand exchange chromatography (HPLEC) with refractometric detection according to the procedure described earlier [30]. A Nexera XR HPLC system (Shimadzu, Kyoto, Japan) which consisted of a DGU-5A vacuum degasser, an LC-20AD chromatographic pump, a SIL-20AC autosampler, a CTO-20AC column thermostat, and an RID-20A refractometric detector was used. The chromatographic separation was carried out at 75 °C on a Rezex RPM-Monosaccharide Pb^{+2} column (Phenomenex, Torrance, CA, USA), 300 × 7.8 mm, using the pure water (flow rate 0.6 mL min^{-1}) as a mobile phase. The injection volume was 10 µL. The system control and quantification of the analytes were performed using LabSolution software ver. 5.71 SP1 (Shimadzu, Kyoto, Japan). The HPLC system was calibrated using the aqueous standard solutions of the monosaccharide mixture with concentrations of 10–1000 mg L^{-1}.

The total monosaccharide content was determined after a preliminary two-stage acid hydrolysis of the extracts according to the following procedure. The dry extract sample (10 mg) was placed in a 4 mL conical glass vial, poured with 100 µL of 72% sulfuric acid, and kept at 30 °C for 60 min in a Reacti-Therm reaction system (Thermo Scientific, Waltham, MA, USA) equipped with a heating block and a magnetic stirring module. Then, 2.5 mL of water was added and the reaction mixture was heated to 100 °C, kept for 3 h under continuous stirring, and allowed to cool down at ambient conditions. After neutralizing the acid by adding an excess of BaCO$_3$ and centrifugation, the obtained solution was injected to the HPLC system. All assays were performed in triplicate.

3.5. FTIR and NMR Spectroscopy

IR spectra were obtained on a Vertex 70 IR Fourier spectrometer (Bruker, Bremen, Germany) equipped with a GladiATR attenuated total reflection (ATR) system (Pike Tech., Madison, WI, USA) with a diamond prism. The spectra were recorded under the following conditions: spectral range 4000–400 cm^{-1}, resolution 4 cm^{-1}, 128 scans. The resulting IR spectrum was subjected to ATR correction with transformation to absorbance units. The instrument was controlled and the spectra were processed with an OPUS software package ver. 8.2.28 (Bruker, Bremen, Germany).

The NMR spectra were recorded on an AVANCE III NMR spectrometer (Bruker, Ettlingen, Germany) with a working frequency for protons of 600 MHz. The ^{1}H-^{13}C HSQC (Heteronuclear Single Quantum Correlation) spectra were recorded using the following parameters: temperature—298 K, spectral window width ~13 ppm for F2 and ~200 ppm for F1 with a number of accumulations—1024 × 256, number of scans—8, delay time between pulses (D1)—2.0 s. To register the ^{1}H-^{13}C HSQC spectra, about 40 mg of sample was dissolved in 0.5 mL of DMSO-d$_6$.

4. Conclusions

1-Butyl-3-methylimidazolium-based ionic liquids possess high dissolution power towards brown algae biomass and can be used for its fractionation into separate polysaccharide and polyphenolic constituents within the biorefinery concept. The effectiveness of ILs as solvents is determined by the nature of anion and increases in the following series corresponding to an increase in the anion basicity: [bmim]MeSO$_4$ < [bmim]Cl < [bmim]OAc. The use of the latter IL ensures dissolution of most bladderwrack (*Fucus vesiculosus*) plant material at temperatures above 120 °C and up to 92% during 24 h treatment at 150 °C. The fractionation strategy involving sequential precipitation of the fractions from IL solution with acetone and water allows for obtainment of the polysaccharide mixture (alginates, cellulose, and fucoidan) and polyphenols (phlorotannins) with the yields of ~40 and ~10%, respectively. Polyphenolic fraction does not contain significant amounts of carbohydrates and can be near-quantitatively extracted from the algal biomass by any of the three studied ILs even under mild conditions (80–100 °C). It contains mainly phlorotannins with a weight-average molecular mass of 10–20 kDa and rather low polydispersity (PDI = 1.6–2.5), as well as an admixture of low-molecular lipophilic substances. Efficient isolation of polysaccharides requires harsh conditions of IL treatment and the use of bmim acetate as a biomass solvent. Higher temperatures facilitate isolation of fucoidan, while alginic acids can be extracted under milder conditions.

Our conceptual study opens up prospects for the development of new approaches for algal biomass valorization for the eco-sustainable production of various products. In this regard, further research should be focused on the detailed characterization of the obtained fractions, optimization of IL treatment conditions, and the development of effective procedures for separation of individual polysaccharides and biologically active low-molecular-weight secondary metabolites.

Author Contributions: Conceptualization, D.S.K. and A.V.B.; methodology, A.V.B.; validation, D.A.L., D.I.F., I.I.P. and A.V.F.; formal analysis, A.V.B.; investigation, D.A.L.; resources, N.V.U.; data curation, A.V.B.; writing—original draft preparation, A.V.B.; writing—review and editing, D.S.K.; visualization, A.V.B. and A.V.F.; supervision, D.S.K.; project administration, A.V.B.; funding acquisition, A.V.L. All authors have read and agreed to the published version of the manuscript.

Funding: The study was supported by the Russian Science Foundation Grant No. 22-23-20047, https://rscf.ru/project/22-23-20047/ (accessed on 1 November 2023), the work of A.V. Belesov on the characterization of polyphenolic fraction was supported by the Ministry of Economic Development, Industry and Science of the Arkhangelsk Region, agreement No. 3 of 28 March 2023.

Institutional Review Board Statement: Not applicable.

Informed Consent Statement: Not applicable.

Data Availability Statement: Data are contained within the article.

Conflicts of Interest: The authors declare no conflict of interest.

References

1. Jerković, I.; Cikoš, A.-M.; Babić, S.; Čižmek, L.; Bojanić, K.; Aladić, K.; Ul'yanovskii, N.V.; Kosyakov, D.S.; Lebedev, A.T.; Čož-Rakovac, R.; et al. Bioprospecting of Less-Polar Constituents from Endemic Brown Macroalga Fucus virsoides J. Agardh from the Adriatic Sea and Targeted Antioxidant Effects In Vitro and In Vivo (Zebrafish Model). *Mar. Drugs* **2021**, *19*, 235. [CrossRef] [PubMed]
2. Li, Y.-X.; Wijesekara, I.; Li, Y.; Kim, S.-K. Phlorotannins as bioactive agents from brown algae. *Process Biochem.* **2011**, *46*, 2219–2224. [CrossRef]
3. Zhao, C.; Yang, C.; Liu, B.; Lin, L.; Sarker, S.D.; Nahar, L.; Yu, H.; Cao, H.; Xiao, J. Bioactive compounds from marine macroalgae and their hypoglycemic benefits. *Trends Food Sci. Technol.* **2018**, *72*, 1–12. [CrossRef]
4. van Alstyne, K.L. Comparison of three methods for quantifying brown algal polyphenolic compounds. *J. Chem. Ecol.* **1995**, *21*, 45–58. [CrossRef]
5. Bogolitsyn, K.G.; Parshina, A.E.; Druzhinina, A.S.; Shulgina, E.V. Comparative Characteristics of the Chemical Composition of Some Brown Algae from the White and Yellow Seas. *Russ. J. Bioorg. Chem.* **2021**, *47*, 1395–1403. [CrossRef]
6. Dobrinčić, A.; Balbino, S.; Zorić, Z.; Pedisić, S.; Bursać Kovačević, D.; Elez Garofulić, I.; Dragović-Uzelac, V. Advanced Technologies for the Extraction of Marine Brown Algal Polysaccharides. *Mar. Drugs* **2020**, *18*, 168. [CrossRef]
7. Stucki, S.; Vogel, F.; Ludwig, C.; Haiduc, A.G.; Brandenberger, M. Catalytic gasification of algae in supercritical water for biofuel production and carbon capture. *Energy Environ. Sci.* **2009**, *2*, 535–541. [CrossRef]

8. Orr, V.C.A.; Rehmann, L. Ionic liquids for the fractionation of microalgae biomass. *Curr. Opin. Green Sustain. Chem.* **2016**, *2*, 22–27. [CrossRef]
9. Obluchinskaya, E.D.; Daurtseva, A.V.; Pozharitskaya, O.N.; Flisyuk, E.V.; Shikov, A.N. Natural Deep Eutectic Solvents as Alternatives for Extracting Phlorotannins from Brown Algae. *Pharm. Chem. J.* **2019**, *53*, 243–247. [CrossRef]
10. Kilpeläinen, I.; Xie, H.; King, A.; Granstrom, M.; Heikkinen, S.; Argyropoulos, D.S. Dissolution of wood in ionic liquids. *J. Agric. Food Chem.* **2007**, *55*, 9142–9148. [CrossRef]
11. Sun, J.; Konda, N.V.S.N.M.; Parthasarathi, R.; Dutta, T.; Valiev, M.; Xu, F.; Simmons, B.A.; Singh, S. One-pot integrated biofuel production using low-cost biocompatible protic ionic liquids. *Green Chem.* **2017**, *19*, 3152–3163. [CrossRef]
12. da Costa Lopes, A.M.; João, K.G.; Morais, A.R.C.; Bogel-Łukasik, E.; Bogel-Łukasik, R. Ionic liquids as a tool for lignocellulosic biomass fractionation. *Sustain. Chem. Process.* **2013**, *1*, 3. [CrossRef]
13. Ma, Z.; Lu, T.; Pan, Y.; Yuan, Y.; Sun, Y. Short alkyl-imidazolium ionic liquids enhanced in-situ transesterification of microalgae. *Fuel* **2024**, *357*, 129828. [CrossRef]
14. Malihan, L.B.; Nisola, G.M.; Chung, W.-J. Brown algae hydrolysis in 1-n-butyl-3-methylimidazolium chloride with mineral acid catalyst system. *Bioresour. Technol.* **2012**, *118*, 545–552. [CrossRef]
15. Park, D.H.; Jeong, G.-T. Production of reducing sugar from macroalgae Saccharina japonica using ionic liquid catalyst. *Korean Chem. Eng. Res.* **2013**, *51*, 106–110. [CrossRef]
16. Orr, V.C.A.; Plechkova, N.V.; Seddon, K.R.; Rehmann, L. Disruption and Wet Extraction of the Microalgae Chlorella vulgaris Using Room-Temperature Ionic Liquids. *ACS Sustain. Chem. Eng.* **2016**, *4*, 591–600. [CrossRef]
17. Ladesov, A.V.; Belesov, A.V.; Kuznetsova, M.V.; Pochtovalova, A.S.; Malkov, A.V.; Shestakov, S.L.; Kosyakov, D.S. Fractionation of Wood with Binary Solvent 1-Butyl-3-methylimidazolium Acetate + Dimethyl Sulfoxide. *Russ. J. Appl. Chem.* **2018**, *91*, 663–670. [CrossRef]
18. Doherty, T.V.; Mora-Pale, M.; Foley, S.E.; Linhardt, R.J.; Dordick, J.S. Ionic liquid solvent properties as predictors of lignocellulose pretreatment efficacy. *Green Chem.* **2010**, *12*, 1967–1975. [CrossRef]
19. Liu, C.; Li, Y.; Hou, Y. Basicity Characterization of Imidazolyl Ionic Liquids and Their Application for Biomass Dissolution. *Int. J. Chem. Eng.* **2018**, *2018*, 7501659. [CrossRef]
20. Belesov, A.V.; Shkaeva, N.V.; Popov, M.S.; Skrebets, T.E.; Faleva, A.V.; Ul'yanovskii, N.V.; Kosyakov, D.S. New Insights into the Thermal Stability of 1-Butyl-3-methylimidazolium-Based Ionic Liquids. *Int. J. Mol. Sci.* **2022**, *23*, 10966. [CrossRef]
21. Kokosa, J.M. Selecting an extraction solvent for a greener liquid phase microextraction (LPME) mode-based analytical method. *TrAC Trends Anal. Chem.* **2019**, *118*, 238–247. [CrossRef]
22. Yilmaz, E.; Soylak, M. Chapter 5—Type of green solvents used in separation and preconcentration methods. In *New Generation Green Solvents for Separation and Preconcentration of Organic and Inorganic Species*; Soylak, M., Yilmaz, E., Eds.; Elsevier: Amsterdam, The Netherlands, 2020; pp. 207–266. [CrossRef]
23. Pereira, L.; Gheda, S.F.; Ribeiro-Claro, P.J.A. Analysis by Vibrational Spectroscopy of Seaweed Polysaccharides with Potential Use in Food, Pharmaceutical, and Cosmetic Industries. *Int. J. Carbohydr. Chem.* **2013**, *2013*, 537202. [CrossRef]
24. Pereira, L.; Sousa, A.; Coelho, H.; Amado, A.M.; Ribeiro-Claro, P.J.A. Use of FTIR, FT-Raman and 13C-NMR spectroscopy for identification of some seaweed phycocolloids. *Biomol. Eng.* **2003**, *20*, 223–228. [CrossRef] [PubMed]
25. Xia, S.; Gao, B.; Li, A.; Xiong, J.; Ao, Z.; Zhang, C. Preliminary Characterization, Antioxidant Properties and Production of Chrysolaminarin from Marine Diatom *Odontella aurita*. *Mar. Drugs* **2014**, *12*, 4883–4897. [CrossRef] [PubMed]
26. Ptak, S.H.; Sanchez, L.; Fretté, X.; Kurouski, D. Complementarity of Raman and Infrared spectroscopy for rapid characterization of fucoidan extracts. *Plant Methods* **2021**, *17*, 130. [CrossRef]
27. Jeon, C.; Park, J.Y.; Yoo, Y.J. Characteristics of metal removal using carboxylated alginic acid. *Water Res.* **2002**, *36*, 1814–1824. [CrossRef]
28. Yeo, M.; Jung, W.-K.; Kim, G. Fabrication, characterisation and biological activity of phlorotannin-conjugated PCL/β-TCP composite scaffolds for bone tissue regeneration. *J. Mater. Chem.* **2012**, *22*, 3568–3577. [CrossRef]
29. Bai, Y.; Chen, X.; Qi, H. Characterization and bioactivity of phlorotannin loaded protein-polysaccharide nanocomplexes. *LWT* **2022**, *155*, 112998. [CrossRef]
30. Ul'yanovskii, N.V.; Falev, D.I.; Kosyakov, D.S. Highly sensitive ligand exchange chromatographic determination of apiose in plant biomass. *Microchem. J.* **2022**, *180*, 107638. [CrossRef]

Article

Functionalized Biochar from the Amazonian Residual Biomass Murici Seed: An Effective and Low-Cost Basic Heterogeneous Catalyst for Biodiesel Synthesis

Thaissa Saraiva Ribeiro, Matheus Arrais Gonçalves, Geraldo Narciso da Rocha Filho and Leyvison Rafael Vieira da Conceição *

Laboratory of Catalysis and Oleochemical, Institute of Exact and Natural Sciences, Federal University of Pará, Belém 66075-110, PA, Brazil; saraivathaissa@gmail.com (T.S.R.); matheusarrais38@gmail.com (M.A.G.); narciso@ufpa.br (G.N.d.R.F.)
* Correspondence: rafaelvieira@ufpa.br

Abstract: This study presents the synthesis of a basic heterogeneous catalyst based on sodium functionalized biochar. The murici biochar (BCAM) support used in the process was obtained through the pyrolysis of the murici seed (*Byrsonimia crassifolia*), followed by impregnation of the active phase in amounts that made it possible to obtain concentrations of 6, 9, 12, 15 and 18% of sodium in the final composition of the catalyst. The best-performing 15Na/BCAM catalyst was characterized by Elemental Composition (CHNS), Thermogravimetric Analysis (TG/DTG), X-ray diffraction (XRD), Fourier Transform Infrared Spectroscopy (FT-IR), Scanning Electron Microscopy (SEM), and Energy Dispersion X-ray Spectroscopy (EDS). The catalyst 15Na/BCAM was applied under optimal reaction conditions: temperature of 75 °C, reaction time of 1.5 h, catalyst concentration of 5% (w/w) and MeOH:oil molar ratio of 20:1, resulting in a biodiesel with ester content of 97.20% $\pm$ 0.31 in the first reaction cycle, and maintenance of catalytic activity for five reaction cycles with ester content above 65%. Furthermore, the study demonstrated an effective catalyst regeneration process, with the synthesized biodiesels maintaining ester content above 75% for another five reaction cycles. Thus, the data indicate a promising alternative to low-cost residual raw materials for the synthesis of basic heterogeneous catalysts.

Keywords: murici seed; biochar; biodiesel; basic heterogeneous catalyst; agroindustrial residue

Citation: Ribeiro, T.S.; Gonçalves, M.A.; da Rocha Filho, G.N.; da Conceição, L.R.V. Functionalized Biochar from the Amazonian Residual Biomass Murici Seed: An Effective and Low-Cost Basic Heterogeneous Catalyst for Biodiesel Synthesis. *Molecules* **2023**, *28*, 7980. https://doi.org/10.3390/molecules28247980

Academic Editors: José A.P. Coelho and Roumiana P. Stateva

Received: 21 October 2023
Revised: 14 November 2023
Accepted: 15 November 2023
Published: 7 December 2023

1. Introduction

In recent decades, emissions of gases that contribute to the greenhouse effect have increased due to the growing demand for energy [1]. The development of a clean, ecological, sustainable alternative that meets the needs of the market has been widely discussed by the scientific community around the world [2,3]. In this scenario, biodiesel has attracted attention because of its properties, which are similar to fossil diesel [4,5], as well as by having advantages, such as the emission of less toxic suspended particles, lower emission of smoke and byproducts of its combustion, and greater safety and production capacity from renewable sources [6].

Chemically, biodiesel is a mixture of alkyl esters of fatty acids derived from renewable sources such as vegetable oils, waste oils, animal fats, and greases [7]. This biofuel can be obtained via esterification of fatty acids or via transesterification of triglycerides with a short-chain alcohol, usually using acidic, basic, or enzymatic catalysts [8–10]. The application of heterogeneous catalysts in the transesterification of oils offers advantages for reducing costs in the production of biofuels because they are easy to separate, noncorrosive, reusable, and can be regenerated [11–14]. In this scenario, several supports were investigated. These supports included ZrO_2 [15], $SrFe_2O_4$ [16], $CuFe_2O_4$ [17], TiO_2 [18], $Na_2Ti_3O_7$ [19], and SiO_2 [20]. The objective was to increase the dispersion of active phases on such supports, as well as increase the reuse capacity of these catalysts. In addition, activated carbon from residual biomass can be used as a potential support [21].

Biochar is a carbonaceous material that can be obtained from the thermochemical degradation of biomass. It can be obtained through several techniques, such as pyrolysis, carbonization, liquefaction, and hydrothermal carbonization and roasting, in which the materials generated will have physicochemical characteristics from the biomass used, as well as the functionalization and activation process [22,23]. Biochar can exhibit indispensable catalytic properties, such as having customizable porous structures, excellent stability in acidic or basic media, and an extensive specific surface area [24]. Several studies of preparation of carbon-based catalysts or biochar from different residual biomasses, functionalized with several active phases, have been reported [25], such as: murumuru kernel shell [26–28] avocado seed [29], date seed [21], rice husk [30], pomelo peel [31], banana peel [32,33], chicken manure [34], citrus fruit peel [35], coconut coir husk [36] and coffee husk [37].

Murici (*Byrsonimia crassifolia*) is a small-sized tree native to South America and widespread throughout the Amazon region. The fruit is of the trilocular drupe type, rounded, about 1.5–2.0 cm in diameter, with the pulp constituting 70.9% of the fruit. It possesses a yellowish coloration when ripe with characteristic aroma and taste. The seed is rounded, rigid, and constitutes 29.1% of the fruit [38,39]. Fruiting occurs usually between November and May, with productivity of around 12.0 kg per tree. The murici fruit is appreciated by local populations, usually raw, and its pulp is used in the manufacturing of juices, sweets, creams, jellies, ice cream, and liqueurs [40].

Bitonto et al. [29] prepared a biochar from the pyrolysis of avocado seeds in a tubular oven at 900 °C for 2 h in an atmosphere of N_2 and then investigated the catalytic activity with different concentrations of CaO (5, 10, and 20%). The authors concluded that the best catalytic activity in the biodiesel synthesis process was the catalyst synthesized with 20% CaO concentration. This catalyst was used in the transesterification of sunflower oil with methanol in the optimal reaction condition of temperature of 99.5 °C, reaction time of 5 h, catalyst concentration of 7.3% *w/w* and MeOH:oil molar ratio of 15.6:1, provided a biodiesel with ester content of 99.5%, and three reaction cycles of the catalyst heat treated in a tubular oven at 550 °C for 3 h under an atmosphere of N_2. Based on the results obtained, the authors concluded that the increase in the concentration of calcium oxide in the biochar positively affected the basicity of the material, which consequently led to an increase in the catalytic activity in the transesterification reaction of triglycerides.

Jamil et al. [21] studied four heterogeneous carbon-based catalysts impregnated with 15% by mass of alkaline metal oxides (CaO, MgO, BaO and SrO). The carbon-based material was obtained from the tubular furnace carbonization of the dates' residue (400 °C/5 h) produced after the extraction of their oil. The biodiesel synthesis was conducted with methanol and using four types of oils, date, coconut, palm, and waste cooking oil, at temperatures of 55–75 °C, reaction times of 0.5–2.5 h, catalyst concentration of 1–5% and MeOH:oil molar ratio of 6:1–18:1. The biodiesel yield obtained was 94.27% under the best reaction conditions and maintenance of high catalytic activity for eight reaction cycles when the catalyst consisting of carbon and strontium oxide was applied. In addition, the authors suggested the use of this catalyst in the transesterification reaction regardless of the oil used in the process.

Thus, the present study aims to synthesize a heterogeneous basic catalyst based on biochar produced from the residual biomass of murici functionalized with sodium. It also aims to evaluate its applicability in the transesterification reaction of soybean oil via the methyl route, as well as to study its ability to be reused and regenerated in the process of biodiesel synthesis.

2. Results and Discussion

2.1. Influence of Sodium Percentage

One of the main factors that can influence the ester content of biodiesel, as well as its other physicochemical properties, is the catalyst used in the transesterification reaction and this, in turn, has a direct relationship with its synthesis process. Thus, the influence of sodium percentages 6, 9, 12, 15 and 18% on the final catalyst composition in the biodiesel production process was evaluated. In this preliminary step, the catalytic tests were performed under fixed reaction parameters: temperature of 120 °C, reaction time of 3 h, catalyst concentration of 8% and MeOH:oil molar ratio of 24:1. Figure 1 presents the results obtained in the study on the influence of the percentage of sodium present in the catalysts, as well as the basicity data of the murici biochar (BCAM) support and catalysts studied.

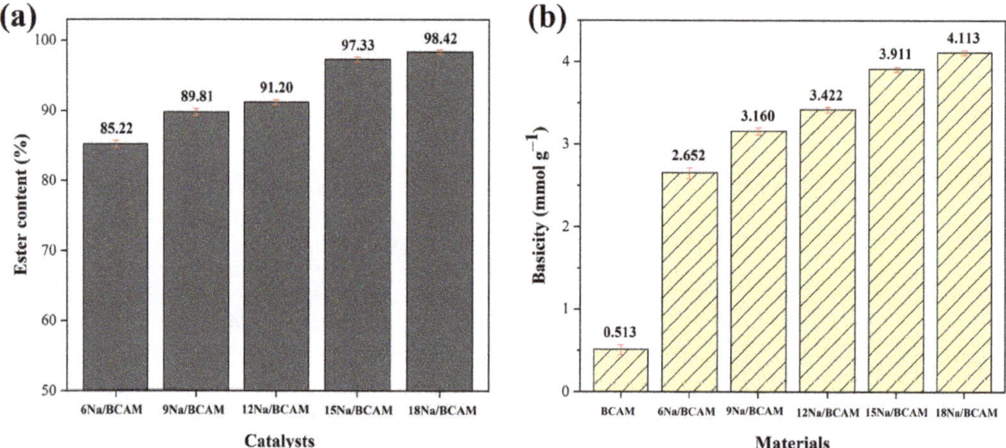

Figure 1. (**a**) Influence of the sodium percentage present in the catalysts and (**b**) basicity of BCAM support and catalysts synthesized.

Based on Figure 1a, it is initially inferred that there is a direct relationship between the percentage of functionalization of biochar and the ester content of biodiesel. During the study, it was observed that 6Na/BCAM did not efficiently conduct the transesterification reaction due to providing the lowest ester content value in biodiesel (85.22% $\pm$ 0.55) among all catalysts used. The low catalytic activity presented may be related to the insufficient amount of active sites responsible for the conversion of triglycerides into esters [41,42]. In addition, it is possible to observe that by tripling the percentage of active phases present in the catalyst composition, that is, by employing the catalyst 18Na/BCAM, an increase of 13.20% in the ester content of biodiesel is observed, reaching the maximum value of 98.42% $\pm$ 0.29 and establishing a linear growth trend of the ester content of biodiesels as the percentage of the active phase impregnated in the BCAM support increases.

It is worth noting that the catalysts 15Na/BCAM and 18Na/BCAM present biodiesels with similar ester contents, with about a 1.09% difference. This may be an indication that impregnations with higher percentages of active phases would not cause a significant impact on the catalytic performance of the catalysts during the conduction of the transesterification reaction. In addition, this fact may be linked to the saturation of these active sites or the high viscosity of the reaction system together with the excess of the active phase, hindering the mass transfer between the MeOH:oil system with the active sites of the catalyst and leading to negligible increases in the conversion of triglycerides to esters [7,18]. Thus, considering the small difference between the synthesized biodiesels and the additional operational cost in the synthesis process to impregnate 3% more active phase in the BCAM support, the percentage corresponding to 15% sodium (the catalyst 15Na/BCAM) was selected for the continuity of the study.

In addition, the selected catalyst leads to a biodiesel with an ester content of 97.3%, 14 times higher than the ester content of biodiesel achieved when only the BCAM support was used (6.95% $\pm$ 0.36 ester content). The increase in ester conversion observed can be attributed to the increase in the basic groups of the catalyst 15Na/BCAM (3.911 mmol g^{-1} $\pm$ 0.031) relative to the BCAM support (0.513 mmol g^{-1} $\pm$ 0.060), caused by the process of impregnation of the active phase, as can be seen in the data presented in Figure 1b. This behavior was also observed by Zhao et al. [31], who reported basicity values of 0.2 mmol g^{-1} for the pomelo peel biochar and 9.0 mmol g^{-1} for the catalyst based on biochar impregnated with 25% K_2CO_3 achieving biodiesel with yields of 5.6% and 98%, respectively. Furthermore, in the study by Zhao et al. [30], the basicity value for rice husk biochar was 1.5 mmol g^{-1}, while the basicity determined for the catalyst impregnated with 30% CaO was 11.4 mmol g^{-1}, resulting in biodiesels with yields of 12.5 and 93.4%, respectively.

It is noteworthy that the 15Na/BCAM catalyst demonstrated high catalytic activity with 15% of the active phase impregnated on the support (ester content of 97.33% $\pm$ 0.38), while the avocado biochar catalyst functionalized with 20% Ca (w/w) led to a biodiesel with methyl ester content (FAME) of only 82.7% [29]. In addition, Zhang et al. [4] synthesized a magnetic catalyst Na_2SiO_3@Ni/C that was used in the optimal reaction condition, resulting in a biodiesel with a yield of 98.1%, a value slightly higher than the ester content obtained in this study when the catalyst 15Na/BCAM was used. However, this yield of 98.1% achieved in the use of the catalyst in the Na_2SiO_3@Ni/C is conditioned on the use of 56% sodium silicate, the equivalent of 21.1% sodium, that is, a 6% more active phase compared to that present in the catalyst 15Na/BCAM developed in this study.

2.2. Characterization of Materials

2.2.1. Elemental Analysis (CHNS)

The results of the elemental analysis for murici seed, BCAM support and catalyst 15Na/BCAM are shown in Table 1. When comparing the results obtained for the murici seed biomass and for the BCAM support, it can be noticed that the carbon (C) content increases from 45.17 to 71.72%, while the oxygen (O) content decreases from 50.87 to 24.81%. This can be attributed to the thermal degradation of the lignocellulosic components during the carbonization process of the murici seed, which results in the elimination of groups that have oxygen, hydrogen, and nitrogen in their structure, resulting in an increase in the carbon content present in the BCAM support [36,43]. In addition, there is a subtle decrease in the hydrogen content (H) from 2.88 to 1.49% and nitrogen (N) from 2.82 to 1.82%. The results obtained for the catalyst 15Na/BCAM suggest an increase from 24.81 to 28.60% in the oxygen (O) content compared to the BCAM support and indicate the presence of silicon (Si) and sodium (Na) percentages of 9.53 and 5.82%, respectively. These elements are assigned to the impregnation step carried out in the catalyst synthesis process.

Table 1. Elemental composition of the murici seed, BCAM support and catalyst 15Na/BCAM.

Sample	C	H	N	O [a]	Na [b]	Si [b]
Murici seed	45.17 $\pm$ 0.04	2.88 $\pm$ 0.02	2.82 $\pm$ 0.01	49.07	-	-
BCAM	71.72 $\pm$ 0.03	1.49 $\pm$ 0.01	1.98 $\pm$ 0.01	24.76	-	-
Catalyst 15Na/BCAM	53.85 $\pm$ 0.03	1.33 $\pm$ 0.01	0.87 $\pm$ 0.01	28.45	9.53	5.82

[a] Determined by difference (C + O + N + H = 100%) per Lim et al. [44]. [b] Not detected, determined by difference and stoichiometric ratio.

2.2.2. TG/DTG Thermogravimetric Analysis

The knowledge of the thermal stability of the synthesized materials at different temperatures is of paramount importance since it helps in optimizing the process of preparing the materials [16]. Thus, TG and DTG thermogravimetric analyses were performed in order to understand the synthesis process of the BCAM support and the catalyst 15Na/BCAM. In addition, the thermal stability of the BCAM support after its formation was also analyzed, as well as the sodium silicate precursor (Na_2SiO_3), as can be seen in Figure 2.

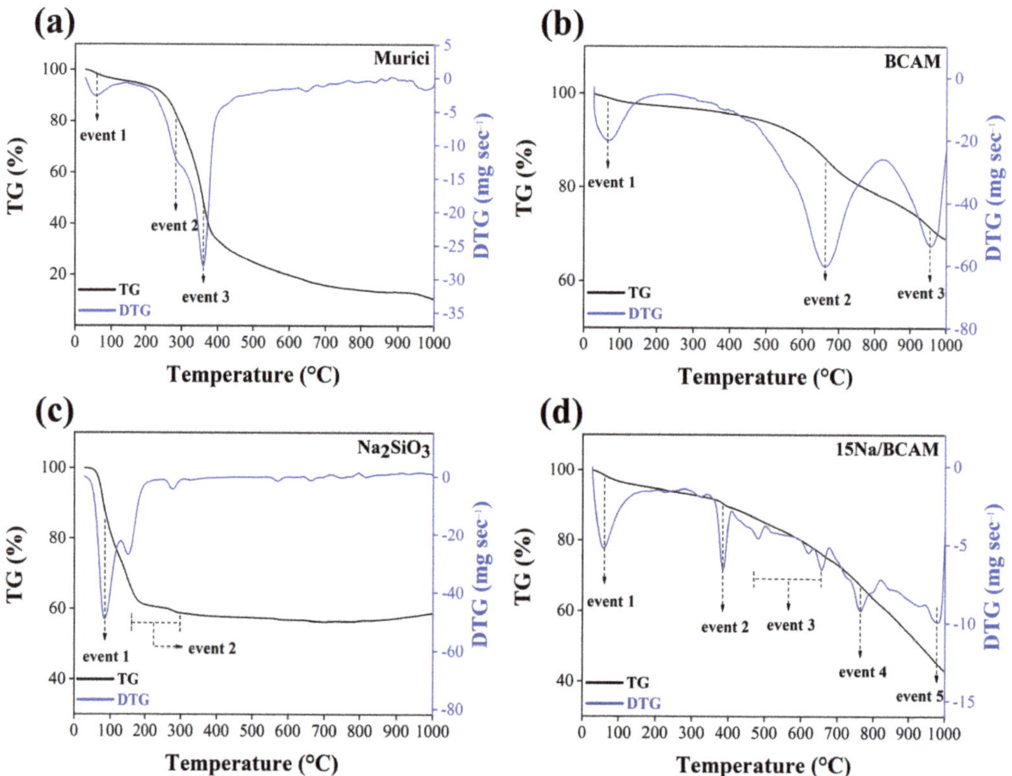

Figure 2. TG/DTG plots of (**a**) murici seed, (**b**) BCAM support, (**c**) Na_2SiO_3 and (**d**) catalyst 15Na/BCAM.

Figure 2a presents the TG and DTG curves for the murici seed, in which the occurrence of three main mass loss events is observed. Event 1, corresponding to a mass loss of approximately 4%, occurs in the 25–100 °C range and can be attributed to the loss of water molecules adsorbed on the surface of the material [31]. Event 2, about 13.5%, temperature range 140–286 °C, is attributed to the onset of pyrolysis of lignocellulosic components of lower thermal stability, i.e., hemicellulose [43,44]. Event 3, with a mass loss of 29%, in the 200–380 °C temperature range, corresponds to the thermal degradation of the other lignocellulosic components, cellulose, and lignin, which have more complex structures and, as a consequence, are more resistant to pyrolysis [44].

From the analysis of the TG and DTG curves referring to the BCAM support, shown in Figure 2b, it is possible to identify three mass loss events. Event 1, approximately 2%, temperature range 25–100 °C, shows a loss of water physically bound to the surface [45], while the mass losses in the temperature range 430–640 °C (event 2) and between 910 and 980 °C (event 3) correspond to pyrolysis of lignocellulosic compounds (hemicellulose, cellulose, and lignin) not volatilized during biochar synthesis by carbonization of the murici seed [27].

Figure 2c,d show the TG and DTG curves for Na_2SiO_3 and catalyst 15Na/BCAM, respectively. The TG and DTG curves present in Figure 2c indicate the occurrence of two main mass loss events. The first, of 13.5% (range 25–85 °C), and the second, 29% (range 100–300 °C), can be attributed to the elimination of water molecules (i) adsorbed on the surface and (ii) structurally bound to the material [46,47]. Based on the analysis of the TG and DTG curves shown in Figure 2d, it is possible to identify five main events of mass loss. Event 1, which corresponds to the temperature range 58–100 °C, refers to a mass loss of 3% and may be attributed to loss of physically bound water and structure [45]. Event 2, in the temperature range 350–400 °C, corresponds to a mass loss of 3% and is related to the process of dihydroxylation of OH groups [31]. The mass loss event 3, with about 10% and occurring in the temperature range 470–650 °C, and event 4, with a mass loss of approximately 2% in the temperature range 740–765 °C, as well as event 5, with mass loss of 4.5% in the range 940–980 °C, can be attributed to the degradation of carbonaceous structures (hemicellulose, cellulose, and lignin) not completely carbonized during the preparation of the BCAM support [26,27,48].

2.2.3. X-ray Diffraction (XRD)

X-ray diffractograms for the BCAM support, in the Na_2SiO_3 and catalyst 15Na/BCAM, are set forth in Figure 3. In the diffractogram referring to the BCAM support (black line), a diffraction peak characteristic of the presence of lignocellulosic materials in the region of $2\theta = 22.5°$ associated with the microcrystalline structure of cellulose can be observed [12]. In addition, the diffractogram showed a halo in the region of $2\theta = 15–30°$ assigned to plane 002 typical of amorphous carbon structures, and a less intense halo in the region of $2\theta = 40–50°$ to plane 101 of amorphous carbonaceous structures [26,27], while the peaks at $2\theta = 28.9°$ and $2\theta = 43.14°$ correspond to planes 002 and 101 in graphitic carbons, respectively [49]. The diffractogram for the BCAM support showed peaks in regions $2\theta = 36°$, $39.3°$, $44.6°$, $47.3°$, $48.4°$, $57.34°$, $61°$ and $65°$ attributed to the $CaCO_3$ structure [50–52].

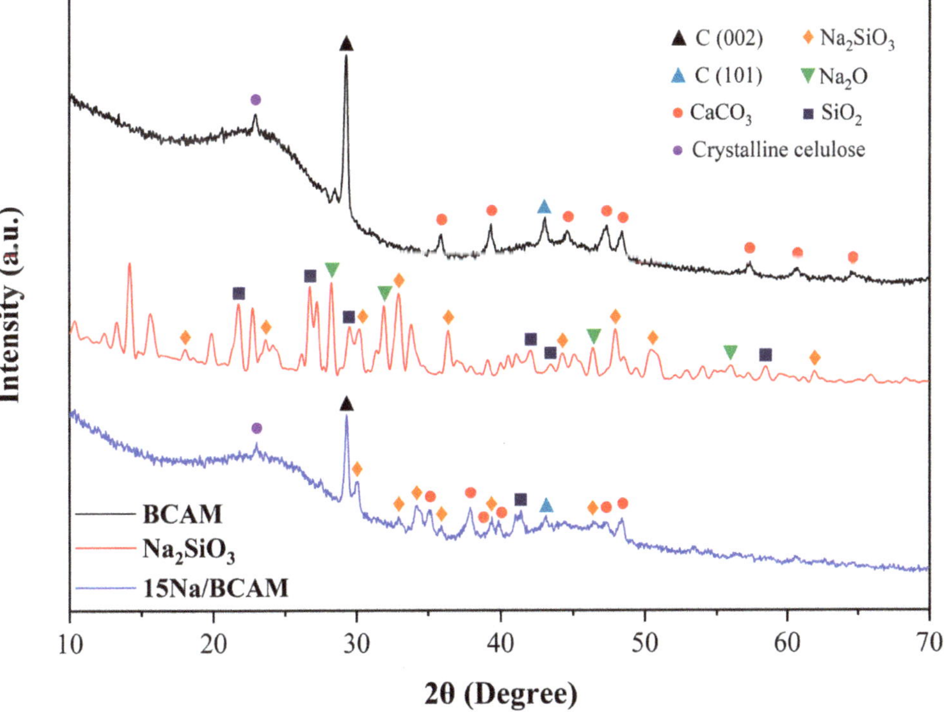

Figure 3. XRD patterns of BCAM support, Na_2SiO_3 and catalyst 15Na/BCAM.

It is worth highlighting that the presence of inorganic materials in the biochar structure is reported in the literature, such as in the studies conducted by Jitjamnong et al. [32], who synthesized a biochar derived from banana peel and observed the presence of potassium and calcium oxides. Conversely, in their study with biochar from banana peel, Patel et al. [33] reported the presence of $CaCO_3$ structures. In the studies reported by Zhao et al. [30], the presence of SiO_2 was observed in the structure of biochar obtained from rice husk; they concluded that the presence of these inorganic structures does not attribute catalytic activity to biochar when applied in the transesterification reaction, for it achieved a biodiesel yield of only 12.5%.

Major peaks present in the diffractogram of Na_2SiO_3 (red line) in the regions of $2\theta = 17.1°, 24°, 32.9°, 36.4°, 44.5°, 47.9°, 50.6°$ and $61.8°$ are mainly attributed to the structure of $Na_2SiO_3.9H_2O$ and crystalline Na_2SiO_3 in the orthorhombic phase [53,54]. In addition, the peaks present in $21.8°, 26.7°, 29.5°, 42.2°, 43.4°$ and $58.4°$ indicate the presence of SiO_2 [8,42,55,56]. The presence of Na_2O is indicated by peaks in the regions of $28.3°$, $32°, 46.3°$ and $55.9°$ [57]. In the diffractogram relative to the catalyst 15Na/BCAM (blue line), the peak was initially observed in $2\theta = 22.5°$ assigned to the microcrystalline phase of cellulose [12], and peaks at $2\theta = 28.9$ and $2\theta = 43.14$ were assigned to planes 002 and 101 in graphitic carbons, respectively [49]. In addition, the typical halos of amorphous carbonaceous materials were observed in the regions of $2\theta = 15–30°$ and $2\theta = 40–50°$ assigned to plane 002 and plane 101, respectively [26,27]. Finally, peaks were observed in $2\theta = 35.1°, 37.8°, 38.8°, 39.84°, 47.1°$ and $48.4°$, attributed to the $CaCO_3$ structure [50,51] and at $2\theta = 30.1°, 32.9°, 34.12°, 35.9°, 39.4°$ and $46.4°$, characteristic of the orthorhombic crystalline phase of Na_2SiO_3 [53,54].

2.2.4. Fourier Transform Infrared Spectroscopy (FT-IR)

FT-IR spectra for the BCAM support in the Na_2SiO_3 and catalyst 15Na/BCAM are set forth in Figure 4. For the spectrum of the BCAM support (black line), typical vibrational bands of carbonized lignocellulosic materials were identified at 1712 and 1570 cm^{-1}, attributed to the C=O and C=C bond stretch, respectively, characteristics of the presence of carboxylic groups and aromatic rings [12,26]. The vibrational bands observed at 1424 cm^{-1} and 873 cm^{-1} correspond to antisymmetric stretching and out-of-plane and in-plane vibrational modes, respectively, assigned to the carbonate group (CO^{-2}) [33]. This reinforces the presence of $CaCO_3$ structures, as indicated by the XRD analysis. In addition, the presence of the vibrational band in the region of 740 cm^{-1} is observed in the spectrum assigned to bond stretch =C–H [26].

The spectrum referring to the Na_2SiO_3 (red line) shows vibrational bands at 1444 cm^{-1} and 1000 cm^{-1} characteristics of symmetric and antisymmetric stretches of Si–O–Si and Si–O–Na bonds, respectively [53,58]. The vibrational band present at 883 cm^{-1} can be attributed to the antisymmetric stretching of the bond Si–O–H characteristic of bonds of silanol groups present in the hydrated form of sodium silicate ($Na_2SiO_3·9H_2O$) [58]. Finally, the spectrum referring to the catalyst 15Na/BCAM (blue line) shows vibrational bands in the regions of 1570 cm^{-1} and 740 cm^{-1} assigned to C=C and =C–H bond stretches, respectively, characteristics of aromatic chains [12,26]. The vibrational bands present at 1444 cm^{-1} and at ~1006 cm^{-1} are attributed to the antisymmetric stretching of the bonds Si–O–Si and antisymmetric Si–O–Na, respectively [53,58]. In addition, a vibrational band is observed in the region of 883 cm^{-1} attributed to the antisymmetric stretching of the Si–O–H bond [58].

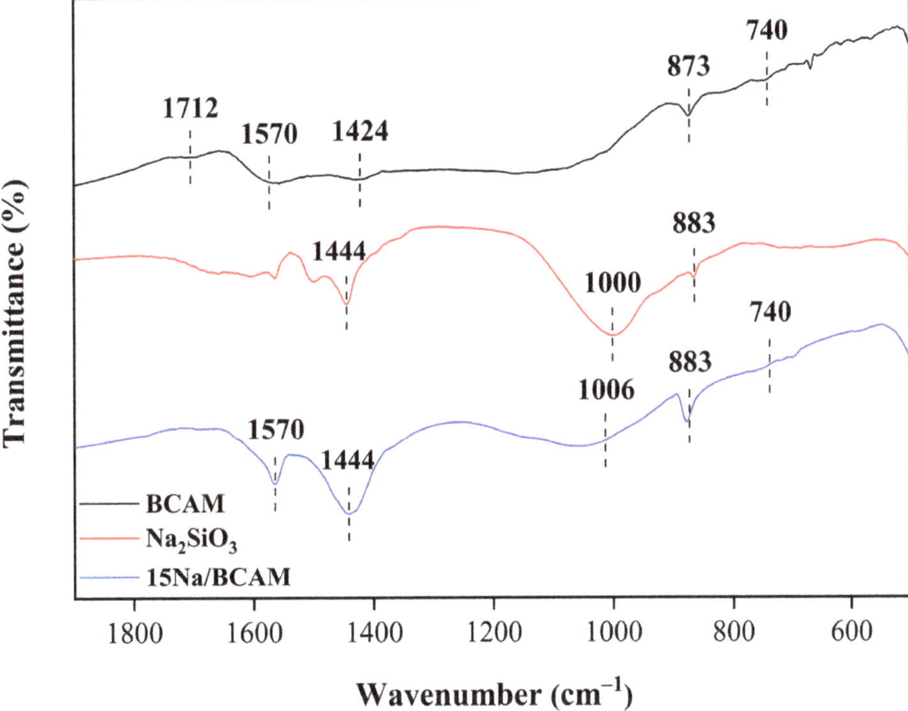

Figure 4. FT-IR spectra of BCAM support, Na_2SiO_3 and catalyst 15Na/BCAM.

2.2.5. Scanning Electron Microscopy (SEM)

Micrographs of the BCAM support and the catalyst 15Na/BCAM are shown in Figure 5. The micrographs referring to the BCAM support are shown in Figure 5a,b and indicate that the BCAM support has an irregular surface morphology with a wide well-developed pore network, formed by the release of gases during the carbonization of organic matter rich in lignocellulosic compounds [26,29,37]. Figure 5c,d show the surface morphology of the catalyst 15Na/BCAM, in which it is possible to observe a marked decrease in the porosity of the support. This is directly related to the impregnation process and, as a consequence, of the anchoring of sodium silicate on the biochar surface. In addition, it can be seen that the deposited sodium silicate does not possess a defined morphology, with small, agglomerated regions of whitish color [46,59].

2.2.6. Energy Dispersion X-ray Spectroscopy (EDS)

The composition and elemental mapping for the BCAM support are shown in Figure 6. It can be observed through the composition (Figure 6a) and elemental mapping (Figure 6b) that the BCAM support has a large carbon content (C), about 86.94% $\pm$ 0.126 distributed homogeneously, and a relatively low oxygen content (O), approximately 11.65% $\pm$ 0.016, homogeneously dispersed in the material with small concentration regions. Both are directly related to the pyrolysis process in nitrogen atmosphere to which the biomass was subjected. Furthermore, these values are close to those indicated in the results of CNHS, 71.72% $\pm$ 0.03 for the carbon content (C) and 24.76% for the oxygen content (O). This small difference between the contents found may be linked to the method of quantification of the elemental composition of the sample used by each technique. Additionally, the composition (Figure 6a) and the elemental mapping (Figure 6b) for the BCAM support indicated the presence of small regions containing calcium (Ca) and oxygen (O) homogeneously and in small clusters. These results are in consensus with the results shown in the XRD analysis, which indicated the presence of calcium carbonate in the crystalline phase present in the BCAM support. However, despite constituting about 1.41% ± 0.024 of the sample, this percentage of calcium carbonate does not attribute catalytic activity to the BCAM support since the ester content value achieved for the biodiesel produced from the reaction using only the BCAM support (blank reaction) was 6.95% ± 0.36.

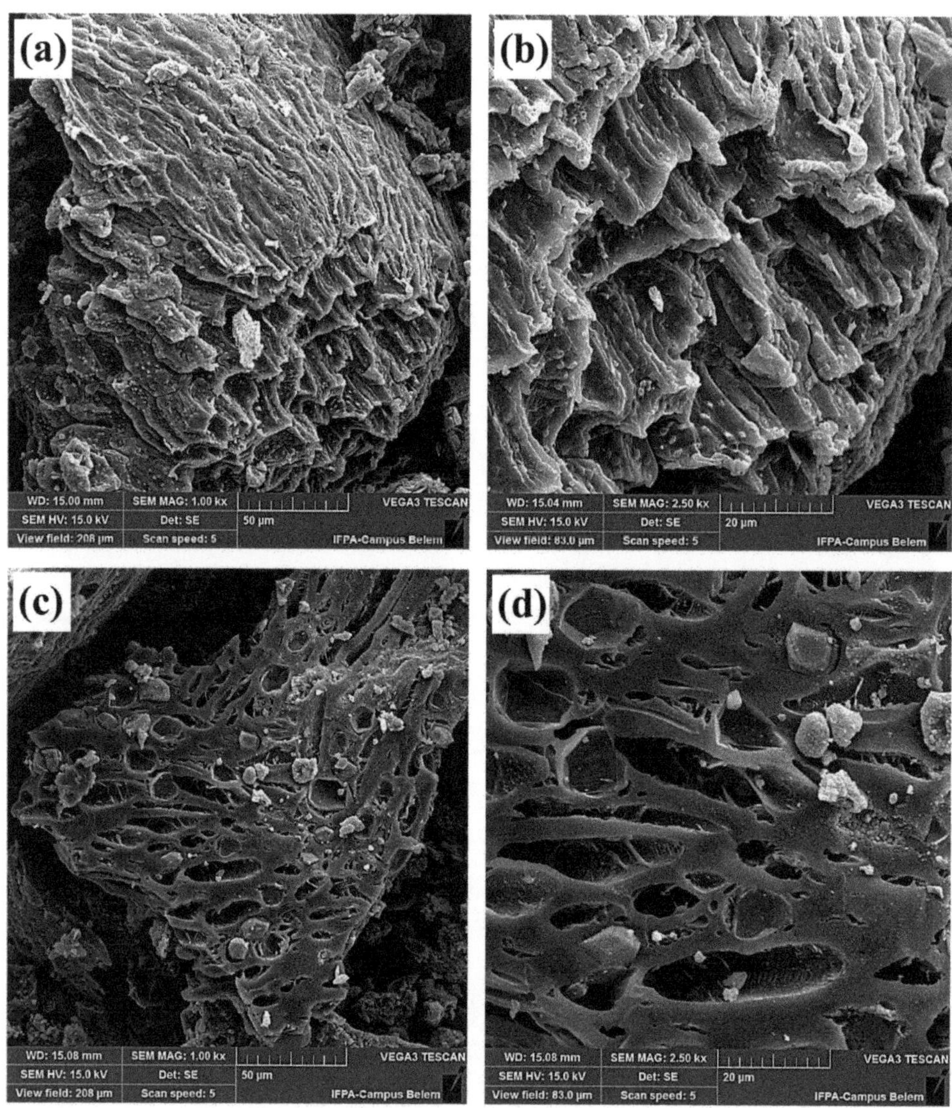

Figure 5. SEM micrographs (**a**) BCAM support 1000× magnification, (**b**) BCAM support 2500× magnification, (**c**) catalyst 15Na/BCAM 1000× magnification and (**d**) catalyst15Na/BCAM 2500× magnification.

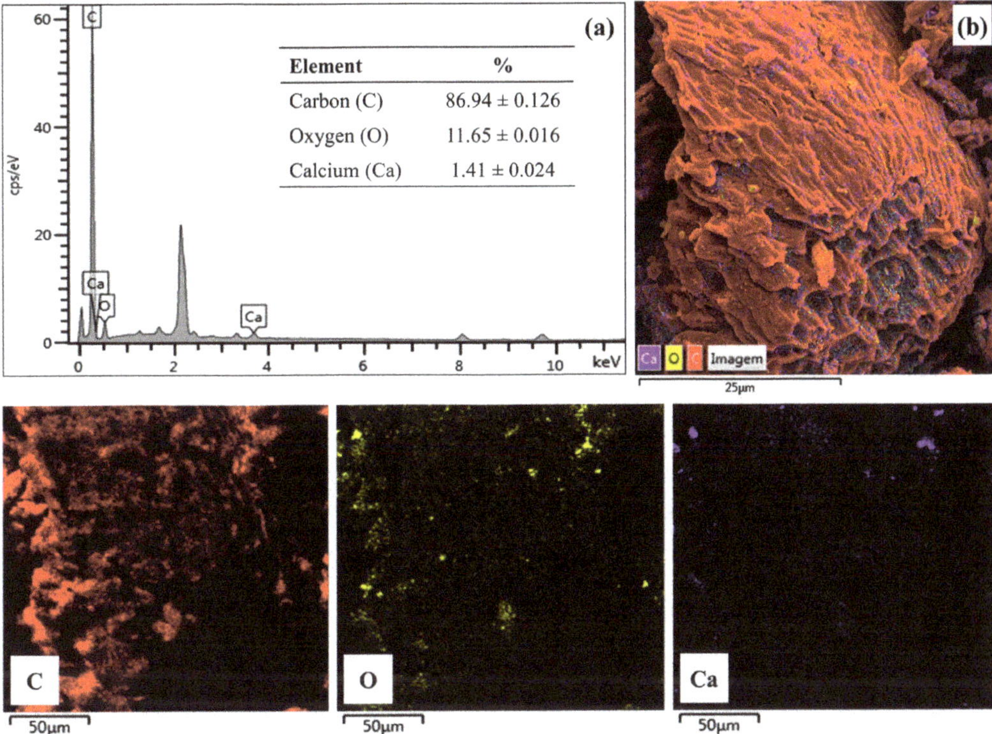

Figure 6. EDS (**a**) chemical composition and (**b**) elementary mapping of the chemical elements on the surface of BCAM support.

Figure 7 shows the composition and elemental mapping for catalyst 15Na/BCAM. The elemental composition shown in Figure 7a initially exposes the presence of 6.32% ± 0.157 sodium in the catalyst, which suggests that the impregnation process was efficient. In addition, the presence of 2.79% ± 0.512 silicon (Si) is also observed, a value close to the value of the 2:1 stoichiometric ratio of Na:Si present in Na_2SiO_3. Moreover, it is possible to perceive the presence of the elements carbon (C), oxygen (O), and calcium (Ca) in the values of 68.68% ± 0.157, 21.11% ± 0.048 and 1.1% ± 0.034, respectively, referring to the elemental composition of the BCAM support. It should be noted that the increase in the percentage of the oxygen element (O) is directly related to the process of impregnation of the BCAM support as a function of the percentage of sodium from the precursor in the Na_2SiO_3. Additionally, the elemental mapping for the 15Na/BCAM catalyst shown in Figure 7b suggests good dispersion of the elements carbon (C), oxygen (O), sodium (Na), silicon (Si), and calcium (Ca) on the catalyst surface.

2.3. Study of the Influence of Reaction Parameters for Biodiesel Synthesis

The study of process optimization is one of the main aspects that must be considered so that the biodiesel production process becomes economically viable for large-scale commercialization [16]. Thus, in order to determine the optimal reaction conditions, the study of the reaction variables temperature, reaction time, catalyst concentration and MeOH:oil molar ratio in the transesterification reaction of soybean oil using the catalyst 15Na/BCAM was performed. Figure 8 shows the results obtained.

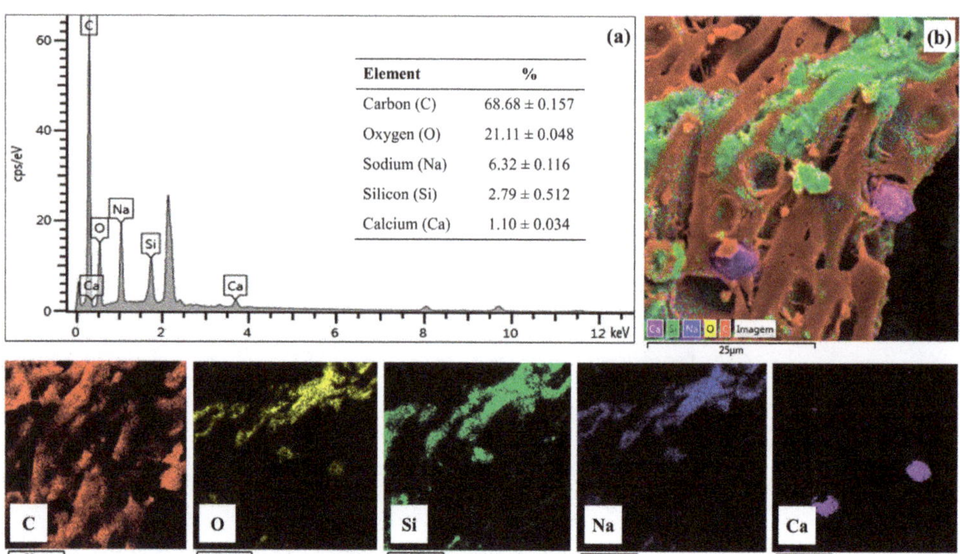

Figure 7. EDS (**a**) elementary composition and (**b**) elemental maps of each constituent of the catalyst 15Na/BCAM.

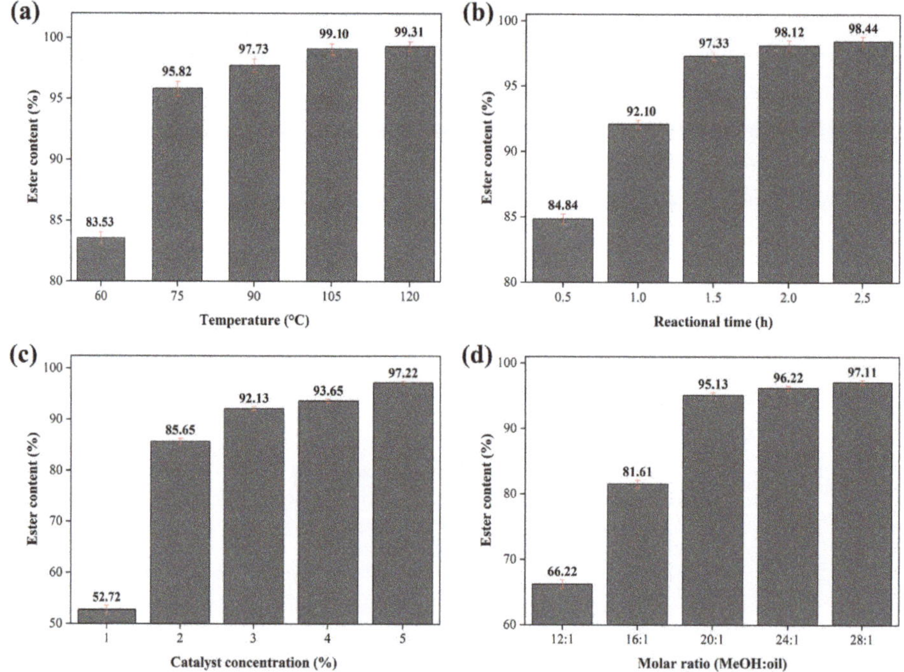

Figure 8. Influence of reactional parameters (**a**) temperature, (**b**) reaction time, (**c**) catalyst concentration and (**d**) MeOH:oil molar ratio in the process of production biodiesel using the catalyst 15Na/BCAM.

The temperature parameter is one of the factors with the greatest influence on the transesterification reaction. The study range used for temperature was 60–120 °C, keeping the other variables fixed: reaction time of 2 h, catalyst concentration 3% (w/w) and MeOH:oil molar ratio of 20:1. It can be inferred from Figure 8a that, as the reaction temperature rises, there is an increase in the conversion of triglycerides into esters, as observed at the reaction temperatures of 60, 75, 90, 105 and 120 °C, in which the synthesized biodiesels obtained ester contents of 83.53 ± 0.51, 95.82 ± 0.61, 97.73 ± 0.55, 99.10 ± 0.44 and 99.31% ± 0.51, respectively. Initially, this may be related to the endothermic character of the transesterification reaction, which requires quantities of heat to proceed in a straight direction [56], that is, as the temperature increases, there is a tendency to produce biodiesels with higher ester contents. In addition, high temperatures in the process promote better interactions between the reagents and, as a consequence, there are greater interactions of these with the active sites of the catalyst, resulting in better conversions [41]. However, it was observed that when the temperature was raised from 105 to 120 °C and from 90 to 105 °C, there were small increases in the ester contents of biodiesels, by about 0.21 and 1.58%, respectively. It is worth noting that the temperatures of 75 and 90 °C led to biodiesels with similar ester contents with a difference of only 1.91%. Thus, considering the energy cost to provide an additional 15 °C, the temperature of 75 °C was selected for the continuity of the study.

The reaction time is also an essential parameter when considering the biodiesel production process, as it significantly influences the catalytic activity of the catalyst used in the process [7]. The time optimization study was performed in the range of 0.5–2.5 h, fixing the other variables: temperature 75 °C, catalyst concentration 3% (w/w) and MeOH:oil molar ratio of 20:1. The data illustrated in Figure 8b suggest that the reaction time of 0.5 h is insufficient to conduct the reaction efficiently, providing a biodiesel with an ester content of 84.84% ± 0.42. This fact can be attributed to insufficient time for the interaction between the reagents and the active sites of the catalyst [42]. It is also possible to suggest, based on Figure 8b, a linear growth relationship between the reaction time and the ester content, as observed more explicitly in the reaction times of 0.5, 1.0 and 1.5 h, which led to biodiesels with ester contents of 84.84 ± 0.42, 92.10 ± 0.33 and 97.33% ± 0.34, respectively. However, when the reaction time was increased from 2 h to 2.5 h, a subtle increase in the ester content of about 0.32% was observed. In addition, the reactions conducted in the reaction times of 1.5 and 2 h provided biodiesels with very close ester contents, with a difference of 1.11%. Thus, in view of the economy of the biodiesel production process, the reaction time chosen was 1.5 h.

The study on the influence of catalyst concentration is illustrated in Figure 8c. The variables temperature of 75 °C, reaction time of 1.5 h and MeOH:oil molar ratio of 20:1 were kept fixed and the concentration of catalyst used in the transesterification reaction varied from 1 to 5%. When the reaction was processed using a 1% catalyst concentration, a biodiesel with ester content of 52.72% ± 0.84 was obtained. This low conversion to esters observed is directly linked to the insufficient number of active sites available for the reaction to process efficiently, as reported by Jamil et al. [21]. In addition, it was observed that, as the concentration of catalysts used in the reaction increased, there was an increase in the content of esters in biodiesels, reaching a maximum value of 97.22% ± 0.31 when the concentration of 5% was used in the process. Thus, the catalyst concentration of greater viability for the reaction process was 5% (w/w).

Finally, the influence of the MeOH:oil molar ratio on the transesterification reaction is shown in Figure 8d. The parameters of temperature 75 °C, reaction time 1.5 h and catalyst concentration of 5% (w/w) defined during the study were kept fixed. Numerous studies based on the stoichiometry of the transesterification reaction report that only 3 moles of alcohol are needed for the reaction to proceed. However, based on the reversible character of this reaction, excess alcohol is used [21]. It can be inferred from Figure 8d that the biodiesels obtained from the molar ratios of 12:1 and 16:1 presented low ester contents of 66.22 ± 0.68 and 81.61% ± 0.61, respectively. This trend may be related to an insufficient amount of alcohol in order to react with the triacylglycerols [42]. However, when the molar ratio was raised from 24:1 to 28:1, there was a small increase in conversion, about 0.89%. This may be attributed to the extremely diluted reaction system caused by excess methanol, which hinders the interaction between the reagents and the active sites of the catalyst [26]. It is worth highlighting that, when using the MeOH:oil molar ratio of 20:1, a biodiesel with an ester content of 95.13% ± 0.45 was obtained, a value very close to the ester contents of 96.22 ± 0.38 and 97.11% ± 0.35 of the biodiesels synthesized using the molar ratios of 24:1 and 28:1, respectively. Thus, the MeOH:oil molar ratio of 20:1 proved to be ideal and economically viable for the reaction process studied.

2.4. Physical and Chemical Properties of Biodiesel

The biodiesel synthesized using the catalyst 15Na/BCAM under optimal reaction conditions was characterized physicochemically according to the ASTM D6751 norm [60], with the results as shown in Table 2. It is observed that the values of 4.47 mm^2 s^{-1} and 0.880 g cm^{-3} regarding the kinematic viscosity and density of biodiesel, respectively, comprise the limits stipulated by ASTM D6751. Such values of these properties for biodiesel are of paramount importance since they are directly related to the fuel flow capacity and atomization during fuel injection into the engine [15,21]. The biodiesel obtained during the study is in accordance with the limit established by the ASTM D6751 norm, since the acidity value determined was 0.2 mg KOH g^{-1}. This data suggest a lower formation of scale in the engine components, since relatively small acidity values in biodiesel lead to an increase in the service life of pumps and filters [21,49].

Table 2. Physicochemical properties of soybean biodiesel synthesized using the catalyst 15Na/BCAM and the limits stipulated by ASTM D6751 standard.

Biodiesel Properties	Unit	Test Methods	ASTM D6751 Limits	Present Study
Kinematic viscosity (at 40 °C)	mm^2 s^{-1}	ASTM D445 [61]	1.9–6.0	4.47
Density (at 20 °C)	g cm^{-3}	ASTM D1298 [62]	0.875–0.900	0.880
Acid value	mg KOH g^{-1}	ASTM D664 [63]	0.5 max	0.20
Cold filtre plugging point	°C	ASTM D6371 [64]	NS	0.0
Flash point	°C	ASTM D93 [65]	130 min	150
Copper strip corrosion	-	ASTM D130 [66]	3 max	1a

NS = Not specified.

The cold plugging point is directly related to the amount of esters derived from long-chain saturated fatty acids, which can lead to clogging of engine filters during fuel cooling [8,67]. The observed value for the cold plugging point was 0.0 °C, indicating that the biodiesel synthesized in this study can be used in engines, even in regions of low temperatures. The flash point is the main aspect responsible for certifying the safety of a fuel because it indicates the amount of residual methanol present in the biodiesel [17,49]. The value obtained for the synthesized soybean biodiesel was 150 °C, a value very similar to that reported by Mares et al. [8], which was 152 °C for biodiesel obtained using the ASA-800/4 catalyst under optimal reaction conditions. The value determined for the copper corrosivity parameter was 1a, which is within the limit stipulated by the ASTM D6751 norm. Thus, the data explained testify that the catalyst 15Na/BCAM is suitable for the synthesis of biodiesel since such a biofuel demonstrates high quality and compliance with the ASTM D6751 norm.

2.5. Catalyst Reuse Study

Among the numerous advantages offered by the application of heterogeneous catalysts in the transesterification of oils, the ability to reuse and the possibility of regeneration are indispensable characteristics when considering the reduction of biodiesel production costs [12]. The reuse study of the catalyst 15Na/BCAM over several reaction cycles using the optimal reaction conditions, temperature of 75 °C, reaction time of 1.5 h, catalyst concentration of 5% (*w/w*) and MeOH:oil molar ratio of 20:1, is shown in Figure 9.

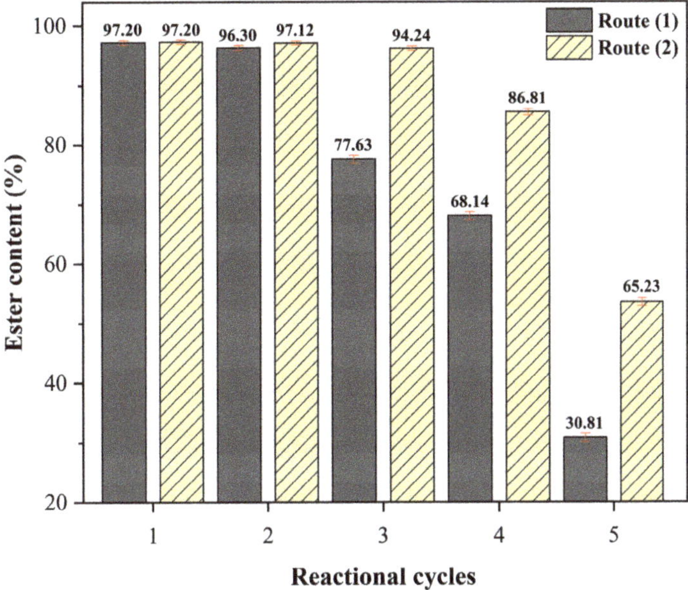

Figure 9. Study of reuse cycles of the catalyst 15Na/BCAM recovered using Route (1) and Route (2). Reaction conditions: temperature of 75 °C, reaction time of 1.5 h, catalyst concentration of 5% (*w/w*) and MeOH:oil molar ratio of 20:1.

From the data shown in the graph of Figure 9, it is possible to observe that the catalyst 15Na/BCAM remained effective in its second reaction cycle because the biodiesel presented an ester content of 96.30% ± 0.42, a decrease of only 0.9% in relation to the biodiesel produced in the first reaction cycle, which was 97.20% ± 0.41. In addition, it can be inferred that the active sites present in the synthesized catalyst were not easily leached into the reaction medium since the biodiesels obtained in the third and fourth reaction cycles showed a reduction in ester content of 19.57 and 29.06%, respectively, in relation to the initial reaction. However, the fifth reaction cycle showed a significant decrease in catalytic activity, providing a product with an ester content of 30.81% ± 0.71. This decrease in ester content may be related to the deposition of unconverted mono-, di-, triglycerides, residual glycerol or esters not removed during the recovery process using Route (1), which consists only of washing with solvents (ethyl alcohol + hexane). This leads to the blocking of active sites on the surface of the catalyst, which makes it difficult for the reaction to proceed efficiently [68].

This hypothesis was tested by performing Route (2), which consists of the combination of the solvent washing process and the thermal reactivation of the catalyst in the tubular oven at 400 °C for 1 h in an atmosphere of N_2 after each reaction cycle. Based on the data shown in Figure 9, it can be inferred that the hypothesis raised is true since the catalyst 15Na/BCAM recovered using Route (2) showed a better catalytic performance throughout the reuse cycles studied, that is, greater stability in the reaction medium since the biodiesels synthesized in the second, third, and fourth reaction cycles resulted in ester contents of 97.12 ± 0.35, 94.24 ± 0.38 and 86.81% ± 0.55, respectively. The biodiesel obtained in the fifth reaction cycle showed a reduction of 31.97% compared to the initial reaction, resulting in an ester content of 65.23% ± 0.65.

This downward trend in the performance of the catalyst 15Na/BCAM may be related to leaching of a significant amount of basic sites. Table 3 shows the results of the basicity analysis of the fresh catalyst, that is, before being used in the reaction and of the catalyst used after the fifth reaction cycle recovered via Route (2). It is observed in Table 3 that there was a decrease in the presence of basic groups between the fresh and the recovered catalysts via Route (2), since they presented basicity values of 3.911 mmol $g^{-1} \pm 0.031$ and 1.420 mmol $g^{-1} \pm 0.011$, respectively. In addition, the data shown in Figure S1 (available in the Supplementary Materials) reinforce the data shown about the basicity of the catalysts, since the percentage of sodium in the catalyst after the fifth reaction cycle was 1.84% $\pm$ 0.056, while in the catalyst before the reaction it was 6.32% $\pm$ 0.116. This behavior was also reported in the study developed by Zhao et al. [30], in which the authors observed a decrease in catalytic activity between the catalyst used in the first use and the catalyst used in the eighth reaction cycle, which provided biodiesels with yields of 98% and 82.4%, respectively. In addition, the authors justify these results by the decrease in basicity over the reuse cycles since the basicity value for the fresh catalyst was 9.0 mmol g^{-1}, while the basicity determined for the catalyst after the 8° reaction cycle was 4.8 mmol g^{-1}.

Table 3. Basicity analysis of the fresh catalysts and catalysts used for 5° cycle reactional recovered using Route (2).

Sample	Basicity (mmol g^{-1})
Catalyst fresh	3.911 ± 0.031
Catalyst after 5° reactional cycle	1.420 ± 0.011

2.6. Catalyst Regeneration Study

During the reuse study of the catalyst 15Na/BCAM recovered via Route (2), a decrease in the ester content of biodiesel produced in the fifth reaction cycle of about 31.97% compared to the first reaction cycle was observed. In general, this reduction in catalyst performance can be attributed to a decrease in the basic active sites responsible for conducting the transesterification reaction efficiently. Given this, the applicability of the catalyst regeneration process using sodium silicate in an amount that would allow obtaining the concentration of 10% (w/w) of sodium was evaluated in the final composition of the catalyst. Figure 10 illustrates the data obtained in the study of the catalyst regeneration process.

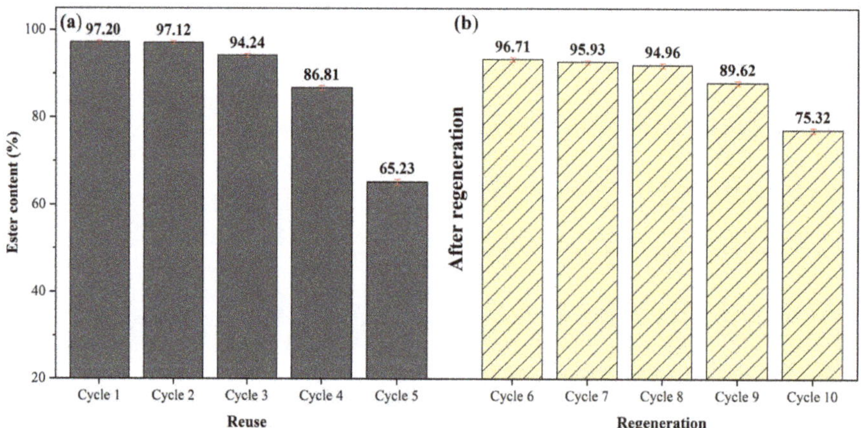

Figure 10. (**a**) Reuse study of the catalyst 15Na/BCAM recovered using Route (2) and (**b**) study of regeneration of the catalyst 15Na/BCAM. Reaction conditions: temperature of 75 °C, reaction time of 1.5 h, catalyst concentration of de 5% (w/w) and MeOH:oil molar ratio of 20:1.

Figure 10a presents the data corresponding to the reuse study of the catalyst 15Na/BCAM recovered Route (2), the starting point for the development of the study of the regeneration process. From the data presented in Figure 10b, it is possible to infer that the catalyst regeneration process was successful since the biodiesel ester content of 96.71% ± 0.65 obtained in the sixth reaction cycle (first after regeneration) represents an increase of 31.48% in relation to the fifth reaction cycle. In addition, the catalyst used in the seventh and eighth reaction cycles led to biodiesels with ester contents of 95.93 ± 0.63 and 94.96% ± 0.73, respectively, which represents insignificant decreases of 0.78 and 1.75% in the ester contents of the biodiesels produced when compared to the sixth reaction cycle. The data suggest that the catalyst 15Na/BCAM exhibits excellent catalytic stability in the reaction medium. However, after the ninth reaction cycle of the catalyst, a decrease in catalytic activity was observed, since the biodiesels synthesized in the ninth and tenth reaction cycles obtained ester contents of 89.62% ± 0.76 and 75.32% ± 0.77, respectively, representing a reduction in catalyst performance by about 7.09 and 21.39% compared to the first use after regeneration process (sixth reaction cycle), as shown in Figure 10b.

The applicability and efficiency and reusability of the catalyst 15Na/BCAM regenerated with 10% (*w/w*) sodium and recovered via Route (2) in the conduction of the transesterification reaction are highlighted due to the fact that it produces biodiesels with ester contents above 89% in the ninth reaction cycle. In addition, the catalyst 15Na/BCAM is highly stable in the reaction medium when compared to the acai seed ash catalyst (ASA-800/4), also submitted to the regeneration study using 20% (*w/w*) of KOH, which showed a decline in its seventh reaction cycle (third reaction cycle after regeneration) resulting in a biodiesel with ester content of 84.9%, and at the end of the ninth reaction cycle (fifth reaction cycle after regeneration) showed significant loss in catalytic activity reaching an ester content of only 36.7% [8], while the catalyst 15Na/BCAM regenerated with 10% (*w/w*) sodium and recovered via Route (2) in its tenth reaction cycle, maintains its catalytic activity producing a biodiesel with an ester content of above 75%.

2.7. Comparison of Catalyst 15Na/BCAM with Different Biochar-Based Catalysts

Table 4 illustrates the performance in the biodiesel production process of the catalyst 15Na/BCAM developed in this study, as well as other heterogeneous basic catalysts based on biochar from different biomasses reported in the literature. Based on the data presented, it can be inferred that the BCAM support has milder synthesis (carbonization) conditions since it is prepared at 600 °C per 1 h when compared to avocado seed biochar and rice husk, which are produced at 900 °C for 2 h and 700 °C for 3 h, respectively [29,30]. In addition, the heat treatment process to which the catalyst 15Na/BCAM was subjected, a temperature of 400 °C for 1 h, which is a lower parameter than the other catalysts presented in Table 4, can be highlighted.

Table 4. Different types of catalysts heterogeneous basic based on carbon and their synthesis parameters for biodiesel production.

| Precursor | Catalyst | Synthesis of Catalyst | | | | Reactional Conditions | | | | Ester Content (%) | Cycles | References |
| | | Carbonization | | Functionalization | | T (°C) | t (h) | Catalyst (wt%) | RM (MeOH:oil) | | | |
		T (°C)	t (h)	T (°C)	t (h)							
Avocado seeds	20 wt%. Ca loaded	900	2	900	2	99.5	5.0	7.3	15.6:1	99.50	3	[29]
Banana peel	30K/BP-600	600	2	600	4	65.0	2.0	4.0	15:1	98.91	5	[32]
Pomelo peel	25K/AP-600	600	2	600	3	65.0	2.5	5.0	8:1	98.00	8	[31]
Date seeds	SrO-carbon	400	5	450	4	65.0	1.5	4.0	15:1	94.27	9	[21]
Sargassum oligocystum	Biochar/CaO/K_2CO_3	350	2	500	3	65.0	3.3	4.0	18:1	98.83	9	[49]
Rice husk	30Ca/RB700	700	3	700	4	65.0	3.0	8.0	9:1	94.40	10	[30]
Urea	CaO-MgO-800-5	800	5	800	5	70.0	4.4	8.0	21:1	91.10	4	[69]
Commercial activated carbon	CaO/AC	—	—	450	1	190.0	1.35	5.5	15:1	80.98	3	[70]
Murici seed	15Na/BCAM	600	1	400	1	75	1.5	5.0	20:1	97.20	10	**This study**

Considering the reaction parameters for the synthesis of biodiesel in which the catalysts listed in Table 4 are used, it is possible to note some similarity with the reaction conditions used for the catalyst 15Na/BCAM. However, the catalyst 15Na/BCAM showed greater catalytic stability throughout the reaction cycles since it provided 10 reaction cycles (five before the regeneration process and five after the process) when used under optimal reaction conditions. This is evidenced when this value is compared with the stability presented by the catalysts 20 wt% Ca loaded and 30K/BP-600, which provided only three and five reaction cycles, respectively [29,32]. In addition, Table 4 shows that the 15Na/BCAM catalyst studied has good catalytic activity, resulting in biodiesel with an ester content of 97.20% ± 0.31 under mild reaction conditions (75 °C, 1.5 h, 5% and 20:1), compared to the catalytic performances of the CaO-MgO-800-5 and CaO/AC catalysts, which produce biodiesels with ester contents of 91.1% and 80.98%, respectively, under relatively abrupt reaction conditions of 190 °C, 1.35 h, 5.5% and 15:1 for the CaO-MgO-800-5 catalyst and 70 °C, 4.4 h, 8% and 21:1 for the CaO/AC catalyst [69,70]. Thus, the data listed in Table 4 reinforce the feasibility of using the residual biomass of the murici seed in the preparation of an efficient basic heterogeneous catalyst for the synthesis of biodiesel.

3. Materials and Methods

3.1. Materials

Murici fruit and refined soybean oil were purchased at a local market in the city of Belém, Pará, Brazil. Pure sodium silicate (Na_2SiO_3) (VERTEC®, Silchester, UK, 1344-09-8) was used in the process of biochar functionalization and catalyst regeneration. Sodium hydroxide (NaOH) 98.0% (Neon®, New York, NY, USA, 310-73-2) and hydrochloric acid (HCl) 37.0% (Qhemis®, Jundiaí, Brazil, 7647-01-0) were used in determining the basic groups present in murici biochar and catalyst. N-heptane UV/HPLC (C_7H_{16}) 99.0% (Neon®, 142-82-5) and methyl heptadecanoate ($C_{18}H_{36}O_2$) 99.0% (Sigma-Aldrich®, St. Louis, MO, USA, 1731-92-6) were used in the quantification of esters by gas chromatography. Methyl alcohol (CH_3OH) 99.8% (ÊXODO CIENTÍFICA®, Sumaré, Brazil, 67-56-1) was used in the transesterification reactions, while ethyl alcohol (C_2H_5OH) (Dinâmica®, Barcelona, Spain, 64-17-5) and n-hexane (C_6H_{14}) 95.0% (Neon®, 110-54-3) were used in catalyst recovery.

3.2. Catalyst Synthesis

Initially, the murici fruit (Figure 11a) was pulped and washed with distilled water to remove dirt and remnants of the pulp. Then, the seeds were dried in an oven at 105 °C for 12 h. Then, the seeds were ground and sieved (10 mesh). Soon after, the material with higher granulometry (>10 mesh) was pyrolyzed in a tubular oven at 600 °C for 1 h with N_2 flow of 80 mL min^{-1} according to the methodology described by Corrêa et al. [26], with minor modifications. Finally, the murici biochar (BCAM) was cooled to room temperature, packed in a hermetically sealed container and stored in a desiccator.

The preparation of catalysts based on sodium functionalized biochar was performed by wet impregnation according to the procedures described by Alsharif et al. [71] and Bitonto et al. [29], with adaptations, as shown in Figure 11b. Initially, the BCAM support mass was dispersed in a beaker with distilled water, to which the sodium silicate mass was added in amounts that made it possible to obtain the concentrations 6, 9, 12, 15 and 18% (w/w) of sodium in the final catalyst composition. In a typical process for the preparation of a catalyst impregnated with 15% (w/w) sodium, the mass of 5.0 g of dry and previously activated BCAM support at 105 °C was dispersed in 60 mL of distilled water and then was added to the mass of 3.35 g of sodium silicate. Then, the suspension was kept in an ultrasonic bath at 25 °C for 2 h, followed by magnetic stirring for 2 h. When it was finished, the mixture was dried in an oven at 105 °C for 12 h, followed by heat treatment in a tubular oven at 400 °C for 1 h under a flow of 80 mL min^{-1} of N_2. The synthesized catalysts were named xNa/BCAM, in which x represents the impregnated sodium concentration.

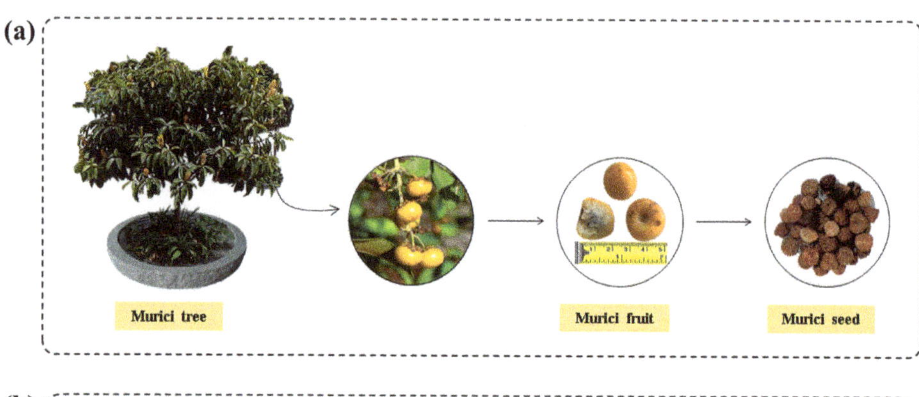

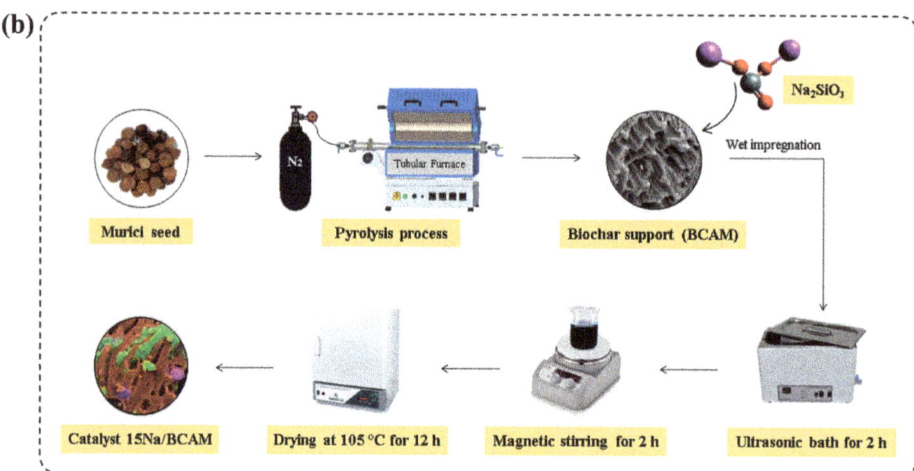

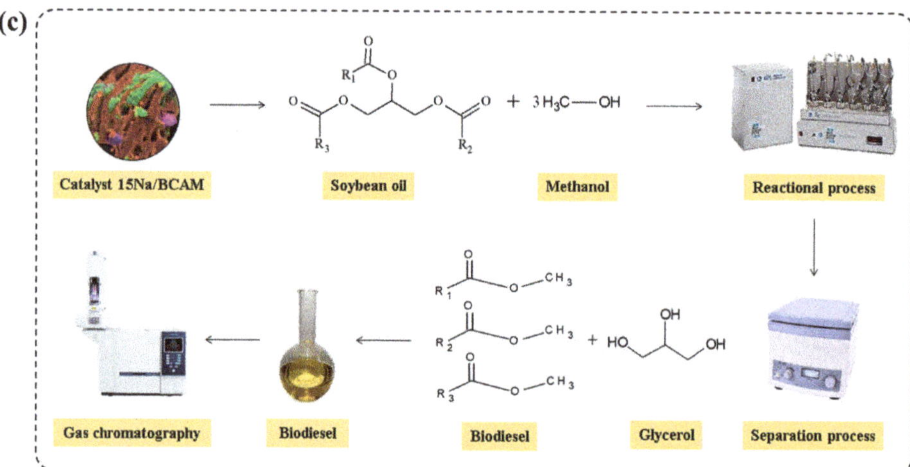

Figure 11. Scheme of synthesis (**a**) BCAM support, (**b**) catalysts xNa/BCAM and (**c**) biodiesel.

3.3. Catalyst Characterization

The determination of the basic groups present in the biochar, catalyst and regenerated catalyst was performed according to the Boehm [72] method, with adaptations. In this method, a mass of 0.25 g of the material (BCAM support and catalyst xNa/BCAM) was dispersed in 30 mL of standard solution HCl 0.10 mol L^{-1} and kept being stirred for 24 h at room temperature. When it was finished, the mixture was centrifuged and an aliquot of 10 mL of the filtrate was mixed with 15 mL of standard NaOH solution 0.10 mol L^{-1}. Then, the excess NaOH was titrated with HCl 0.10 mol L^{-1} solution, using phenolphthalein as an indicator. The Elemental Composition (CHNS) of the materials (murici seed, BCAM support, and xNa/BCAM) was performed on a Fisons Instruments EA1108 equipment with the aid of the EAGER 200-RESULTS software. The masses of approximately 2.0 mg of samples were weighed on a Sartorius microscale (XM-1000 P). Thermogravimetric Analysis (TG/DTG) for the murici seed, BCAM support, Na_2SiO_3, and xNa/BCAM was performed using Shimadzu equipment, model DTG-60H, between the temperature range of 25 °C to 1000 °C (heating rate of 10 °C min^{-1}), in a flow of N_2 of 50 mL min^{-1}, using an alumina crucible. The X-ray Diffraction (XRD) patterns of the BCAM support, in the Na_2SiO_3 and xNa/BCAM, were obtained using the powder method in a PANalytical diffractometer, EMPYREAN model, with radiation employed and scanning range of Cu Kα (1.541874 Å) at 40 KV and 30 mA and $10° < 2\theta < 70°$, respectively. The analysis of Fourier Transform Infrared Spectroscopy (FT-IR) for the BCAM support, Na_2SiO_3, and xNa/BCAM, was performed using a Shimadzu spectrometer model IRPrestige-21 in the spectral range of analysis of 1900–500 cm^{-1} with 4 cm^{-1} resolution and 32 scans. The surface morphology of the BCAM support and xNa/BCAM was determined by Scanning Electron Microscopy (SEM) using a TESCAN microscope, model VEGA 3 LMU, operating with acceleration voltage of 20 kV. The elemental composition of the BCAM support and xNa/BCAM surface was determined by X-ray Spectroscopy by Energy Dispersion (EDS) using an Oxford microanalysis system, model AZTec Energy X-Act, with resolution of 129 eV.

3.4. Transesterification Reaction

The catalytic tests of methyl transesterification of soybean oil, using functionalized biochar as a catalyst, was conducted in a 80 mL capacity closed steel reactor of the Parr Series 5000 Multiple Reactor System under fixed agitation of 700 rpm. The reaction parameters were optimized from the evaluation of reaction temperature (60–120 °C), reaction time (0.5–2.5 h), catalyst concentration (1–5%), and MeOH:oil molar ratio (12:1–28:1). In a typical reaction procedure, 12 g of soybean oil was added to the reaction vessel, followed by the addition of 11.25 mL of methanol (referring to a MeOH:oil molar ratio of 20:1) and 0.6 g of catalyst (considering a catalyst concentration of 5%). Then, temperature and reaction time were set to 75 °C and 1.5 h, respectively. The catalyst concentrations used in catalytic tests were calculated as a function of the mass of oil employed in the reaction process.

At the end of each reaction experiment, the products were centrifuged to recover the catalyst, transferred to a settling funnel, and washed with portions of hot distilled water ~80 °C. Later, the remnants of washing water present were removed by drying in an oven at 60 °C for 12 h. Finally, the synthesized biodiesel was stored for the ester content analysis. A test was also performed using only the BCAM support (control) in order to evaluate its contribution to the catalytic activity of the catalyst in the reaction process. The biodiesel synthesis procedure is illustrated in Figure 11c.

3.5. Characterization of Biodiesel

3.5.1. Ester Content

The determination of the ester content of biodiesels was performed according to the methodology adapted by Silva et al. [73] from the European standard EN 14103, using a Varian gas chromatograph (GC), model CP 3800, equipped with flame ionization detector (FID), capillary column CP WAX 52 CB (30 m length, 0.32 mm diameter and 0.25 μm film), with initial oven temperature programming of 170 °C, with heating rate of 10 °C min^{-1}

up to 250 °C (same temperature as FID and injector). Helium gas was used in the flow of 1 mL min^{-1} as mobile phase, heptane as solvent, methyl heptadecanoate as internal standard, and an injection volume of 1.0 µL. The ester content (EC) was calculated using Equation (1).

$$\text{Ester content}(\%) = \frac{\sum A_T - A_{P.I}}{A_{P.I}} \times \frac{C_{P.I}}{C_{B100}} \times 100 \tag{1}$$

where: $\sum A_T$ is the sum of the total area of the ester peaks; A_{PI} is the area of the internal pattern; C_{PI} is the concentration of the internal standard solution (mg L^{-1}); and C_{B100} is the final concentration of biodiesel after dilution (mg L^{-1}).

3.5.2. Physicochemical Properties of Biodiesel

The biodiesel produced was analyzed physicochemically using the American Society for Testing and Materials (ASTM) standard methodology [60]. The kinematic viscosity at 40 °C was determined according to ASTM D445 method [61], using a cannon-Fenske viscometer (SCHOTT GERATE, Model No. 520 23). The density was performed at 20 °C using the ASTM D1298 method [62], with an automatic KEM DAS-500 densimeter. The acid value of biodiesel was determined according to the ASTM D664 method [63]. The cold filter plugging point was determined according to the ASTM D6371 method [64] in a TANAKA equipment, model AFP-102. The flash point was determined according to the ASTM D93 method [65] on a TANAKA APM 7 Pensky-Martens automatic equipment. Finally, the corrosiveness of copper was determined according to the ASTM D130 method [66] in a Koehler corrosion bath.

3.6. Catalyst Recovery

After the catalytic tests, the catalyst was recovered by centrifugation. Then, the catalyst was subjected to two recovery routes. Route (1) consisted only of washing with portions of ethyl alcohol and hexane, and then drying at 60 °C for 12 h. In Route (2), the catalyst was subjected to the combination of the washing process used in Route (1) with thermal reactivation in a tubular oven at 400 °C for 1 h in an atmosphere of N$_2$. The catalyst recovered by both routes was reused in successive transesterification reactions of soybean oil under optimal conditions.

3.7. Catalyst Regeneration

As the catalyst demonstrated a reduction in its catalytic activity during the reuse study, the feasibility of employing the regeneration process using sodium silicate in an amount that would allow the concentration of 10% (w/w) of sodium in the final composition of the catalyst was evaluated. The regeneration process was performed by wet impregnation according to the procedures described by Alsharif et al. [71] and Bitonto et al. [29], with minor adaptations. Initially, 0.23 g of sodium silicate was dissolved in 20 mL of distilled water and then the catalyst previously used, activated at 105 °C, was slowly added. Then, the suspension was kept in an ultrasonic bath at 25 °C for 2 h, followed by magnetic stirring for 2 h. Then, the mixture was dried in an oven at 105 °C for 12 h, followed by heat treatment in a tubular oven at 400 °C for 1 h under flow of 80 mL min^{-1} of N$_2$. The regenerated catalyst was evaluated for its reusability.

4. Conclusions

The present study proposed the synthesis of a heterogeneous basic catalyst based on biochar produced from the residual biomass of murici functionalized with sodium for application in the synthesis of biodiesel. The characterizations performed by CHNS, TG/DTG, XRD, FTIR, SEM, EDS and basicity confirmed that the impregnation process used in the synthesis of the catalyst was effective. The synthesized catalyst 15Na/BCAM showed the best results during the study, obtaining a biodiesel with an ester content of 97.20% ± 0.31, as well as in accordance with the ASTM D6751 norm, using optimal reaction conditions: temperature of 75 °C, reaction time of 1.5 h, catalyst concentration of 5% (w/w)

and MeOH:oil molar ratio of 20:1. The catalyst 15Na/BCAM was used in five reaction cycles, maintaining biodiesels with ester contents above 65%. In addition, the catalyst was subjected to the regeneration process resulting in increased catalytic activity, providing five more reaction cycles and obtaining biodiesels with ester contents above 75%. Thus, the heterogeneous catalyst based on biochar functionalized with sodium proved to be suitable for the application in the transesterification reaction of soybean oil via methyl route, and applicable for reuse and regeneration processes, indicating a promising alternative to low-cost waste raw materials for the synthesis of basic heterogeneous catalysts.

Supplementary Materials: The following supporting information can be downloaded at: https://www.mdpi.com/article/10.3390/molecules28247980/s1, Figure S1: EDS chemical composition and elementary mapping of the chemical elements on the surface Catalyst after 5° reactional cycle.

Author Contributions: T.S.R.: Conceptualization, Methodology, Investigation, Visualization, Formal analysis, Writing—original draft. M.A.G.: Methodology, Resources, Investigation. G.N.d.R.F.: Supervision. L.R.V.d.C.: Project administration, Conceptualization, Visualization, Supervision, Project administration, Writing—Review and Editing. All authors have read and agreed to the published version of the manuscript.

Funding: This research was funded by Coordination for the Improvement of Higher Education Personnel—Brazil (CAPES)—funding code 001 and was supported by the Pró-Reitoria de Pesquisa da UFPA PROPESP/UFPA (PAPQ).

Institutional Review Board Statement: Not applicable.

Informed Consent Statement: Not applicable.

Data Availability Statement: Data are contained within the article and Supplementary Materials.

Acknowledgments: The authors thank the Graduate Program in Chemistry of the Federal University of Pará (PPGQ/UFPA) and the Laboratory of Catalysis and Oleochemical (LCO/UFPA), Research Laboratory and Fuel Analysis (LAPAC/UFPA), X-ray Diffraction Laboratory (PPGF/UFPA), Metallurgy Laboratory of Federal Institute of Technical Education of Pará (IFPA) for structural support.

Conflicts of Interest: The authors declare no conflict of interest.

References

1. Lin, K.S.; Mdlovu, N.V.; Chan, H.Y.; Wu, K.C.W.; Wu, J.C.S.; Huang, Y.T. Preparation and characterization of mesoporous polymer-based solid acid catalysts for biodiesel production via transesterification of palmitic oils. *Catal. Today* **2022**, *397–399*, 145–154. [CrossRef]
2. Al-Saadi, A.; Mathan, B.; He, Y. Esterification and transesterification over SrO–ZnO/Al$_2$O$_3$ as a novel bifunctional catalyst for biodiesel production. *Renew. Energy* **2020**, *158*, 388–399. [CrossRef]
3. Soares, S.; Fernandes, G.M.; Moraes, L.M.B.; Batista, A.D.; Rocha, F.R.P. Single-phase determination of calcium and magnesium in biodiesel using smartphone-based digital images. *Fuel* **2022**, *307*, 121837. [CrossRef]
4. Zhang, F.; Wu, X.H.; Yao, M.; Fang, Z.; Wang, Y.T. Production of biodiesel and hydrogen from plant oil catalyzed by magnetic carbon-supported nickel and sodium silicate. *Green Chem.* **2016**, *18*, 3302–3314. [CrossRef]
5. Bambase, M.E.; Almazan, R.A.R.; Demafelis, R.B.; Sobremisana, M.J.; Dizon, L.S.H. Biodiesel production from refined coconut oil using hydroxide-impregnated calcium oxide by cosolvent method. *Renew. Energy* **2021**, *163*, 571–578. [CrossRef]
6. Sabzevar, A.M.; Ghahramaninezhad, M.; Shahrak, M.N. Enhanced biodiesel production from oleic acid using TiO$_2$-decorated magnetic ZIF-8 nanocomposite catalyst and its utilization for used frying oil conversion to valuable product. *Fuel* **2021**, *288*, 119586. [CrossRef]
7. Rezania, S.; Mahdinia, S.; Oryani, B.; Cho, J.; Kwon, E.E.; Bozorgian, A.; Nodeh, H.R.; Darajeh, N.; Mehranzamir, K. Biodiesel production from wild mustard (*Sinapis arvensis*) seed oil using a novel heterogeneous catalyst of LaTiO$_3$ nanoparticles. *Fuel* **2022**, *307*, 121759. [CrossRef]
8. Mares, E.K.L.; Gonçalves, M.A.; da Luz, P.T.S.; Rocha Filho, G.N.; Zamian, J.R.; Conceição, L.R.V. Acai seed ash as a novel basic heterogeneous catalyst for biodiesel synthesis: Optimization of the biodiesel production process. *Fuel* **2021**, *299*, 120887. [CrossRef]
9. Zhang, S.; Wang, L.; Wang, J.; Yang, J.; Fu, L.; Cai, Z.; Fu, Y. A novel in situ transesterification of yellow horn seeds to biodiesel using a dual function switchable solvent. *Fuel* **2022**, *312*, 122974. [CrossRef]
10. Conceição, L.R.V.; Carneiro, L.M.; Giordani, D.S.; Castro, H.F. Synthesis of biodiesel from macaw palm oil using mesoporous solid catalyst comprising 12-molybdophosphoric acid and niobia. *Renew. Energy* **2017**, *113*, 119–128. [CrossRef]

11. Conceição, L.R.V.; Carneiro, L.M.; Rivaldi, J.D.; Castro, H.F. Solid acid as catalyst for biodiesel production via simultaneous esterification and transesterification of macaw palm oil. *Ind. Crop. Prod.* **2016**, *89*, 416–424. [CrossRef]

12. Quevedo-Amador, R.A.; Reynel-Avila, H.E.; Mendoza-Castillo, D.I.; Badawi, M.; Bonilla-Petriciolet, A. Functionalized hydrochar-based catalysts for biodiesel production via oil transesterification: Optimum preparation conditions and performance assessment. *Fuel* **2022**, *312*, 122731. [CrossRef]

13. Li, D.; Feng, W.; Chen, C.; Chen, S.; Fan, G.; Liao, S.; Wu, G.; Wang, Z. Transesterification of Litsea cubeba kernel oil to biodiesel over zinc supported on zirconia heterogeneous catalysts. *Renew. Energy* **2021**, *177*, 13–22. [CrossRef]

14. Ning, Y.; Niu, S.; Wang, Y.; Zhao, J.; Lu, C. Sono-modified halloysite nanotube with $NaAlO_2$ as novel heterogeneous catalyst for biodiesel production: Optimization via GA_BP neural network. *Renew. Energy* **2021**, *175*, 391–404. [CrossRef]

15. Munir, M.; Ahmad, M.; Saeed, M.; Waseem, A.; Nizami, A.S.; Sultana, S.; Zafar, M.; Rehan, M.; Srinivasan, G.R.; Ali, A.M.; et al. Biodiesel production from novel non-edible caper (*Capparis spinosa* L.) seeds oil employing Cu–Ni doped ZrO_2 catalyst. *Renew. Sustain. Energy Rev.* **2021**, *138*, 110558. [CrossRef]

16. Gonçalves, M.A.; Mares, E.K.L.; da Luz, P.T.S.; Zamian, J.R.; Rocha Filho, G.N.; Conceição, L.R.V. Statistical optimization of biodiesel production from waste cooking oil using magnetic acid heterogeneous catalyst $MoO_3/SrFe_2O_4$. *Fuel* **2021**, *304*, 121463. [CrossRef]

17. Santos, H.C.L.; Gonçalves, M.A.; Viegas, A.C.; Figueira, B.A.M.; da Luz, P.T.S.; Rocha Filho, G.N.; Conceição, L.R.V. Tungsten oxide supported on copper ferrite: A novel magnetic acid heterogeneous catalyst for biodiesel production from low quality feedstock. *RSC Adv.* **2022**, *12*, 34614–34626. [CrossRef]

18. De, A.; Boxi, S.S. Application of Cu impregnated TiO_2 as a heterogeneous nanocatalyst for the production of biodiesel from palm oil. *Fuel* **2020**, *265*, 117019. [CrossRef]

19. Martínez-Klimov, M.E.; Ramírez-Vidal, P.; Tejeda, P.R.; Klimova, T.E. Synergy between sodium carbonate and sodium titanate nanotubes in the transesterification of soybean oil with methanol. *Catal. Today* **2020**, *353*, 119–125. [CrossRef]

20. Boonphayak, P.; Khansumled, S.; Yatongchai, C. Synthesis of $CaO-SiO_2$ catalyst from lime mud and kaolin residue for biodiesel production. *Mater. Lett.* **2021**, *283*, 128759. [CrossRef]

21. Jamil, F.; Kumar, P.S.M.; Al-Haj, L.; Myint, M.T.Z.; Al-Muhtaseb, A.H. Heterogeneous carbon-based catalyst modified by alkaline earth metal oxides for biodiesel production: Parametric and kinetic study. *Energy Convers. Manag. X* **2021**, *10*, 100047. [CrossRef]

22. Cao, X.; Sunab, S.; Sun, R. Application of biochar-based catalysts in biomass upgrading: A review. *RSC Adv.* **2017**, *7*, 48793–48805. [CrossRef]

23. Low, Y.W.; Yee, K.F. A review on lignocellulosic biomass waste into biochar-derived catalyst: Current conversion techniques, sustainable applications and challenges. *Biomass Bioenerg.* **2021**, *154*, 106245. [CrossRef]

24. Sudarsanam, P.; Zhong, R.; Van den Bosch, S.; Coman, S.M.; Parvulescu, V.I.; Bert, F.S. Functionalised heterogeneous catalysts for sustainable biomass valorisation. *Chem. Soc. Rev.* **2018**, *47*, 8349–8402. [CrossRef] [PubMed]

25. Anto, S.; Sudhakar, M.P.; Ahamed, T.S.; Samuel, M.S.; Mathimani, T.; Brindhadevi, K.; Pugazhendhi, A. Activation strategies for biochar to use as an efficient catalyst in various applications. *Fuel* **2021**, *285*, 119205. [CrossRef]

26. Corrêa, A.P.L.; Bastos, R.R.C.; Rocha Filho, G.N.; Zamian, J.R.; Conceição, L.R.V. Preparation of sulfonated carbon-based catalysts from murumuru kernel shell and their performance in the esterification reaction. *RSC Adv.* **2020**, *10*, 20245–20256. [CrossRef] [PubMed]

27. Bastos, R.R.C.; Corrêa, A.P.L.; da Luz, P.T.S.; Rocha Filho, G.N.; Zamian, J.R.; Conceição, L.R.V. Optimization of biodiesel production using sulfonated carbon-based catalyst from an amazon agro-industrial waste. *Energy Convers. Manag.* **2020**, *205*, 112457. [CrossRef]

28. Lima, R.P.; da Luz da Luz, P.T.S.; Braga, M.; dos Santos Batista, P.R.; da Costa, C.E.F.; Zamian, J.R.; do Nascimento, L.A.S.; Rocha Filho, G.N. Murumuru (*Astrocaryum murumuru* Mart.) butter and oils of buriti (*Mauritia flexuosa* Mart.) and pracaxi (*Pentaclethra macroloba* (Willd.) Kuntze) can be used for biodiesel production: Physico-chemical properties and thermal and kinetic studies. *Ind. Crop. Prod.* **2017**, *97*, 536–544. [CrossRef]

29. Bitonto, L.; Reynel-Ávila, H.E.; Mendoza-Castillo, D.I.; Bonilla-Petriciolet, A.; Durán-Valle, C.J.; Pastore, C. Synthesis and characterization of nanostructured calcium oxides supported onto biochar and their application as catalysts for biodiesel production. *Renew. Energy* **2020**, *160*, 52–66. [CrossRef]

30. Zhao, C.; Yang, L.; Xing, S.; Luo, W.; Wang, Z.; Pengmei, L.V. Biodiesel production by a highly effective renewable catalyst from pyrolytic rice husk. *J. Clean. Prod.* **2018**, *199*, 772–780. [CrossRef]

31. Zhao CPengmei Lv Yang, L.; Xing, S.; Luo, W.; Wang, Z. Biodiesel synthesis over biochar-based catalyst from biomass waste pomelo peel. *Energy Convers. Manag.* **2018**, *160*, 477–485. [CrossRef]

32. Jitjamnong, J.; Thunyaratchatanon, C.; Luengnaruemitchai, A.; Kongrit, N.; Kasetsomboon, N.; Sopajarn, A.; Chuaykarn, N.; Khantikulanon, N. Response surface optimization of biodiesel synthesis over a novel biochar-based heterogeneous catalyst from cultivated (*Musa sapientum*) banana peels. *Biomass Conv. Bioref.* **2021**, *11*, 2795–2811. [CrossRef]

33. Patel, M.; Kumar, R.; Pittman, J.C.U.; Mohan, D. Ciprofloxacin and acetaminophen sorption onto banana peel biochars: Environmental and process parameter influences. *Environ. Res.* **2021**, *201*, 111218. [CrossRef] [PubMed]

34. Jung, J.M.; Oh, J.I.; Baek, K.; Lee, J.; Kwon, E.E. Biodiesel production from waste cooking oil using biochar derived from chicken manure as a porous media and catalyst. *Energy Convers. Manag.* **2018**, *165*, 628–633. [CrossRef]

35. Rajendran, N.; Kang, D.; Han, J.; Gurunathan, B. Process optimization, economic and environmental analysis of biodiesel production from food waste using a citrus fruit peel biochar catalyst. *J. Clean. Prod.* **2022**, *365*, 132712. [CrossRef]

36. Thushari, I.; Babel, S.; Samart, C. Biodiesel production in an autoclave reactor using waste palm oil and coconut coir husk derived catalyst. *Renew. Energy* **2019**, *134*, 125–134. [CrossRef]

37. Fernández, J.V.; Faria, D.N.; Santoro, M.C.; Mantovaneli, R.; Cipriano, D.F.; Brito, G.M.; Carneiro, M.T.W.D.; Schettino, M.A.; Gonzalez, J.L.; Freitas, J.C.C. Use of Unmodifed Cofee Husk Biochar and Ashes as Heterogeneous Catalysts in Biodiesel Synthesis. *Bioenerg. Res.* **2022**, *16*, 1746–1757. [CrossRef]

38. Pires, F.C.S.; Silva, A.P.S.; Salazar, M.A.R.; Costa, W.A.; Costa, H.S.C.; Lopes, A.S.; Rogez, H.; Junior, R.N.C. Determination of process parameters and bioactive properties of the murici pulp (*Byrsonima crassifolia*) extracts obtained by supercritical extraction. *J. Supercrit. Fluids* **2019**, *146*, 128–135. [CrossRef]

39. Carvalho, A.V.; Paracampo, N.E.N.P.; Mattietto, R.A.; Nascimento, W.M.O.; Junior, R.A.G.; Resende, M.D.V. *Avaliação Nutricional da Polpa de Frutos Provenientes de Clones de Muricizeiro*; Boletim de Pesquisa e Desenvolvimento; Embrapa Amazônia Oriental: Belém, Brazil, 2020.

40. Ferreira, M.G.R. Murici (*Byrsonima crassifolia* (L.) Rich.), Embrapa. 2005. Available online: https://www.embrapa.br/busca-de-publicacoes/-/publicacao/859538/murici-byrsonima-crassifolia-l-rich (accessed on 15 May 2023).

41. Rabie, A.M.; Shaban, M.; Abukhadra, M.R.; Hosny, R.; Ahmed, S.A.; Negm, N.A. Diatomite supported by CaO/MgO nanocomposite as heterogeneous catalyst for biodiesel production from waste cooking oil. *J. Mol. Liq.* **2019**, *279*, 224–231. [CrossRef]

42. Daimary, N.; Eldiehy, K.S.H.; Boruah, P.; Deka, D.; Bora, U.; Kakati, B.K. Potato peels as a sustainable source for biochar, bio-oil and a green heterogeneous catalyst for biodiesel production. *J. Environ. Chem. Eng.* **2022**, *10*, 107108. [CrossRef]

43. Lim, X.Y.; Yek, P.N.Y.; Liew, R.K.; Chiong, M.C.; Mahari, W.A.W.; Peng, W.; Chong, C.T.; Lin, C.Y.; Aghbashlo, M.; Tabatabaei, M.; et al. Engineered biochar produced through microwave pyrolysis as a fuel additive in biodiesel combustion. *Fuel* **2022**, *312*, 122839. [CrossRef]

44. Lam, S.S.; Liew, R.K.; Lim, X.Y.; Ani, F.N.; Jusoh, A. Fruit waste as feedstock for recovery by pyrolysis technique. *Int. Biodeterior. Biodegrad.* **2016**, *13*, 325–333. [CrossRef]

45. Liew, R.K.; Nam, W.L.; Chong, M.Y.; Phang, X.Y.; Su, M.H.; Yek, P.N.Y.; Ma, N.L.; Cheng, C.K.; Chong, C.T.; Lam, S.S. Oil palm waste: An abundant and promising feedstock for microwave pyrolysis conversion into good quality biochar with potential multi-applications. *Process Saf. Environ. Protect.* **2018**, *115*, 57–69. [CrossRef]

46. Guo, F.; Peng, Z.G.; Dai, J.Y.; Xiu, Z.L. Calcined sodium silicate as solid base catalyst for biodiesel production. *Fuel Process. Technol.* **2010**, *91*, 322–328. [CrossRef]

47. Oliveira, K.G.; Lima, R.R.S.; Longe, C.; Bicudo, T.C.; Sales, R.V.; Carvalho, L.S. Sodium and potassium silicate-based catalysts prepared using sand silica concerning biodiesel production from waste oil. *Arab. J. Chem.* **2022**, *15*, 103603. [CrossRef]

48. Ogbu, I.M.; Ajiwe, V.I.E.; Okoli, C.P. Performance evaluation of carbon-based heterogeneous acid catalyst derived from *Hura crepitans* seed pod for esterification of high FFA vegetable oil. *Bioenerg. Res.* **2018**, *11*, 772–783. [CrossRef]

49. Foroutan, R.; Mohammadi, R.; Razeghi, J.; Ramavandi, B. Biodiesel production from edible oils using algal biochar/CaO/K_2CO_3 as a heterogeneous and recyclable catalyst. *Renew. Energy* **2021**, *168*, 1207–1216. [CrossRef]

50. Maneerung, T.; Kawi, S.; Wang, C.H. Biomass gasification bottom ash as a source of CaO catalyst for biodiesel production via transesterification of palm oil. *Energy Convers. Manag.* **2015**, *92*, 234–243. [CrossRef]

51. Syazwani, O.N.; Teo, S.H.; Islam, A.; Taufiq-Yapa, Y.H. Transesterification activity and characterization of natural CaO derived from waste venus clam (*Tapes belcheri* S.) material for enhancement of biodiesel production. *Process Saf. Environ. Protect.* **2017**, *105*, 303–315. [CrossRef]

52. Nabgan, W.; Nabgan, B.; Ikram, M.; Jadhav, A.H.; Ali, M.W.; Ul-Hamid, A.; Nam, H.; Lakshminarayana, P.; Kumar, A.; Bahari, M.B.; et al. Synthesis and catalytic properties of calcium oxide obtained from organic ash over a titanium nanocatalyst for biodiesel production from dairy scum. *Chemosphere* **2021**, *290*, 133296. [CrossRef]

53. Wang, S.; Hao, P.; Li, S.; Zhang, A.; Guan, Y.; Zhang, L. Synthesis of glycerol carbonate from glycerol and dimethyl carbonate catalyzed by calcined silicates. *Appl. Catal. A Gen.* **2017**, *542*, 174–181. [CrossRef]

54. Gui, X.; Chen, S.; Yun, Z. Continuous production of biodiesel from cottonseed oil and methanol using a column reactor packed with calcined sodium silicate base catalyst. *Chin. J. Chem. Eng.* **2016**, *24*, 499–505. [CrossRef]

55. Das, A.; Shi, D.; Halder, G.; Rokhum, S.L. Microwave-assisted synthesis of glycerol carbonate by transesterification of glycerol using Mangifera indica peel calcined ash as catalyst. *Fuel* **2022**, *330*, 125511. [CrossRef]

56. Chen, G.; Shan, R.; Li, S.; Shi, J. A biomimetic silicification approach to synthesize CaO–SiO_2 catalyst for the transesterification of palm oil into biodiesel. *Fuel* **2015**, *153*, 48–55. [CrossRef]

57. Ibrahim, M.L.; Khalil, N.N.A.N.A.; Islam, A.; Rashid, U.; Ibrahim, S.F.; Mashuri, S.I.S.; Taufiq-Yap, Y.H. Preparation of Na_2O supported CNTs nanocatalyst for efficient biodiesel production from waste-oil. *Energy Convers. Manag.* **2020**, *205*, 112445. [CrossRef]

58. Daramola, M.O.; Mtshali, K.; Senokoane, L.; Fayemiwo, O.M. Influence of operating variables on the transesterification of waste cooking oil to biodiesel over sodium silicate catalyst: A statistical approach. *J. Taibah Univ. Sci.* **2016**, *10*, 675–684. [CrossRef]

59. Roschat, W.; Siritanon, T.; Yoosuk, B.; Promarak, V. Rice husk-derived sodium silicate as a highly efficient and low-cost basic heterogeneous catalyst for biodiesel production. *Energy Convers. Manag.* **2016**, *119*, 453–462. [CrossRef]

60. *ASTM International D6751*; Standard Specification for Biodiesel Fuel Blend Stock (B100) for Middle Distillate Fuels. ASTM: West Conshohocken, PA, USA, 2015.

61. *ASTM D445*; Standard Test Method for Kinematic Viscosity of Transparent and Opaque Liquids (and Calculation of Dynamic Viscosity). ASTM: West Conshohocken, PA, USA, 2006.

62. *ASTM D1298*; Standard Test Method for Density, Relative Density, or API Gravity of Crude Petroleum and Liquid Petroleum Products by Hydrometer Method. ASTM: West Conshohocken, PA, USA, 2005.

63. *ASTM D664*; Standard Test Method for Acid Number of Petroleum Products by Potentiometric Titration. ASTM: West Conshohocken, PA, USA, 2018.

64. *ASTM D6371*; Standard Test Method for Cold Filter Plugging Point of Diesel and Heating Fuels. ASTM: West Conshohocken, PA, USA, 2016.

65. *ASTM D93*; Standard Test Methods for Flash Point by Pensky-Martens Closed CupTester. ASTM: West Conshohocken, PA, USA, 2013.

66. *ASTM D130*; Standard Test Method for Corrosiveness to Copper from Petroleum Products by Copper Strip Test. ASTM: West Conshohocken, PA, USA, 2018.

67. Gonçalves, M.A.; Mares, E.K.L.; da Luz, P.T.S.; Zamian, J.R.; Rocha Filho, G.N.; Castro, H.F.; Conceição, L.R.V. Biodiesel synthesis from waste cooking oil using heterogeneous acid catalyst: Statistical optimization using linear regression model. *J. Renew. Sustain. Energy* **2021**, *13*, 04310. [CrossRef]

68. Borah, M.J.; Sarmah, H.J.; Bhuyan, N.; Mohanta, D.; Deka, D. Application of Box-Behnken design in optimization of biodiesel yield using WO$_3$/graphene quantum dot (GQD) system and its kinetics analysis. *Biomass Conv. Bioref.* **2022**, *12*, 221–232. [CrossRef]

69. Liu, K.; Zhang, L.; Wei, G.; Yuan, Y.; Huang, Z. Synthesis, characterization and application of a novel carbon-doped mix metal oxide catalyst for production of castor oil biodiesel. *J. Clean. Prod.* **2022**, *373*, 133768. [CrossRef]

70. Wan, Z.; Hameed, B.H. Transesterification of palm oil to methyl ester on activated carbon supported calcium oxide catalyst. *Biores. Technol.* **2011**, *102*, 2659–2664. [CrossRef]

71. Alsharifi, M.; Znad, H.; Hena, S.; Ang, M. Biodiesel production from canola oil using novel Li/TiO$_2$ as a heterogeneous catalyst prepared via impregnation method. *Renew. Energy* **2017**, *114*, 1077–1089. [CrossRef]

72. Boehm, H.P. Some aspects of the surface chemistry of carbon blacks and other carbons. *Carbon* **1994**, *32*, 759–769. [CrossRef]

73. Silva, C.; Weschenfelder, T.A.; Rovani, S.; Corazza, F.C.; Corazza, M.L.; Dariva, C.; Oliveira, J.V. Continuous production of fatty acid ethyl esters from soybean oil in compressed ethanol. *Ind. Eng. Chem. Res.* **2007**, *46*, 5304–5309. [CrossRef]

Article

Biological Valorization of Lignin-Derived Aromatics in Hydrolysate to Protocatechuic Acid by Engineered *Pseudomonas putida* KT2440

Xinzhu Jin, Xiaoxia Li, Lihua Zou, Zhaojuan Zheng and Jia Ouyang *

Jiangsu Co-Innovation Center of Efficient Processing and Utilization of Forest Resources, College of Chemical Engineering, Nanjing Forestry University, Nanjing 210037, China; jxznjfu_1010@163.com (X.J.); fqqdxy2022@163.com (X.L.); lihuazou@njfu.edu.cn (L.Z.); zhengzj@njfu.edu.cn (Z.Z.)
* Correspondence: hgouyj@njfu.edu.cn

Abstract: Alongside fermentable sugars, weak acids, and furan derivatives, lignocellulosic hydrolysates contain non-negligible amounts of lignin-derived aromatic compounds. The biological funnel of lignin offers a new strategy for the "natural" production of protocatechuic acid (PCA). Herein, *Pseudomonas putida* KT2440 was engineered to produce PCA from lignin-derived monomers in hydrolysates by knocking out protocatechuate 3,4-dioxygenase and overexpressing vanillate-*O*-demethylase endogenously, while acetic acid was used for cell growth. The sugar catabolism was further blocked to prevent the loss of fermentable sugar. Using the engineered strain, a total of 253.88 mg/L of PCA was obtained with a yield of 70.85% from corncob hydrolysate 1. The highest titer of 433.72 mg/L of PCA was achieved using corncob hydrolysate 2 without any additional nutrients. This study highlights the potential ability of engineered strains to address the challenges of PCA production from lignocellulosic hydrolysate, providing novel insights into the utilization of hydrolysates.

Keywords: *Pseudomonas putida* KT2440; sugar loss; protocatechuic acid; biological funnel; lignocellulosic hydrolysate

Citation: Jin, X.; Li, X.; Zou, L.; Zheng, Z.; Ouyang, J. Biological Valorization of Lignin-Derived Aromatics in Hydrolysate to Protocatechuic Acid by Engineered *Pseudomonas putida* KT2440. *Molecules* **2024**, 29, 1555. https://doi.org/10.3390/molecules29071555

Academic Editors: José A.P. Coelho and Roumiana P. Stateva

Received: 14 February 2024
Revised: 25 March 2024
Accepted: 27 March 2024
Published: 30 March 2024

1. Introduction

Lignocellulosic biomass represents a vast and renewable resource for humanity [1]. Its low cost and abundance make it an attractive alternative for the production of fuel and chemicals. Lignocellulosic biomass mainly consists of polysaccharides and lignin [2]. The efficient utilization of both polysaccharides and lignin is critical for the economic efficiency of biomass biorefinery [3]. Over the years, numerous researchers have dedicated their efforts to developing various technologies for the valorization of polysaccharides, including hemicellulose and cellulose [4,5]. The process of separating lignin from a lignocellulosic matrix is well-established and has been used for many years [6–8]. However, the utilization of lignin remains challenging in conversion owing to its recalcitrant and diversified structure [9]. Meanwhile, the limited progress in industrial-scale lignin valorization is mainly attributed to the emergence of new products that replace petroleum-derived alternatives but are not cost-effective. To decrease the product cost, there is a growing urgency to develop an efficient method for utilizing waste lignin for the commercialization of lignocellulose biorefinery.

Lignin is the most abundant aromatic feedstock on earth [10]. Every year, approximately 100 million tons of lignin are produced as by-products from various sources [11]. In general, achieving industrial-scale implementation of lignin valorization cannot be accomplished solely through straightforward physicochemical or biological approaches. Combining both of them is a promising solution. Through the physicochemical treatment, a significant amount of monomeric aromatic compounds are released into the pretreated liquor. However, the diverse lignin degradation products are difficult to separate and purify during the chemical upgrading [12]. To prevent the wastage of aromatic resources, some microorganisms are selected and engineered to utilize the monomers to produce high-value-added products. As more and more pathways involved in aromatic metabolism were identified, the biological funnel for the conversion of lignin-related aromatic compound mixtures to bioproducts has been well demonstrated [13,14]. Using alkaline pretreated liquor, Linger et al. achieved 0.252 g/L polyhydroxyalkanoic acid (PHA) by *P. putida* KT2440 [15]. Similarly, 1.8 g/L *cis, cis*-muconic acid was obtained using lignin hydrolysate from softwood as a substrate and recombinant *Corynebacterium glutamicum* MA-2 [16].

Protocatechuic acid (PCA, 3,4-dihydroxybenzoic acid) is a natural catabolic intermediate of lignin and lignin-derived aromatics found in plants and fruits [17]. It has been reported to possess multiple biological activities, such as antioxidation, anti-aging, antibacterial, antiviral, and anti-inflammatory, as reviewed elsewhere [18–20]. Given its versatility in the application of pharmaceutical, health food, and cosmetic industries, the production of "natural" PCA has attracted more attention and holds greater business value than that of "artificial" chemical PCA. This is due to its higher price and highly desirable market. However, the preparation of PCA through solvent extraction from plants and fruits often comes with high cost and low yield [21,22]. In recent years, biological approaches for PCA synthesis have been studied using some genetically engineered microorganisms such as *C. glutamicum, Saccharomyces cerevisiae, Escherichia coli,* and *P. putida.* To date, PCA has been biosynthesized from glucose fermentation through the shikimate pathway and 4.27 g/L and 0.64 g/L of PCA were obtained, respectively, in *E. coli* and *C. glutamicum* [23,24]. Additionally, the highest titer of 12.5 g/L with a yield of 0.313 (g/g) was achieved in *P. putida* KT2440, which implied that *P. putida* KT2440 is an ideal host strain. Apart from shikimate pathways, PCA can also be synthesized from aromatic compounds using the lignin biological funnel at a higher theoretical yield and atom efficiency [16]. Recently, a lignin-derived biosensor in *P. putida* KT2440 was developed for PCA production from a mixture of *p*-coumaric acid (*p*-CA) and ferulic acid (FA) [25]. With vanillin as the substrate, 0.146 g/L of PCA was produced from 0.15 g/L of vanillin by engineered *E. coli* [26]. The biological funnel of lignin offers a new strategy for "natural" PCA production. However, more detailed studies on PCA production using lignin are still limited and need to be further developed.

The hydrolysate derived from lignocellulose following dilute acid pretreatment comprises fermentable sugars and weak acids originating from cellulose and hemicellulose, as well as phenolic compounds derived from lignin. In our previous work, *P. putida* KT2440 exhibited high resistance and tolerance to such hydrolysate [27]. In this work, the degradation abilities of *P. putida* KT2440 for diverse phenolic compounds in hydrolysate were assessed. Based on the biological funnel pathway, an engineered strain capable of producing PCA from phenolic compounds and hydrolysates was constructed. In addition, the genetically modified strain used acetic acid as a carbon source and no longer possessed the capacity to metabolize fermentable sugars, resulting in the retention of sugars in the hydrolysate where they could serve as substrates for other microorganisms to generate novel products.

2. Results and Discussion

2.1. Evaluation of the Degradation Abilities of P. putida KT2440 towards Phenolic Compounds

Here, two types of corncob hydrolysates obtained from the factory were chosen as representatives for analyzing their composition. Hydrolysate 1 was the liquid fraction directly generated during the dilute acid pretreatment of corncobs, while hydrolysate 2 was obtained by concentrating the liquid fraction after pretreatment using vacuum concentration. As shown in Figure 1a,b, both hydrolysates 1 and 2 are xylose-rich liquors due to degradation of hemicellulose in corncob during pretreatment. Besides xylose, there glucose, arabinose, and cellobiose and their degradation products exist, including weak acids and furans. Their amount is basically consistent with previous similar studies [28]. The phenolic compounds generated from lignin in hydrolysates were at low concentrations of mg/L. Lignin mainly consists of H-type, G-type, and S-type phenyl propane units, which correspond to three different phenolic acids: *p*-CA, FA, and syringic acid (SA) respectively. A variety of low-molecular-weight phenolic compounds, including six phenolic acids and three phenolic aldehydes, was observed in hydrolysates and their proportion of total phenolic is illustrated in Figure S1. Among these, FA and *p*-CA were the main lignin monomers, which account for over 160, 120 mg/L and 530, 250 mg/L, respectively, in hydrolysates 1 and 2.

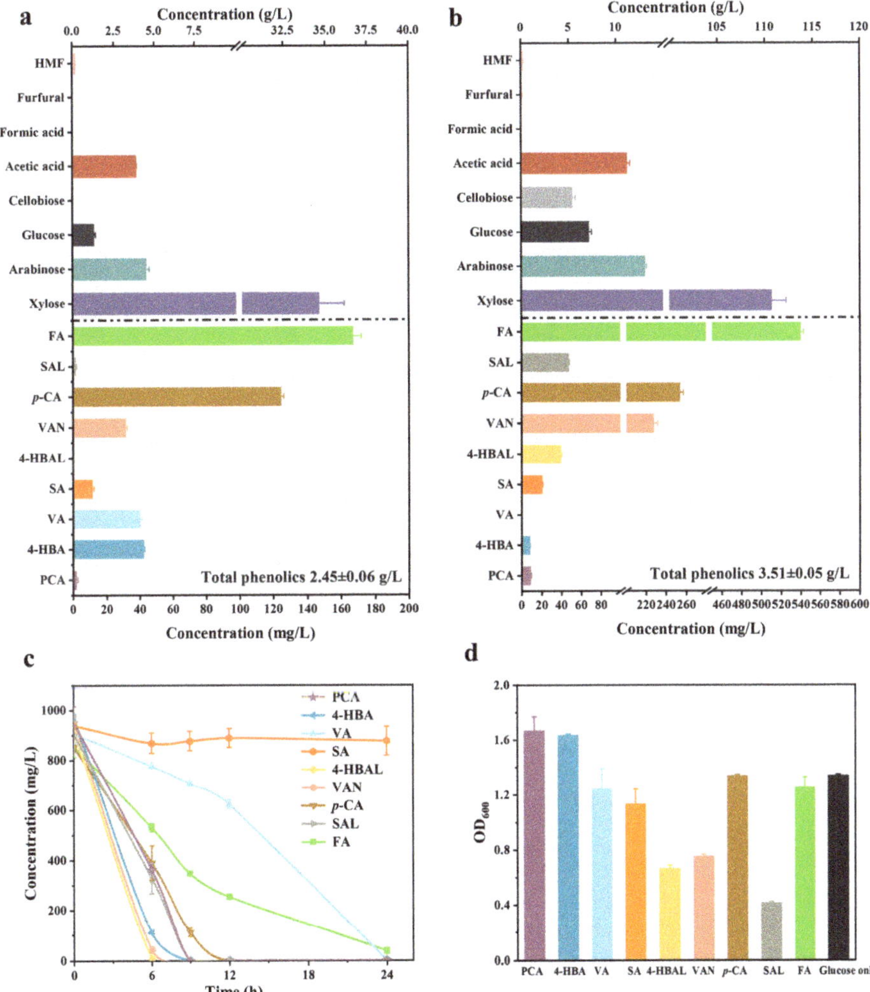

Figure 1. Composition of fermentable sugars, weak acids, furans, and phenolic compounds in (**a**) hydrolysate 1 and (**b**) hydrolysate 2. (**c**) Conversion of 1 g/L of lignin-related monomers present in hydrolysates by KT2440 within 24 h. (**d**) The cell growth profiles of KT2440 after 6 h fermentation in the presence of a simple phenolic compound. Abbreviations: Protocatechuic acid (PCA), ferulic acid (FA), vanillin (VAN), vanillic acid (VA), *p*-coumaric acid (*p*-CA), 4-hydroxybenzaldehyde (4-HBAL), 4-hydroxybenzoic acid (4-HBA), syringaldehyde (SAL), syringic acid (SA). Error bars indicate the standard deviation in two replicates.

In order to evaluate the bio-degradation ability of *P. putida* KT2440 for phenolic compounds in corncob hydrolysate, the above-determined lignin degradation products were individually applied to fermentation by *P. putida* KT2440. The concentration of phenolics was 1 g/L, and 10 mM glucose was added to the M9 minimal medium for keeping the cell growth. As illustrated in Figure 1c, *P. putida* KT2440 degraded most phenolics within 24 h except for SA. Despite *P. putida* KT2440 did not consume SA, it was seen that glucose in the medium was completely used up (Figure S2a) and the cell still showed significant growth (Figure 1d). This implied that *P. putida* KT2440 lacks the degradation metabolism for SA, but still demonstrates good tolerance for SA. Compared with phenolic acids, the results of cell growth in Figure 1d showed that phenolic aldehydes exhibit higher toxicity than phenolic acids for *P. putida* KT2440. Among three phenolic aldehydes, syringaldehyde (SAL) has the strongest toxicity, followed by vanillin (VAN) and 4-hydroxybenzaldehyde (4-HBAL). After 24 h fermentation, all of these phenolic aldehydes were found to be converted to the corresponding phenolic acids through aldehyde oxidation reactions [29]. As shown in Figure S2b,c, *P. putida* KT2440 could convert SAL into SA, which was accumulated due to lack of further degradation pathway during fermentation, while the amount of VA was observed in the first 36 h and completely consumed at the end during fermentation of VAN. Moreover, VA also served as an intermediate product during FA degradation (Figure S2d). This suggests that VA is located downstream point of FA and VAN pathways and that the production of VA is faster than its consumption [30].

Overall, *P. putida* KT2440 demonstrated good abilities for the degradation of a variety of low-weight phenolic compounds at 1 g/L. All of the H-type lignin-derived compounds, such as *p*-CA, 4-HBAL, and 4-HBA, are quickly consumed within 12 h. In comparison, VAN, VA, and FA from G-type lignin exhibited a relatively slower degradation rate. Even so, a complete degradation also occurred within a 24 h period. These results suggest that *P. putida* KT2440 is capable of efficiently utilizing the majority of aromatic monomers in hydrolysates. Hence, it is a suitable starting strain for efficiently converting aromatic monomers derived from lignin into products.

2.2. Effects of Sugar and Weak Acid in Hydrolysates on Phenolic Compound Metabolism

According to the component analysis of hydrolysates, there are many other chemicals that existed in the liquor, mainly sugar and weak acids except for phenolic compounds. The metabolism by which these compounds interfere with the degradation of phenolic compounds is unknown when using *P. putida* KT2440. Thus, Figure 2 shows the fermentation performance of *P. putida* KT2440 using sugar and a weak acid, either the sole or the mixture.

When using *p*-CA as the sole carbon source, *P. putida* KT2440 completely metabolized *p*-CA within 9 h and there was no obvious product accumulated. Meanwhile, significant cell growth was observed in the first 9 h. It suggested that *P. putida* KT2440 could directly use *p*-CA as the sole carbon source for cell growth. As supplemented with 10 mM glucose, *P. putida* KT2440 promoted better cell growth throughout the process than that of *p*-CA as the sole carbon source. However, glucose supplementation delayed *p*-CA metabolism to 12 h (Figure 2b). Both glucose and *p*-CA were used up within 12 h. The accumulation of gluconic acid and 2-ketogluconic acid was observed momently at 3 h and 6 h, respectively. Previously, it was reported that glucose is first oxidized to gluconate or 2-ketogluconic acid, then transported to the cytosol and finally enters the TCA cycle in *P. putida* KT2440 [31]. Hence, glucose is proposed as the preferred carbon and energy source over *p*-CA, and the presence of glucose inhibits the metabolism of *p*-CA. In Figure 2c, the conversion of xylose was lagged behind *p*-CA, but it did not have a significant effect on cell growth. The consumption of xylose was basically consistent with the production of xylonic acid. *p*-CA is still the main carbon and energy source and xylose is mainly oxidized to xylonic acid [32]. Acetic acid is a kind of typical byproduct of hydrolysate. In *P. putida* KT2440, acetic acid was consumed within 12 h (Figure 2d), indicating that acetic acid could serve as a potential carbon source.

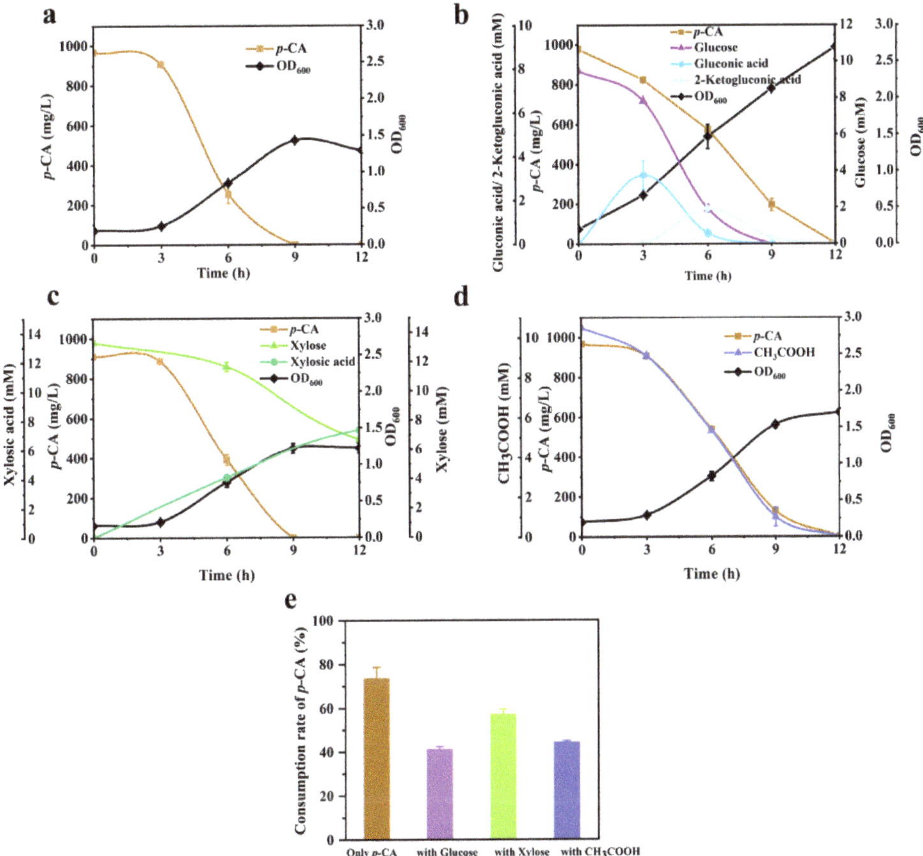

Figure 2. Fermentation profiles of KT2440 with additional supplement of different carbon sources in the presence of *p*-CA. (**a**) *p*-CA only, (**b**) 10 mM glucose, (**c**) 10 mM xylose, and (**d**) 10 mM acetic acid in 12 h. The consumption rate of *p*-CA at 6 h under four environments was shown by (**e**). Error bars indicate the standard deviation in two replicates.

Figure 2e summarizes the consumption rate of *p*-CA at 6 h fermentation by *P. putida* KT2440 using four different types of carbon sources. It was found that the presence of glucose, xylose, and acetic acid all slowed down the degradation rate of *p*-CA. The consumption rates decreased to 41.19%, 57.12%, and 44.23%, respectively, lower than the control rate of 73.62%. When hydrolysate was directly used for the bioconversion of lignin-derived compounds, this inhibition effect would be a major obstacle. However, it should be noted that once *p*-CA is designed to be converted into one product by engineering *P. putida* KT2440, its effect as a carbon and energy supplier needs to be weakened due to its main flow distribution towards conversion. By that time, glucose and acetic acid in hydrolysates would be used as alternative carbon and energy sources for cell growth. Furthermore, considering the significant inhibition of *p*-CA metabolism in *P. putida* KT2440 by glucose, acetic acid appears to be a suitable and effective carbon source for maintaining cell activation in real hydrolysate applications.

2.3. Construction of an Engineered P. putida Strain for Protocatechuic Acid Production

Preliminary tests indicated that *P. putida* KT2440 was capable of efficiently utilizing most aromatic monomers commonly found in hydrolysates; however, it was unable to produce any valuable product. In general, the majority of aromatic compounds undergo complete degradation via biological funneling and the β-ketoadipate pathway before entering the TCA cycle [33]. Biological funneling is the process by which various aromatic compounds in hydrolysates converge to PCA which serves as their common intermediate metabolite. As mentioned in Figure 3, all six phenolic compounds in hydrolysates could be converted into PCA and then split by *pcaGH*. Specifically, these phenolic compounds can be divided into two metabolic pathways: G-lignin-derived compounds (FA, VAN, and VA) are funneled into VA and further oxidized to PCA by vanillic acid O-demethylase oxygenase (VanAB); *p*-CA, 4-HBAL, and 4-HBA from H-lignin-derived compounds are funneled into 4-HBA and further degraded to PCA by *p*-hydroxybenzoate hydroxylase (PobA). Regardless of their origin, it has been found that knocking out of the gene *pcaGH* is necessary for the accumulation of PCA.

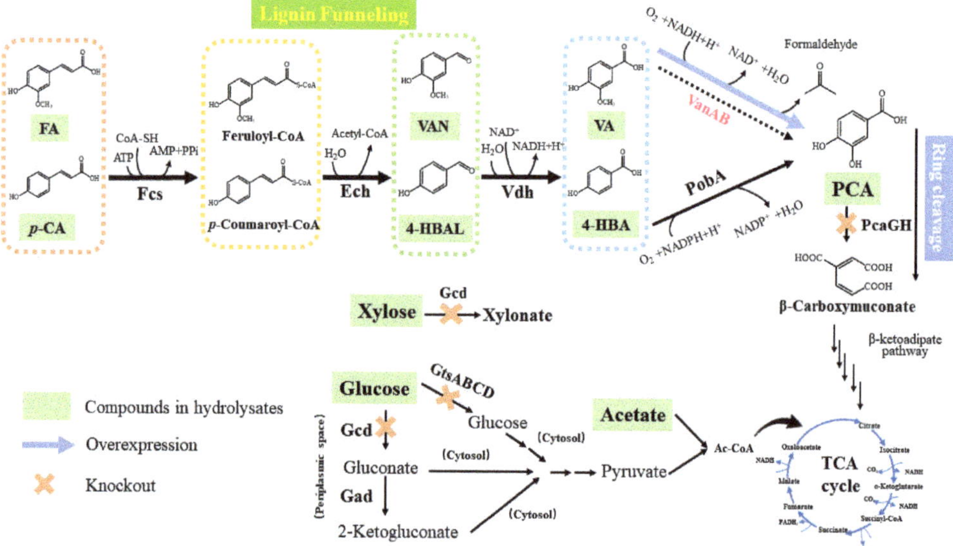

Figure 3. Metabolic pathways of substances present in hydrolysate in *P. putida* KT2440 with engineered pathways are shown. Abbreviations: *p*-coumarate-CoA ligase (Fcs), enoyl-CoA hydratase/aldolase (Ech), vanillin dehydrogenase (Vdh), *p*-hydroxybenzoate hydroxylase (PobA), vanillic acid O-demethylase oxygenase (VanAB), glucose dehydrogenase (Gcd), glucose ABC transporter (GtsABCD), gluconate dehydrogenase (Gad), tricarboxylic acid cycle (TCA cycle).

By a homologous recombination knockout technology [27] shown in Figure 4b, the engineered strain KT1 was obtained by knockout of *pcaGH* (Figure S3). Using the KT1, we investigated the production of PCA using six different phenolic compounds at a concentration of 1 g/L supplemented with 10 mM glucose as a carbon source. As shown in Figure 4c, the knockout of *pcaGH* did not significantly influence the consumption of phenolic compounds, with the majority being used within 24 h except for FA and VA. however, only three H-lignin-derived monomers, *p*-CA, 4-HBAL, and 4-HBA, demonstrated exceptional abilities for PCA production, whose yields of PCA were 97.7%, 98.5%, and 93.1%, respectively. The complete conversion and production demonstrated the uninterrupted flow of this pathway to PCA. Meanwhile, the block of the PCA to TCA cycle did not exert any discernible impact on cellular proliferation. One plausible explanation is that the presence of glucose potentially furnishes an abundant supply of carbon and energy. In a previous study, the knockout of *pcaGH* in the KT2440 strain resulted in a production of only 5.2 mg/L of PCA from glucose [34]. Our results suggested that the production of PCA from *p*-CA may be more efficient compared to glucose due to a shorter metabolic pathway and higher atom efficiency.

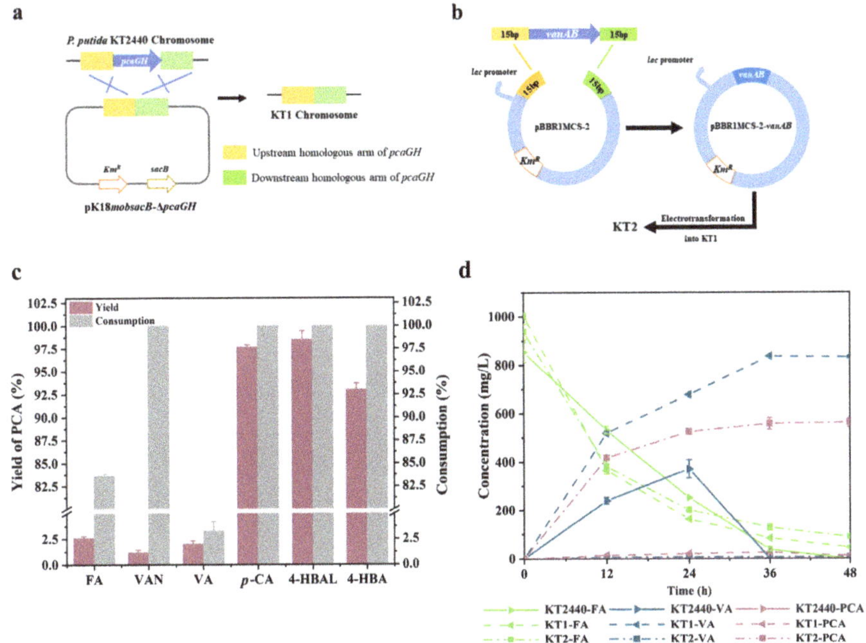

Figure 4. Construction of PCA-producing strains and removal of restriction factors for PCA production. (**a**) Overview of *pcaGH* knockout by pK18*mobsacB* and (**b**) *vanAB* overexpression by pBBR1MCS-2, (**c**) Histogram of PCA yield and consumption of monophenols by KT01. (**d**) Shake-flask evaluation of the effect of knockout and overexpression on PCA production from FA by KT2440, KT1, and KT2. Error bars indicate the standard deviation in two replicates.

Regarding H-lignin-derived monomers, the consumptions of FA, VAN, and VA were 83.7%, 100%, and 3.3% after 24 h fermentation. However, all yields of PCA were less than 3% (Figure 4c). Meanwhile, a large amount of VA was produced or maintained during the fermentation process (Figure 4d). The result suggested that the oxidation of VA into PCA by the *vanAB* encoding enzymes was seriously inhibited owing to the block out of the downstream pathway. Thus, it was speculated that the expression of the *vanAB* gene in the *P. putida* KT2440 genome was regulated by the downstream pathway. Therefore, to enhance the expression of *vanAB*, the endogenous *vanAB* gene was introduced into KT1 by overexpression, generating KT2. pBBR1MCS-2 containing a lac promoter was used as the expression vector to overexpress *vanAB*. Endogenous *vanAB* was ever reported to be a key rate-limiting step in the FA pathway [34]. Upadhyay and Lali overexpressed several different sources of *vanAB* genes in KT2440 Δ*pcaGH* and found that the conversion rate of VA to PCA could increase to 51.33% from 10.54% by overexpressing *vanAB* from *Acinetobacter* sp. ADP1 [35]. In this work, both KT1 and KT2 were able to completely consume 1 g/L of FA almost totally within 36 h, but the overexpression of the *vanAB* gene in KT1 led to a significant increase in PCA yield from 2.61% to 75.63%, while only a minimal accumulation of VA was observed (Figure 4d). The construction of an engineered strain of *P. putida* KT2440 capable of efficiently producing PCA from *p*-CA and FA has been successfully accomplished so far.

2.4. PCA Production from Single Phenolic Acid in the Engineered P. putida

Firstly, the production boundaries of PCA by KT2 were investigated using *p*-CA and FA at three different concentration levels (1, 2, 5, and 10 g/L). Figure 5a shows a complete conversion of 10 g/L *p*-CA within 72 h, resulting in a maximum PCA production of 6.11 g/L. Despite the PCA production from *p*-CA being far from that from glucose [24], its high yield by KT2 confirmed its potential to form lignin into PCA. In terms of the production of PCA from FA (Figure 5b), the assimilation of FA exhibited significant delays compared to the *p*-CA pathway. Only 2.5 g/L of FA could be consumed within 72 h and the maximum concentration of PCA was at only 1.9 g/L. This comparison suggested that further enhancements were required for efficient PCA production from FA in the future.

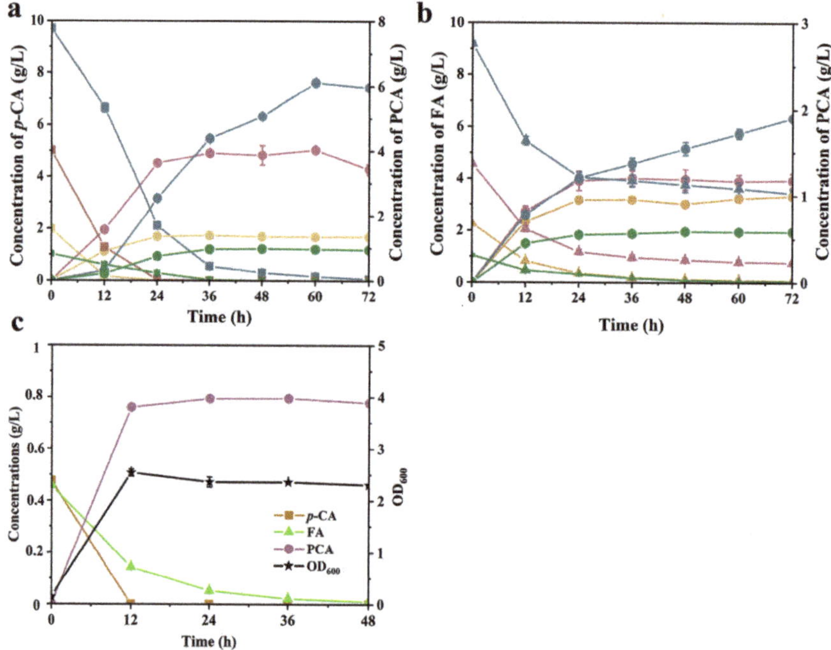

Figure 5. Assessment of PCA production by KT2 in the presence of glucose using different concentrations of (**a**) *p*-CA and (**b**) FA and their production of PCA. (**c**) PCA production using mixed phenolic acid (*p*-CA:FA = 1:1) by KT2 in the presence of glucose. Squares represent *p*-CA, circles represent PCA, and triangles represent FA. The blue line corresponds to 10 g/L substrate concentration, the pink line corresponds to 5 G/L, the yellow line corresponds to 2 g/L, and the green line corresponds to 1 g/L. Error bars indicate the standard deviation in two replicates.

A comparison of the PCA production pathways from *p*-CA and FA revealed a shared partial pathway (Figure 3). Their difference in degradation ability was mainly dependent on the types of substrates, implying that *p*-CA utilization was preferred. As the substrate loading increased, the PCA yield decreased whether for *p*-CA or for FA. At a high substrate loading (10 g/L), the PCA yield from *p*-CA and FA decreased to 67.0% and 22%, respectively, accompanied by the accumulation of the intermediate products of 4-HBA and VA (Figure S4a). The accumulation of intermediate products indicates an inadequate conversion in the final step of PCA production. For *p*-CA and FA, *pobA* and *vanAB* were, respectively, responsible for the PCA biosynthesis from 4-HBA and VA, which required distinct cofactors, NADPH and NADH. By comparison, the accumulation of VA was more serious than 4-HBA (Figure S4b). The decrease in PCA yield might be attributed to the inadequate activity of *pobA* with increasing concentration of *p*-CA. For FA metabolism, it might be due to an excess unbalance of redox metabolism [36]. Additionally, the detrimental impact of formaldehyde as a by-product should also be taken into account due to its inherent toxicity.

2.5. PCA Production from Hydrolysates

As mentioned above, although the PCA yield decreased with increasing substrate loading, KT2 was still capable of meeting the demand for biosynthesis of PCA from lignin hydrolysates due to the presence of a low concentration of phenolic acid. However, several factors still influenced the conversion. A series of aromatics, mainly including *p*-CA, FA, and their related phenolic compounds, coexist in the hydrolysates (Figure 1). Hence it was necessary to assess the co-utilization efficiency of *p*-CA and FA. In Figure 5c, a mixture of *p*-CA and FA in a 1:1 (*w/w*) ratio was used at a total of 1 g/L loading in the presence of 10 mM glucose. The strain KT2 depleted the two phenolic acids simultaneously, achieving a maximum yield of 98.1% corresponding to 0.8 g/L PCA.

Besides phenolic compounds, the hydrolysates also contained sugars and acetic acids. Previous research indicated that the presence of glucose contradicted the mechanism of phenolic compound degradation due to the existence of a global regulator (crc), whose deletion promoted the expression of *pobA* and *vanAB* [37]. Additionally, xylose was present at a relatively high level in two hydrolysates. Although xylose has little influence on *p*-CA metabolism and cell growth, its consumption would result in a massive accumulation of xylonic acid, which would affect the separation of the product during the downstream process. In light of the aforementioned drawbacks of sugar metabolism, it was imperative to further impede the sugar utilization pathway. As an alternative approach, harnessing acetic acid from hydrolysates as a carbon source held promise for sustaining cell growth.

As mentioned in Section 2.2, the genes *gcd* and *gtsABCD* were knockout and the strain KT3 was obtained based on KT2. Due to the lack of these genes, KT3 lost the ability to catabolize xylose and glucose for cell growth (Figure S5). When using 1 g/L acetic acid as an additional carbon source instead of glucose, the PCA-producing ability of KT3 remained largely unaffected by the knockout of *gcd* and *gtsABCD* (Figure 6a). The strain KT3 synchronously converted 94.2% of the two phenolic acids with a maximum PCA yield of 78.4%, equivalent to 0.6 g/L PCA. There was no significant intermediate metabolite accumulation during the whole fermentation process. Meanwhile, 2.75-fold cell growth was observed within 24 h. Furthermore, two hydrolysates as complex mixtures were used to investigate the effect on KT3. Interestingly KT3 demonstrated a robust performance for a real, heterogeneous, lignin-enriched, hydrolysate-derived pilot-scale biomass pretreatment. Furthermore, 100% of hydrolysate 1 could be mechanized well without any dilution or other additional substances. Meanwhile, 253.88 mg/L PCA was achieved from hydrolysate1 with a yield of 70.85% (Figure 6b). The value of yield was nearly the same as that of mixed phenolic acid (Figure 5a). The amount of all five phenolic compounds in hydrolysate 1 decreased by a maximum of 52–100%, indicating that they were converted into PCA through the bio-funneling pathway. Among them, FA degradation was the worst one with only 52%. This outcome was likely attributed to the limited efficiency of the FA metabolism in *P. putida*. Hydrolysate 2 exhibited more serious toxicity than hydrolysate 1, probably due to the presence of a higher concentration of phenolic acid. Since KT3 could not survive in hydrolysate 2, a diluted solution of hydrolysate 2 mixed with sterilized water at an 80% concentration was used for PCA production. As shown in Figure 6c, using 80% hydrolysate 2, the optimal production of 433.72 mg/L PCA was observed by KT3 with a yield of 56.74%. 4-HBA as intermediates were observed shortly and completely depleted at the end. All glucose and xylose were retained, and the residual acetic acid concentration of 1.9 g/L corresponded to a degradation rate of 76.31%. However, KT2 only produced 111.77 mg/L of PCA, corresponding to a yield of 11.34%. Additionally, not only FA but also the metabolic process of 4-HBA was suppressed, highlighting the significance of disrupting the *gcd* and *gtsABCD* genes (Figure S6).

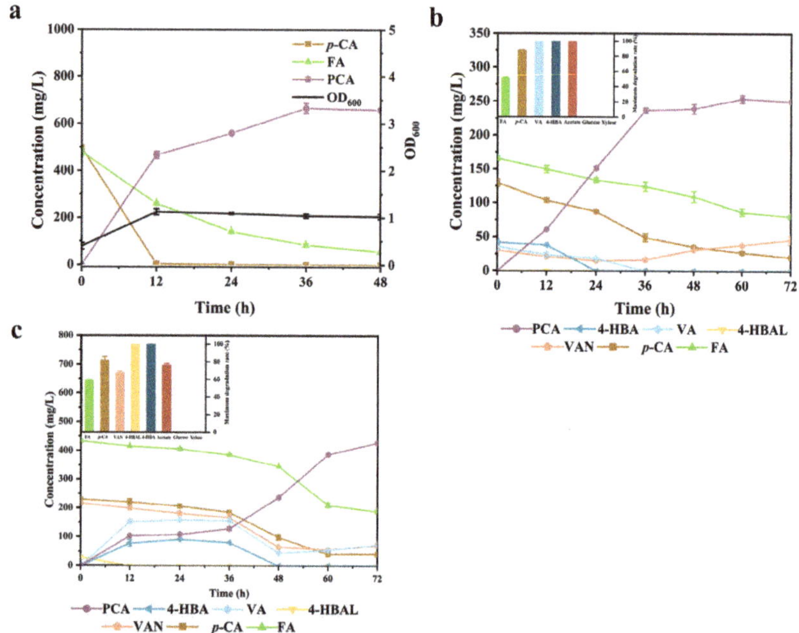

Figure 6. Production of PCA from mixed phenolic acid and real hydrolysates by KT3. (**a**) PCA production using mixed phenolic acid (*p*-CA and FA) in the presence of acetic acid. The fermentation of KT3 for PCA production and the maximum degradation rate of each substance from (**b**) hydrolysate 1 and (**c**) 80% hydrolysate 2. Error bars indicate the standard deviation in two replicates.

Table 1 summarizes the production of PCA from aromatic compounds and lignocellulose hydrolysate. Previous studies were mainly limited to the conversion of two phenolic acids (*p*-CA and FA). Hydrolysate is a complex system, which inevitably contains a large amount of xylose and weak acid. The utilization of hydrolysate in the past has relied on sugar as a carbon source, resulting in increased sugar consumption and the production of xylonic acid as a by-product. For the first time, acetic acid was used as the carbon source for the transformation, and the constructed strain could obtain a single product by co-conversion of multiple substances at the same time. Furthermore, the absence of sugar loss makes it suitable for subsequent biorefining processes.

Table 1. Comparative PCA production reported from aromatic compounds and lignocellulose hydrolysate.

Strain	Substrate	Fermentation Duration (h)	Titer (g/L)	Yield	Reference
Engineered *P. putida* KT2440	1 g/L VA	n. a.	0.51	0.51	[35]
P. putida KT14	3.28 g/L *p*-CA	48	3.1	1.00	[25]
S. Cerevisiae yPCA12	1.6 g/L *p*-CA	96	0.72	0.48	[38]
C. glutamicum ATCC 21420 FVan	0.89 g/L FA	48	0.44	0.63	[39]
P. putida KT2440 KT16	3.88 g/L FA	48	2.1	0.68	[25]
P. putida KT2	1.02 g/L *p*-CA	72	0.95	0.99	This work
P. putida KT2	1 g/L FA	72	0.76	0.76	This work
P. putida KT2	*p*-CA:FA = 1:1 (mol/mol)	48	0.8	0.98	This work
S. cerevisiae yPCA12	0.5×Corn stover APL	96	0.81	0.74 [a]	[38]
P. putida KT3	Corncob hydrolysate 1	72	0.25	0.70 [b]	This work
P. putida KT3	Corncob hydrolysate 2	72	0.43	0.57 [b]	This work

[a] PCA yield was based on the total mol of FA and *p*-CA; [b] PCA yield was based on the total mol of FA, *p*-CA, VAN, 4-HBAL, VA, and 4-HBA; n. a.: not available; APL: alkaline pretreatment liquor.

3. Materials and Methods

3.1. Materials

Corncob was composed of 34.5% cellulose, 30.9% hemicellulose, and 22.7% lignin. After pretreatment with dilute sulfuric acid at 160 °C for 60 min, the remaining solid residue was composed of 54.6% cellulose, 3.1% hemicellulose, and 32.3% lignin. The hydrolysate pretreatment was kindly provided by a biomass factory in China. Syringaldehyde and ferulic acid were purchased from TCI (Shiga, Japan). 4-hydroxybenzoic acid, vanillic acid, syringic acid, 4-hydroxybenzaldehyde, vanillin, and *p*-coumaric acid were purchased from Sigma-Aldrich (Saint Louis, MO, USA).

3.2. Strains and Plasmids

All bacterial strains and plasmids used in this study are listed in Table 2. *E. coli* Trans1-T1 was purchased from TransGen (Beijing, China). *P. putida* KT2440 was obtained from the Biochemical Engineering Research Institute of Nanjing Forestry University. *P. putida* KT1, KT2, and KT3 were independently constructed in this study.

Table 2. Strains and plasmids used in this study.

Strains/Plasmids	Characteristics	Source
Escherichia coli		
Tans1-T1	Cloning host	TransGen Biotech
Pseudomonas putida		
KT2440	Wild-type strain	Lab stock
KT1	KT2440 with scarless deletion of *pcaGH* (PP_4655-4656) encoding protocatechuate 3,4-dioxygenase dehydrogenase	This study
KT2	KT01 harboring plasmid pBBR1MCS2-*vanAB* (PP_3736-3737) encoding vanillate O-demethylase oxygenase	This study
KT3	KT02 with scarless deletion of glucose dehydrogenase encoding *gcd* (PP_1444) and glucose ABC transporter *gtsABCD* (PP_1015-1018)	This study
Plasmids		
pBBR1MCS2	Broad-host-range cloning vector; Km^R	Lab stock
pK18*mobsacB*	The suicide vector containing the *sacB* gene; Km^R	Lab stock
pBBR1MCS2-*vanAB*	pBBR1MCS2 with *vanAB* from *P. putida* KT2440	This study
pK18*mobsacB*-Δ*pcaGH*	pK18*mobsacB* containing the homology arms of *pcaGH* from *P. putida* KT2440	This study

3.3. Genetic Manipulation and Plasmid Construction

The genome of *P. putida* KT2440 was extracted by TaKaRa MiniBEST Bacterial Genomic DNA Extraction Kit Ver.3.0 (Takara, Shiga, Japan).

As for the construction of a *pcaGH* (PP_4655-4656) disruption mutant, all procedures were conducted using the protocol previously [27]. Briefly, the 500 bp upstream and downstream homology arms of *pcaGH* were amplified from the *P. putida* KT2440 genome by primers *pcaGH*-up-f/r and *pcaGH*-down-f/r respectively. After gel purification by Wizard® SV Gel and PCR Clean-Up System (Promega, Wisconsin, USA), overlap extension PCR was performed, using primers *pcaGH*-up-f and *pcaGH*-down-r and thus the upstream and downstream regions were fused. The purified 1000 bp PCR products were then ligated with the pEASY-Blunt cloning vector to form a new plasmid. The plasmid and pK18*mobsacB* were digested using *Eco*R I and *Bam*H I to generate cohesive ends and then ligated to obtain *pcaGH* knockout plasmid code as pK18*mobsacB*-Δ*pcaGH*. The plasmid was introduced into *P. putida* KT2440 by electro-transformation. Single crossovers in which the plasmid had recombined into the genome were selected on LB plates supplemented with 50 μg/mL kanamycin. After two rounds of passaging in LB broth with 15% (*w/v*) sucrose without antibiotics, the correct double-crossover transformants were selected using LB agar plates supplemented with 15% (*w/v*) sucrose, referred to as KT1. The deletion was found successful on the basis of the shortened PCR product of 1000 bp using the primers *pcaGH*-up-f and *pcaGH*-down-r (Figure S3). KT1 was used as the starting strain for further construction of *gcd* and *gtsABCD* knockout mutant generating KT1 Δ*gcd*Δ*gtsABCD*. The process of knockout was the same as described above.

Regarding the overexpression of the gene *vanAB* (PP_4655-4656) from *P. putida* KT2440, pBBR1MCS-2 was used to express *vanAB*. The gene *vanAB* was amplified from *P. putida* KT2440 genome with primers *vanAB*-f/r. Then using pBBR1MCS-2-f/r to obtain a linearized vector. At last, the target gene was recombined with the linearized vector using the pEASY®-Basic Seamless Cloning and Assembly Kit (TransGen Biotech Co., Ltd., Beijing, China), generating plasmid pBBR1MCS-2-*vanAB*. The plasmid was electroporated into KT1 and KT1 Δ*gcd*Δ*gtsABCD*, after screening single colonies on LB agar plates with 50 µg/mL kanamycin, the plasmid extraction was verified and the correct strains were marked as KT2 and KT3.

3.4. P. putida Cultivations

P. putida KT2440 and its derived strains were all pre-cultured overnight in 5 mL Luria-Bertani (LB) medium (10 g/L tryptone, 5 g/L yeast extract, 10 g/L NaCl) at 30 °C and 200 rpm. If required, 50 µg/mL kanamycin was added to the medium to avoid loss of plasmid.

For the fermentation of lignin-derived monomers, the overnight cultures were harvested by centrifugation at 8000 rpm for 5 min, after discarding the supernatant, the cells were resuspended twice with 0.9 (*w/v*) NaCl. The initial OD_{600} was set to 0.2 in a 250 mL shake flask containing 10% filling volume of M9 minimal medium (2 mM $MgSO_4$, 0.1 mM $CaCl_2$, M9 salts (12.8 g/L $Na_2HPO_4 \cdot 7H_2O$, 3 g/L KH_2PO_4, 0.5 g/L NaCl, 1.0 g/L NH_4Cl), trace element solution (0.3 g/L H_3BO_3, 0.05 g/L $ZnCl_2$, 0.03 g/L $MnCl_2 \cdot 4H_2O$, 0.2 g/L $CoCl_2$, 0.01 g/L $CuCl_2 \cdot 2H_2O$, 0.02 g/L $NiCl_2 \cdot 6H_2O$, 0.03 g/L $NaMoO_4 \cdot 2H_2O$) then the volume was made up to 1 L with distilled water) supplemented with carbon sources and aromatic compounds. The pH of the medium was adjusted by 10 M NaOH. 10 mM glucose was the carbon source for *P. putida* KT2440, KT1, and KT2, while KT3 utilized 16 mM acetate as its carbon source.

For the fermentation of the hydrolysates, CaO was used to adjust the pH of the hydrolysate to 7.2. KT3 was pre-activated overnight, and then 1% seed cultures were inoculated into 25 mL of LB medium in a 250 mL shake flask for 12 h. The initial OD_{600} was set to 5 in a 100 mL shake flask containing 10% filling volume of hydrolysates. Hydrolysate 2 was diluted with sterilized distilled water. All fermentations were carried out at 200 rpm and 30 °C. All samples were centrifuged at 12,000 rpm for 5 min. The supernatant was diluted to the appropriate concentration using deionized water in preparation for further analysis.

3.5. Analytical Methods

The consumption of glucose and acetate acid was detected by an Agilent 1260 HPLC system equipped with an Aminex HPX-87H column (Bio-Rad, Hercules, CA, USA) which had a refractive index, the above components were separated at 55°C using 5 mM H_2SO_4 as eluent at a flow rate of 0.6 mL/min [28].

The concentrations of phenolic compounds were identified by using an HPLC system with a Zorbax SB-C18 column (150 × 4.6 mm, Alto, CA, USA) with a UV detector. The gradient elution was performed using 100% acetonitrile as mobile phase A and 1.5% acetic acid as mobile phase B at a flow rate of 0.8 mL/min [40].

Xylose, xylonic acid, gluconate, and 2-ketogluconate were analyzed by high-performance anion exchange chromatography (Thermo, ICS-3000, Waltham, MA, USA) equipped with a CarboPac™ PA10 column and pulsed amperometric detection set at 30 °C. The mobile phase was 100 mM NaOH and 500 mM NaAc at a flow rate of 0.3 mL/min [41]. All experiments were performed at least in duplicate, and values were expressed as mean ± standard deviation.

Cell growth was monitored turbidimetrically at an optical density of 600 nm (OD_{600}).

The yield of PCA is calculated as the ratio of PCA (mol) to the sum of the initial monophenols (mol).

4. Conclusions

In this work, *P. putida* KT2440 exhibited remarkable degradation abilities toward diverse phenolic acids and aldehydes in hydrolysates. By knockout of *pcaGH* and overexpression of the endogenous *vanAB* in *P. putida* KT2440, the maximum yield of PCA from FA and *p*-CA increased to 75.6% and 97.7%, respectively. Furthermore, by blocking sugar metabolism, the engineered strain KT3 did not consume the sugar, and only utilized the residual acetic acid in hydrolysates for cell growth and converted at least five phenolic compounds to PCA. Additionally, 253.88 mg/L of PCA was achieved by hydrolysate 1 with a yield of 70.85% and the highest titer of PCA was obtained at 433.72 mg/L PCA using hydrolysate 2. In addition to the common chemicals from sugars, the production of PCA further enhances the utilization value of the dilute acid pretreatment hydrolysate.

Supplementary Materials: The following supporting information can be downloaded at: https://www.mdpi.com/article/10.3390/molecules29071555/s1, Figure S1: Proportion of nine monophenols in total phenols in two hydrolysates; Figure S2: Metabolic process and growth of SAL, SA, VAN, and FA by *P. putida* KT2440; Figure S3: Confirmation of gene deletion and overexpression; Figure S4: The intermediate concentration (4-HBA and VA) curves; Figure S5: The growth on glucose and xylose by engineered *P. putida* KT2440; Figure S6: Production of PCA from 80% (*v/v*) hydrolysate 2 by KT2; Figure S7: HPLC analysis of PCA production in hydrolysates 1 and 2. Table S1: Primers used in this study.

Author Contributions: Methodology, formal analysis, investigation, writing—original draft, X.J.; data curation, resources, X.L.; writing—review and editing, resources, L.Z.; writing—review and editing, Z.Z.; conceptualization, funding acquisition, supervision, writing—review and editing, J.O. All authors have read and agreed to the published version of the manuscript.

Funding: This work was financially supported by The National Key Research and Development Program of China (2021YFC2101603).

Data Availability Statement: The data presented in this study are available on request from the corresponding author.

Conflicts of Interest: The authors declare no conflicts of interest.

References

1. Liu, Y.; Nie, Y.; Lu, X.; Zhang, X.; He, H.; Pan, F.; Zhou, L.; Liu, X.; Ji, X.; Zhang, S. Cascade Utilization of Lignocellulosic Biomass to High-Value Products. *Green Chem.* **2019**, *21*, 3499–3535. [CrossRef]
2. Kant, S.; Sadashiv, S.; Ashok, A.; Kant, R. Bioresource Technology Recent Developments in Pretreatment Technologies on Lignocellulosic Biomass: Effect of Key Parameters, Technological Improvements, and Challenges. *Bioresour. Technol.* **2020**, *300*, 122724. [CrossRef]
3. Hassan, S.S.; Williams, G.A.; Jaiswal, A.K. Lignocellulosic Biorefineries in Europe: Current State and Prospects. *Trends Biotechnol.* **2019**, *37*, 231–234. [CrossRef] [PubMed]
4. Chen, S.S.; Maneerung, T.; Tsang, D.C.W.; Ok, Y.S.; Wang, C.H. Valorization of Biomass to Hydroxymethylfurfural, Levulinic Acid, and Fatty Acid Methyl Ester by Heterogeneous Catalysts. *Chem. Eng. J.* **2017**, *328*, 246–273. [CrossRef]
5. Toor, M.; Kumar, S.S.; Malyan, S.K.; Bishnoi, N.R.; Mathimani, T.; Rajendran, K.; Pugazhendhi, A. An Overview on Bioethanol Production from Lignocellulosic Feedstocks. *Chemosphere* **2020**, *242*, 125080. [CrossRef] [PubMed]
6. Gogoi, G.; Hazarika, S. Coupling of Ionic Liquid Treatment and Membrane Filtration for Recovery of Lignin from Lignocellulosic Biomass. *Sep. Purif. Technol.* **2017**, *173*, 113–120. [CrossRef]
7. Zhou, M.; Fakayode, O.A.; Ahmed Yagoub, A.E.G.; Ji, Q.; Zhou, C. Lignin Fractionation from Lignocellulosic Biomass Using Deep Eutectic Solvents and Its Valorization. *Renew. Sustain. Energy Rev.* **2022**, *156*, 111986. [CrossRef]
8. Wang, Z.; Deuss, P.J. The Isolation of Lignin with Native-like Structure. *Biotechnol. Adv.* **2023**, *68*, 108230. [CrossRef] [PubMed]
9. Rinaldi, R.; Jastrzebski, R.; Clough, M.T.; Ralph, J.; Kennema, M.; Bruijnincx, P.C.A.; Weckhuysen, B.M. Paving the Way for Lignin Valorisation: Recent Advances in Bioengineering, Biorefining and Catalysis. *Angew. Chemie -Int. Ed.* **2016**, *55*, 8164–8215. [CrossRef]
10. Gillet, S.; Aguedo, M.; Petitjean, L.; Morais, A.R.C.; Da Costa Lopes, A.M.; Łukasik, R.M.; Anastas, P.T. Lignin Transformations for High Value Applications: Towards Targeted Modifications Using Green Chemistry. *Green Chem.* **2017**, *19*, 4200–4233. [CrossRef]

11. Bajwa, D.S.; Pourhashem, G.; Ullah, A.H.; Bajwa, S.G. A Concise Review of Current Lignin Production, Applications, Products and Their Environment Impact. *Ind. Crops Prod.* **2019**, *139*, 111526. [CrossRef]
12. Kim, J.Y.; Park, S.Y.; Lee, J.H.; Choi, I.G.; Choi, J.W. Sequential Solvent Fractionation of Lignin for Selective Production of Monoaromatics by Ru Catalyzed Ethanolysis. *RSC Adv.* **2017**, *7*, 53117–53125. [CrossRef]
13. Liu, H.; Liu, Z.H.; Zhang, R.K.; Yuan, J.S.; Li, B.Z.; Yuan, Y.J. Bacterial Conversion Routes for Lignin Valorization. *Biotechnol. Adv.* **2022**, *60*, 108000. [CrossRef] [PubMed]
14. Reshmy, R.; Athiyaman Balakumaran, P.; Divakar, K.; Philip, E.; Madhavan, A.; Pugazhendhi, A.; Sirohi, R.; Binod, P.; Kumar Awasthi, M.; Sindhu, R. Microbial Valorization of Lignin: Prospects and Challenges. *Bioresour. Technol.* **2022**, *344*, 126240. [CrossRef] [PubMed]
15. Linger, J.G.; Vardon, D.R.; Guarnieri, M.T.; Karp, E.M.; Hunsinger, G.B.; Franden, M.A.; Johnson, C.W.; Chupka, G.; Strathmann, T.J.; Pienkos, P.T.; et al. Lignin Valorization through Integrated Biological Funneling and Chemical Catalysis. *Proc. Natl. Acad. Sci. USA* **2014**, *111*, 12013–12018. [CrossRef] [PubMed]
16. Becker, J.; Kuhl, M.; Kohlstedt, M.; Starck, S.; Wittmann, C. Metabolic Engineering of Corynebacterium Glutamicum for the Production of Cis, Cis-Muconic Acid from Lignin. *Microb. Cell Fact.* **2018**, *17*, 115. [CrossRef] [PubMed]
17. Antony, F.M.; Wasewar, K. Effect of Temperature on Equilibria for Physical and Reactive Extraction of Protocatechuic Acid. *Heliyon* **2020**, *6*, e03664. [CrossRef] [PubMed]
18. Song, J.; He, Y.; Luo, C.; Feng, B.; Ran, F.; Xu, H.; Ci, Z.; Xu, R.; Han, L.; Zhang, D. New Progress in the Pharmacology of Protocatechuic Acid: A Compound Ingested in Daily Foods and Herbs Frequently and Heavily. *Pharmacol. Res.* **2020**, *161*, 105109. [CrossRef] [PubMed]
19. Muthukumaran, J.; Srinivasan, S.; Venkatesan, R.S.; Ramachandran, V.; Muruganathan, U. Syringic Acid, a Novel Natural Phenolic Acid, Normalizes Hyperglycemia with Special Reference to Glycoprotein Components in Experimental Diabetic Rats. *J. Acute Dis.* **2013**, *2*, 304–309. [CrossRef]
20. Semaming, Y.; Pannengpetch, P.; Chattipakorn, S.C.; Chattipakorn, N. Pharmacological Properties of Protocatechuic Acid and Its Potential Roles as Complementary Medicine. *Evid. -Based Complement. Altern. Med.* **2015**, *2015*, 593902. [CrossRef]
21. Yang, Y.C.; Wei, M.C.; Huang, T.C.; Lee, S.Z. Extraction of Protocatechuic Acid from Scutellaria Barbata D. Don Using Supercritical Carbon Dioxide. *J. Supercrit. Fluids* **2013**, *81*, 55–66. [CrossRef]
22. Liu, Q.; Tang, G.Y.; Zhao, C.N.; Feng, X.L.; Xu, X.Y.; Cao, S.Y.; Meng, X.; Li, S.; Gan, R.Y.; Li, H. Bin Comparison of Antioxidant Activities of Different Grape Varieties. *Molecules* **2018**, *23*, 2432. [CrossRef] [PubMed]
23. Guo, X.; Wang, X.; Chen, T.; Lu, Y.; Zhang, H. Comparing E. Coli Mono-Cultures and Co-Cultures for Biosynthesis of Protocatechuic Acid and Hydroquinone. *Biochem. Eng. J.* **2020**, *156*, 107518. [CrossRef]
24. Kogure, T.; Suda, M.; Hiraga, K.; Inui, M. Protocatechuate Overproduction by *Corynebacterium Glutamicum* via Simultaneous Engineering of Native and Heterologous Biosynthetic Pathways. *Metab. Eng.* **2021**, *65*, 232–242. [CrossRef] [PubMed]
25. Li, J.; Yue, C.; Wei, W.; Shang, Y.; Zhang, P.; Ye, B.C. Construction of a P-Coumaric and Ferulic Acid Auto-Regulatory System in *Pseudomonas Putida* KT2440 for Protocatechuate Production from Lignin-Derived Aromatics. *Bioresour. Technol.* **2022**, *344*, 126221. [CrossRef]
26. Vignali, E.; Pollegioni, L.; Di Nardo, G.; Valetti, F.; Gazzola, S.; Gilardi, G.; Rosini, E. Multi-Enzymatic Cascade Reactions for the Synthesis of *Cis,Cis*-Muconic Acid. *Adv. Synth. Catal.* **2022**, *364*, 114–123. [CrossRef]
27. Zou, L.; Ouyang, S.; Hu, Y.; Zheng, Z.; Ouyang, J. Efficient Lactic Acid Production from Dilute Acid-Pretreated Lignocellulosic Biomass by a Synthetic Consortium of Engineered *Pseudomonas Putida* and *Bacillus Coagulans*. *Biotechnol. Biofuels* **2021**, *14*, 227. [CrossRef]
28. Fu, J.; Wang, Z.; Miao, H.; Yu, C.; Zheng, Z.; Ouyang, J. Rapid Adaptive Evolution of *Bacillus Coagulans* to Undetoxified Corncob Hydrolysates for Lactic Acid Production and New Insights into Its High Phenolic Degradation. *Bioresour. Technol.* **2023**, *383*, 129246. [CrossRef]
29. Overhage, J.; Priefert, H.; Rabenhorst, J.; Steinbüchel, A. Biotransformation of Eugenol to Vanillin by a Mutant of *Pseudomonas* Sp. Strain HR199 Constructed by Disruption of the Vanillin Dehydrogenase (Vdh) Gene. *Appl. Microbiol. Biotechnol.* **1999**, *52*, 820–828. [CrossRef]
30. Ravi, K.; García-hidalgo, J.; Gorwa-grauslund, M.F.; Lidén, G. Conversion of Lignin Model Compounds by *Pseudomonas Putida* KT2440 and Isolates from Compost. *Appl. Microb. Cell Physiol.* **2017**, *101*, 5059–5070. [CrossRef]
31. Nikel, P.I.; Chavarría, M.; Fuhrer, T.; Sauer, U.; De Lorenzo, V. Pseudomonas Putida KT2440 Strain Metabolizes Glucose through a Cycle Formed by Enzymes of the Entner-Doudoroff, Embden-Meyerhof-Parnas, and Pentose Phosphate Pathways. *J. Biol. Chem.* **2015**, *290*, 25920–25932. [CrossRef]
32. Dvořák, P.; de Lorenzo, V. Refactoring the Upper Sugar Metabolism of *Pseudomonas Putida* for Co-Utilization of Cellobiose, Xylose, and Glucose. *Metab. Eng.* **2018**, *48*, 94–108. [CrossRef]
33. Erickson, E.; Bleem, A.; Kuatsjah, E.; Werner, A.Z.; DuBois, J.L.; McGeehan, J.E.; Eltis, L.D.; Beckham, G.T. Critical Enzyme Reactions in Aromatic Catabolism for Microbial Lignin Conversion. *Nat. Catal.* **2022**, *5*, 86–98. [CrossRef]
34. Li, J.; Ye, B.C. Metabolic Engineering of *Pseudomonas Putida* KT2440 for High-Yield Production of Protocatechuic Acid. *Bioresour. Technol.* **2021**, *319*, 124239. [CrossRef]
35. Upadhyay, P.; Lali, A. Protocatechuic Acid Production from Lignin-Associated Phenolics. *Prep. Biochem. Biotechnol.* **2021**, *51*, 979–984. [CrossRef]

36. Graf, N.; Altenbuchner, J. Genetic Engineering of Pseudomonas Putida KT2440 for Rapid and High-Yield Production of Vanillin from Ferulic Acid. *Appl. Microbiol. Biotechnol.* **2014**, *98*, 137–149. [CrossRef]
37. Johnson, C.W.; Salvachúa, D.; Khanna, P.; Smith, H.; Peterson, D.J.; Beckham, G.T. Enhancing Muconic Acid Production from Glucose and Lignin-Derived Aromatic Compounds via Increased Protocatechuate Decarboxylase Activity. *Metab. Eng. Commun.* **2016**, *3*, 111–119. [CrossRef]
38. Zhang, R.K.; Tan, Y.S.; Cui, Y.Z.; Xin, X.; Liu, Z.H.; Li, B.Z.; Yuan, Y.J. Lignin Valorization for Protocatechuic Acid Production in Engineered: Saccharomyces Cerevisiae. *Green Chem.* **2021**, *23*, 6515–6526. [CrossRef]
39. Okai, N.; Masuda, T.; Takeshima, Y.; Tanaka, K.; Yoshida, K.-I.; Miyamoto, M.; Ogino, C.; Kondo, A. Biotransformation of Ferulic Acid to Protocatechuic Acid by Corynebacterium Glutamicum ATCC 21420 Engineered to Express Vanillate O-Demethylase. *AMB Express* **2017**, *7*, 130. [CrossRef] [PubMed]
40. Jiang, T.; Qiao, H.; Zheng, Z.; Chu, Q.; Li, X.; Yong, Q.; Ouyang, J. Lactic Acid Production from Pretreated Hydrolysates of Corn Stover by a Newly Developed *Bacillus Coagulans* Strain. *PLoS ONE* **2016**, *11*, e0149101. [CrossRef] [PubMed]
41. Dai, L.; Jiang, W.; Jia, R.; Zhou, X.; Xu, Y. Directional Enhancement of 2-Keto-Gluconic Acid Production from Enzymatic Hydrolysate by Acetic Acid-Mediated Bio-Oxidation with *Gluconobacter Oxydans. Bioresour. Technol.* **2022**, *348*, 126811. [CrossRef] [PubMed]

Article

Sustainable Coating Based on Zwitterionic Functionalized Polyurushiol with Antifouling and Antibacterial Properties

Kaiyue Xu [†], Huimin Xie [†], Chenyi Sun, Wenyan Lin, Zixuan You, Guocai Zheng, Xiaoxiao Zheng, Yanlian Xu, Jipeng Chen * and Fengcai Lin *

Fujian Engineering and Research Center of New Chinese Lacquer Materials, College of Materials and Chemical Engineering, Minjiang University, Fuzhou 350108, China; xukaiyue@stu.mju.edu.cn (K.X.); xiehuimin@stu.mju.edu.cn (H.X.); sunchenyi@stu.mju.edu.cn (C.S.); linwenyan@stu.mju.edu.cn (W.L.); youzixuan@stu.mju.edu.cn (Z.Y.); 2231@mju.edu.cn (G.Z.); xxzheng@mju.edu.cn (X.Z.); ylxu@mju.edu.cn (Y.X.)
* Correspondence: jpchen@mju.edu.cn (J.C.); fengcailin@mju.edu.cn (F.L.)
[†] These authors contributed equally to this work.

Abstract: Zwitterionic polymer coatings facilitate the formation of hydration layers via electrostatic interactions on their surfaces and have demonstrated efficacy in preventing biofouling. They have emerged as a promising class of marine antifouling materials. However, designing multifunctional, environmentally friendly, and natural products-derived zwitterionic polymer coatings that simultaneously resist biofouling, inhibit protein adhesion, exhibit strong antibacterial properties, and reduce algal adhesion is a significant challenge. This study employed two diisocyanates as crosslinkers and natural urushiol and ethanolamine as raw materials. The coupling reaction of diisocyanates with hydroxyl groups was employed to synthesize urushiol-based precursors. Subsequently, sulfobetaine moieties were introduced into the urushiol-based precursors, developing two environmentally friendly and high-performance zwitterionic-functionalized polyurushiol antifouling coatings, denoted as HUDM-SB and IPUDM-SB. The sulfobetaine-functionalized polyurushiol coating exhibited significantly enhanced hydrophilicity, with the static water contact angle reduced to less than 60°, and demonstrated excellent resistance to protein adhesion. IPUDM-SB exhibited antibacterial efficacy up to 99.9% against common Gram-negative bacteria (*E. coli* and *V. alginolyticus*) and Gram-positive bacteria (*S. aureus* and *Bacillus*. sp.). HUDM-SB achieved antibacterial efficacy exceeding 95.0% against four bacterial species. Furthermore, the sulfobetaine moieties on the surfaces of the IPUDM-SB and HUDM-SB coatings effectively inhibited the growth and reproduction of algal cells by preventing microalgae adhesion. This zwitterionic-functionalized polyurushiol coating does not contain antifouling agents, making it a green, environmentally friendly, and high-performance biomaterial-based solution for marine antifouling.

Keywords: urushiol; zwitterionic polymers; marine antifouling coating; biomass-based coating

---

Citation: Xu, K.; Xie, H.; Sun, C.; Lin, W.; You, Z.; Zheng, G.; Zheng, X.; Xu, Y.; Chen, J.; Lin, F. Sustainable Coating Based on Zwitterionic Functionalized Polyurushiol with Antifouling and Antibacterial Properties. *Molecules* **2023**, *28*, 8040. https://doi.org/10.3390/molecules28248040

Academic Editors: José A.P. Coelho, Roumiana P. Stateva and Joannis K. Kallitsis

Received: 16 November 2023
Revised: 4 December 2023
Accepted: 9 December 2023
Published: 11 December 2023

1. Introduction

The attachment and growth of microorganisms, algae, barnacles, and other aquatic organisms on the surfaces of ships, marine facilities, or other underwater structures in the marine environment is referred to as marine biofouling [1,2]. A globally recognized issue, biofouling leads to surface contamination of vessels and underwater equipment, increased hydrodynamic resistance, corrosion of underwater structures, threats to safe operations, economic losses, and environmental concerns [3]. These challenges significantly impede the development and utilization of marine resources worldwide. Protective coatings are the most direct and effective means of mitigating marine biofouling. However, the marine antifouling coatings used domestically and internationally often rely on petroleum-based resin materials containing toxic metal ions and bioactive additives, such as Cu_2O [4,5], zinc pyrithione [6,7], and copper/zinc pyrithione sulfate [8]. Although they combat marine biofouling, the self-degradation products of these coatings can severely harm non-target organisms and the marine environment. Consequently, within the framework of achieving global "carbon neutrality", there is a pressing need to design and synthesize environmentally friendly, non-toxic, renewable, and high-performance marine antifouling coatings based on natural extracts (such as cellulose [9,10], urushiol [11], tannic acid [12], and tung oil [13]).

Hydrophilic materials that are resistant to protein adhesion represent a prominent category of fouling-resistant coatings. This property is often achieved through the incorporation of polyethylene glycol (PEG) and zwitterions to enhance the hydrophilicity of the material [14,15]. PEG is a non-toxic material that forms hydration layers through extensive hydrogen bonding with water molecules and effectively inhibits protein adhesion [14,16]. However, PEG lacks inherent biocidal properties and has insufficient mechanical performance to support its use as a long-term antifouling coating. Zwitterionic compounds are an alternative class of hydrophilic antifouling material featuring chemical structures that contain positively and negatively charged groups [17]. These compounds undergo electrostatic interactions to form hydration layers and display strong hydrophilicity, deterring the adhesion of fouling organisms via inhibition of protein adhesion [18].Typical cationic zwitterions include quaternary ammonium ions, while anionic zwitterions include sulfobetaines, carboxybetaines, and phosphocholines [18,19]. Zwitterions can be utilized to modify surface grafting to tailor the surface properties of polymers and achieve high hydrophilicity in their coatings. Additionally, the quaternary ammonium ions in zwitterionic compounds impart effective antibacterial properties [20,21]. Therefore, zwitterion-modified polymer surfaces form hydration layers on the coating surfaces to prevent adhesion by fouling organisms and also exhibit biocidal effects due to the quaternary ammonium groups. The synergistic action of these two mechanisms gives these coatings exceptional antifouling characteristics. Zhang et al. [22] employed a one-pot reaction to graft zwitterionic esters and capsaicin polymer on a polydimethylsiloxane (PDMS) network, resulting in a multifunctional marine antifouling coating with excellent adhesion resistance for protein, bacteria, and diatoms. Guan et al. [23] significantly enhanced the surface hydrophilicity of modified membranes using layer-by-layer interfacial polymerization of zwitterionic polymers, concurrently increasing water permeability to create a material with outstanding antifouling properties. The coatings do not release other antifouling agents because the quaternary ammonium groups are grafted onto them. Therefore, zwitterion surface functionalization offers an environmentally friendly and synergistic antifouling strategy. Although the exceptional performance of zwitterionic compounds in marine antifouling has been confirmed, their independent application in marine antifouling is challenging. Firm coupling with substrate materials through chemical or physical methods is required to fully leverage their excellent antibacterial and antifouling properties.

Raw lacquer, a distinctive and high-quality natural resource in Asia, is a natural resin coating with excellent adhesion, film-forming properties, renewability, and environmental friendliness [24,25]. It has been used as an adhesive and coating for furniture, wooden structures, and other items for thousands of years. Urushiol is the primary component of raw lacquer and has a catechol-like ortho-quinone structure with alkyl side chains of C_{15}–C_{17} on the benzene ring [26]. Urushiol is a premium natural resource for constructing green marine antifouling coatings because the dual-active hydroxyl and unsaturated double bonds within its molecular structure render it highly chemically reactive and modifiable [11]. The active phenolic hydroxyl group of urushiol can readily form a stable interface with the functional groups of zwitterionic compounds through covalent or non-covalent bonds. This interaction creates active sites with high potential for adhesion in zwitterionic compounds. The flexible, unsaturated long side-chain structure of urushiol can undergo a self-polymerization reaction, leading to the formation of a mesh polymer cross-linking network with favorable mechanical properties for use in zwitterionic polymer coatings. Urushiol polymer coatings have a smooth and compact texture, high hardness, excellent stability, and resistance to organic solvents and chemical corrosion. These properties may collectively enhance the underwater stability of zwitterionic polymer coatings. Furthermore, the structure and properties of urushiol-based polymer coatings can be effectively adjusted by altering the ratio of the components. This flexibility enables the creation of a novel, green, bio-based marine antifouling coating with superior performance.

In the present study, environmentally friendly sulfobetaine-functionalized polyurushiol coatings, specifically HUDM-SB and IPUDM-SB, were synthesized using natural urushiol and ethanolamine as raw materials. Two types of diisocyanates were selected as cross-linking agents, and the coupling reaction of diisocyanates with the hydroxyl groups of urushiol was employed to synthesize urushiol-based precursors. Subsequently, sulfobetaine groups were introduced into these precursors, resulting in two environmentally friendly and high-performance zwitterionic-functionalized urushiol-polymer antifouling coatings. The impact of zwitterionic groups on the wettability, thermal stability, and physical–mechanical properties of the coatings was investigated using contact angle measurements, liquid-droplet adhesion tests, thermogravimetric analysis, and tests of mechanical properties. Furthermore, bovine serum albumin (BSA) and γ-globulin were used to systematically evaluate the ability of these coatings to resist protein adhesion. Additionally, the antibacterial and anti-diatom-fouling performances of the coatings were comprehensively examined using the Gram-negative bacteria *Escherichia coli* (*E. coli*) and *Vibrio alginolyticus* (*V. alginolyticus*), the Gram-positive bacteria *Staphylococcus aureus* (*S. aureus*) and *Bacillus species* (*Bacillus. Sp.*), and the algae *Nitzschia closterium* (*N. closterium*) and *Phaeodactylum tricornutum* (*P. tricornutum*). These coatings were primarily sourced from urushiol, a natural product, and sulfobetaine zwitterionic groups and contained no additional fouling agents. The coatings offer a novel approach to developing green, environmentally friendly, and high-performance biomaterial-based marine antifouling materials.

2. Results and Discussions

2.1. Synthesis and Structural Characterization of the HUDM-SB and IPUDM-SB Monomers

As shown in Figure 1a, the reaction between DMEA and HMDI formed PHDE containing aliphatic isocyanate (NCO) groups. Subsequently, the NCO groups in PHDE reacted with OH groups in urushiol to produce a urushiol monomer with quaternary amine structures (HUDM). Next, zwitterionic sulfobetaine glycine monomers were covalently linked to HUDM molecules through an opening reaction with 1,3-propane sulfone and the tertiary amine, yielding HUDM-SB. The chemical structure of HUDM-SB was characterized using FTIR spectroscopy, as shown in Figure 1c. The FTIR spectrum of HMDI exhibited characteristic peaks near 2270 cm^{-1} corresponding to the NCO groups [27]. Additionally, NCO peaks were observed around 2270 cm^{-1} in the PHDE spectrum, albeit with reduced intensity compared to HMDI, indicating a reaction between a few NCO groups in HMDI and DMEA. However, the characteristic NCO peaks disappeared in the FTIR spectrum of HUDM, suggesting the reaction of the remaining isocyanate groups in HMDI with urushiol [28]. Furthermore, a series of characteristic absorption peaks corresponding to the phenolic amide ester linkages were observed, confirming the successful coupling of NCO and OH groups. In the PHDE and HUDM spectra, peaks at 3330 and 3299 cm^{-1} were attributed to the N-H stretching vibrations in amide esters and phenolic amide ester bonds [29], respectively. The peaks at 1721 and 1706 cm^{-1} corresponded to the C=O stretching vibrations [30], and overlapping peaks at 1249–1259 cm^{-1} indicated the stretching and bending vibrations of the C-N bond in the amide and N-H bonds [31]. Therefore, these results verified the successful synthesis of HUDM, the precursor containing aminoethyl methacrylate and phenolic amide ester bonds. Moreover, the FTIR spectrum of HUDM-SB displayed strong characteristic absorption peaks at 1170 and 1067 cm^{-1}, corresponding to the vibrational absorption peaks of the –SO$_3^-$ groups, confirming the successful coupling of sulfobetaine glycine groups to the urushiol molecules [32].

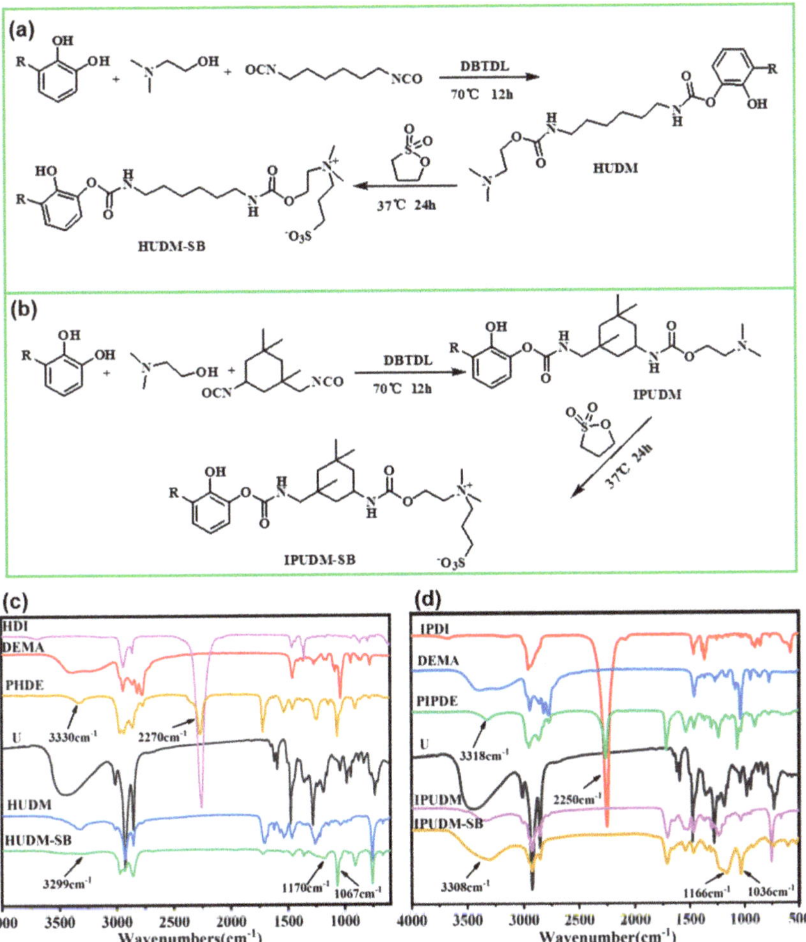

Figure 1. (**a**,**b**) Synthesis of HUDM, HUDM−SB, IPUDM and IPUDM−SB. ATR−FTIR spectra of (**c**) HUDM−SB and (**d**) IPUDM−SB.

Similarly, the reaction between DMEA and IPDI formed PIPDE containing NCO groups, as shown in Figure 1b. The NCO groups in PIPDE readily coupled with OH groups in urushiol molecules to yield IPUDM, which is urushiol with tertiary amine structures. Subsequently, zwitterionic sulfobetaine glycine monomers were covalently linked to urushiol through an opening reaction with 1,3-propane sulfone and the tertiary amine to form IPUDM-SB. The chemical structure of IPUDM-SB was characterized using FTIR spectroscopy, as shown in Figure 1d. Compared to the FTIR spectrum of IPDI shown in Figure 1d, the absorption peak intensity of the NCO characteristic peak around 2250 cm^{-1} in PIPDE was significantly reduced, indicating a reaction between the NCO groups in IPDI and DMEA [33]. The NCO characteristic peaks disappeared in the FTIR spectrum of IPUDM, suggesting a complete reaction of the two isocyanates in IPDI with DMEA and urushiol. The peaks at 3308 and 3318 cm^{-1} in the PIPDE and IPUDM spectra corresponded to the N-H stretching vibrations in amide ester and phenolic amide ester bonds, respectively. The peaks at 1719 and 1713 cm^{-1} represented the C=O stretching vibrations [34]. The overlapping peaks at 1237–1238 cm^{-1} indicated the stretching and bending vibrations of the C-N bond in the amide and N-H bonds. Therefore, these results confirmed the successful synthesis of IPUDM, the precursor containing aminoethyl methacrylate and phenolic amide ester bonds. Furthermore, the FTIR spectrum of IPUDM-SB showed strong characteristic absorption peaks at 1166 and 1036 cm^{-1}, corresponding to the vibrational absorption peaks of the –SO$_3^-$ groups, confirming the successful coupling of sulfobetaine glycine groups with the urushiol molecules [32].

2.2. Surface Elemental Composition of the HUDM-SB and IPUDM-SB Coatings

As illustrated in Figure 2, XPS was employed to characterize the surface elemental composition of the zwitterion-functionalized urushiol polymer coatings. The XPS spectra of the IPUDM and HUDM coatings showed only three peaks, representing C1s (284.5 eV), N1s (400.1 eV), and O1s (544.9 eV). However, as shown in Figure 2a,e, additional characteristic peaks for S elements, S2p and S2s, were evident in the XPS spectra of IPUDM-SB and HUDM-SB coatings where sulfobetaine moieties were introduced, indicating the successful grafting of sulfobetaine to urushiol molecules [35]. As shown in the high-resolution N1s spectra in Figure 2b,f, IPUDM and HUDM coatings had a single primary peak at ~400.1 eV. As shown in Figure 2c,g, the N1s spectra of the IPUDM-SB coatings had two distinct peaks at 402.3 and 399.9 eV, and the N1s spectra of the HUDM-SB coatings had two distinct peaks at 402.5 and 399.8 eV. The peaks at 399.9 and 399.8 eV corresponded to the unreacted tertiary amines on the IPUDM-SB and HUDM-SB coatings, respectively. The other peaks at 402.5 and 402.3 eV represented the quaternary ammonium ions (+NR$_4$) formed after the successful chemical grafting of 1,3-propane sulfone [32]. The quaternary ammonium ions required higher energy to displace electrons from their N1s orbitals due to their electron-deficient and more stable nature. Additionally, as shown in Figure 2d,h, the surfaces of the IPUDM-SB coatings had two sulfur (S) elemental peaks at 169.0 and 167.7 eV, and the surfaces of the HUDM-SB coatings had two S elemental peaks at 169.1 and 167.8 eV, corresponding to the S2p$_{1/2}$ and S2p$_{3/2}$ signal absorption peaks of the –SO$_3^-$ groups [36], respectively. Therefore, the results confirmed that the reaction of the IPUDM and HUDM precursors with 1,3-propane sulfone and tertiary amines through an opening reaction covalently linked sulfobetaine glycine to the urushiol molecules. Additionally, this synthesis resulted in zwitterionic urushiol-based polymeric antifouling coatings, rich in cationic (+NR$_4$) and anionic (–SO$_3^-$) functional groups.

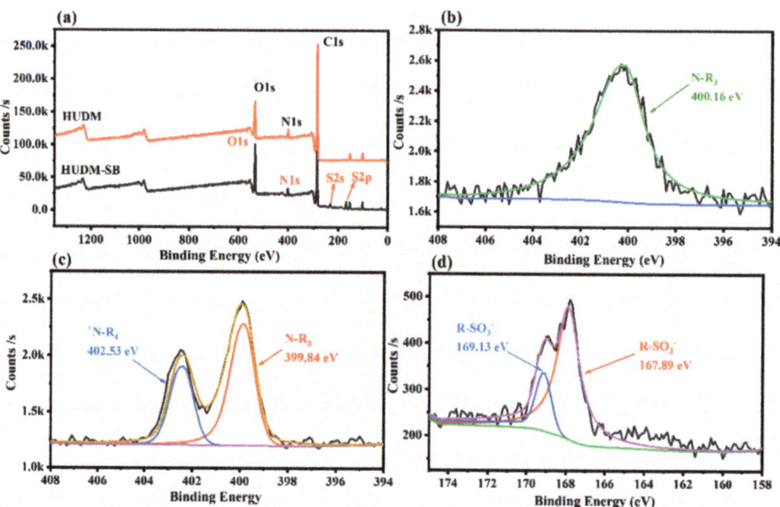

Figure 2. *Cont.*

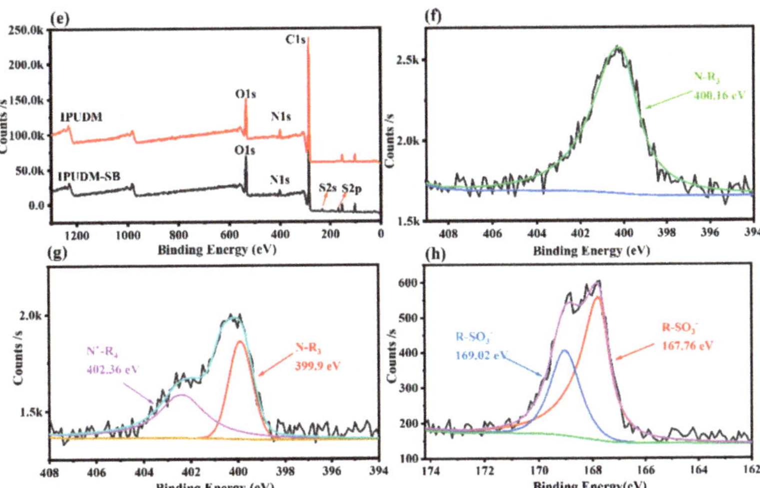

Figure 2. (**a**,**e**) XPS survey spectra stack graph of IPUDM, HUDM, IPUDM−SB and HUDM−SB. N1s core-grade XPS spectra of (**b**) HUDM, (**c**) HUDM−SB, (**f**) IPUDM, and (**g**) IPUDM−SB. S2p core-grade XPS spectra of (**d**) HUDM−SB and (**h**) IPUDM−SB.

2.3. Surface and Physicochemical Properties of the Coatings

The physical and mechanical properties of zwitterionic urushiol-polymer coatings, which were synthesized by coupling urushiol molecules with different types of isocyanates, are presented in Table 1. The urushiol coating coupled with cycloaliphatic isocyanates (IPUDM-SB) exhibited superior adhesion, pencil hardness, and gloss compared to the urushiol coating coupled with aliphatic isocyanates (HUDM-SB). This observation suggests that coatings with a higher proportion of rigid structures within the urushiol polymer have enhanced hardness. Importantly, both IPUDM-SB and HUDM-SB coatings demonstrated higher adhesion and hardness than the unmodified zwitterionic urushiol-based polymer coatings, indicating that the introduction of sulfobetaine moieties did not compromise the original physical properties of the coatings.

Table 1. Physical and mechanical properties of different coatings.

Samples	Adhesion (Grade)	Pencil Hardness	Glossiness (%)
IPUDM	7	1H	93.7
IPUDM-SB	3	2H	88.7
HUDM	6	4B	65.8
HUDM-SB	4	HB	44.6

In order to evaluate the hydrophilic properties of the coated surfaces, static water contact angles (WCAs) were measured at room temperature for each of the coatings, namely, IPUDM, IPUDM-SB, HUDM, and HUDM-SB, as depicted in Figure 3. Contact-angle measurements revealed that the WCAs for IPUDM and HUDM coatings were $95.5 \pm 0.9°$ and $88.7 \pm 5.8°$, respectively. The hydrophobic nature of the IPUDM coating was attributed to the presence of rigid structures originating from cycloaliphatic isocyanates. In contrast, the HUDM coating, which contained aliphatic isocyanates and saturated long-chain flexible structures, exhibited reduced hydrophobicity. The introduction of zwitterionic functionalities in the IPUDM-SB and HUDM-SB coatings led to a significant reduction in WCAs, with values of $56.2 \pm 2.19°$ and $57.0 \pm 5.77°$, respectively. This finding represents respective decreases of 41.12% and 35.74% in contact angles, indicating notably enhanced hydrophilicity. This enhancement can be attributed to the abundance of anions and cations on the coating surfaces, which, through electrostatic hydration, formed a dense hydration layer without disrupting hydrogen bonding between water molecules [37]. Remarkably, IPUDM-SB coatings demonstrated the greatest increase in hydrophilicity, possibly due to the isophorone diisocyanate structure and the presence of two different NCO reactivity sites. Generally, coatings with higher hydrophilicity tend to exhibit stronger binding with water droplets. As shown in Figure 3b, the unmodified IPUDM and HUDM coatings displayed significantly lower binding with water droplets compared to the zwitterionic functionalized IPUDM-SB and HUDM-SB coatings. The adhesion forces between the IPUDM-SB and HUDM-SB coatings and water droplets measured 0.51 mN and 0.43 mN, respectively, with the IPUDM-SB coating demonstrating a higher binding force than the HUDM-SB coating, consistent with the WCA test results.

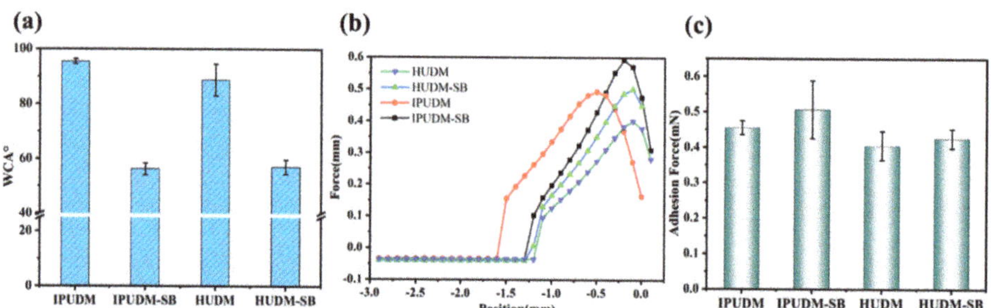

Figure 3. Basic performance tests for IPUDM, IPUDM−SB, HUDM and HUDM−SB coatings: (**a**) WCA, (**b**) adhesion force measurements, and (**c**) statistical adhesion force measurements. Each sample was measured thrice and averaged.

The thermal stability of room-temperature-cured IPUDM-SB and HUDM-SB coatings was examined using thermogravimetric analysis (TGA), as depicted in Figure 4a,b. The results indicated that IPUDM-SB and HUDM-SB coatings began to decompose at temperatures of 199.17 °C and 222.50 °C, with residual carbon contents of 7.47% and 8.26%, respectively. The thermal decomposition can be described as follows. In the initial step, the carbonyl bonds of the carbamate groups on the polymer's main chain undergo thermal cleavage, resulting in isocyanates, polyols, or polyphenols. In the subsequent step, the hard segments of isocyanates break down into diisocyanates. The final step involves polyol or polyphenol decomposition. This finding indicates that the breakdown of the urea bonds in the isocyanates and the carbamate groups in the hard segments occurs earlier than the decomposition of the soft segments. IPUDM-SB, which contains cycloaliphatic isocyanates, begins to decompose at a slightly lower temperature than HUDM-SB, which includes aliphatic isocyanates, indicating that IPUDM-SB has lower thermal stability.

The thermal performance of the room-temperature-cured coatings, namely IPUDM, IPUDM-SB, HUDM, and HUDM-SB, was further investigated through differential scanning calorimetry (DSC) analysis. As illustrated in Figure 4c,d, distinct glass transition temperatures (T_g) were observed during the second heating scan in the DSC test: 40.72 °C, 93.56 °C, 24.51 °C, and 74.12 °C, respectively. It is evident that HUDM and HUDM-SB coatings have slightly lower T_g values compared to IPUDM and IPUDM-SB coatings. This difference can be attributed to the use of 1,6-hexamethylene diisocyanate as the crosslinker in HUDM and HUDM-SB coatings, which introduces more flexible structures and thus results in lower T_g values. Additionally, the DSC results reveal that IPUDM-SB and HUDM-SB coatings, which are functionalized with sulfobetaine, exhibit significantly higher T_g values compared to IPUDM and HUDM coatings. This significant change in T_g values suggests that the introduction of sulfobetaine moieties may increase cross-linking density and reduce the free volume fraction of the main chains of IPUDM-SB and HUDM-SB coatings. Furthermore, IPUDM-SB, which contains cycloaliphatic isocyanates, exhibits a slightly higher cross-linking density compared to HUDM-SB, which contains aliphatic isocyanates.

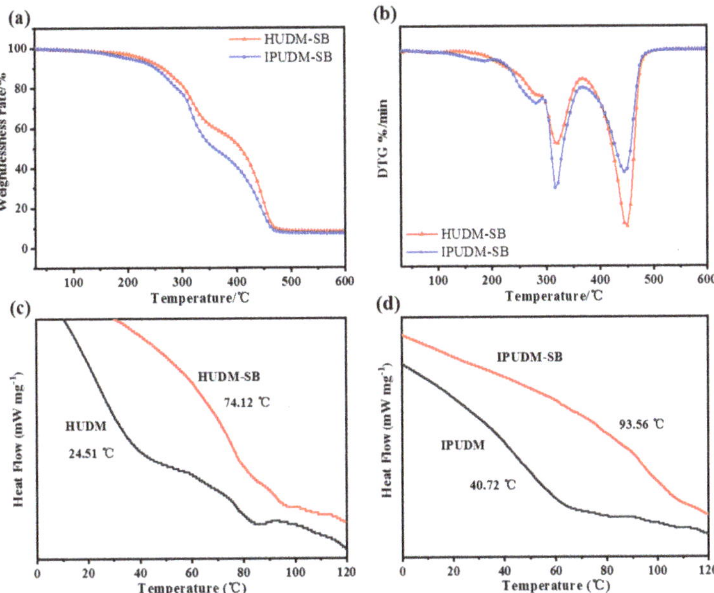

Figure 4. Thermal stability tests for HUDM−SB and IPUDM−SB coatings: (**a**) DTG (**b**) TGA. Glass transition temperature tests for coatings: (**c**) HUDM and HUDM−SB (**d**) IPUDM and IPUDM−SB.

2.4. Surface and Physicochemical Properties of the Coatings

The occurrence of biofouling on the surfaces of underwater equipment involves several stages, including formation of an organic monolayer, development of a biofilm, adhesion of algal spores and protozoa, and attachment of macrofouling organisms. A crucial initial step in macrofouling is the successful adsorption of organic compounds, such as proteins, polysaccharides, and nucleic acids, onto underwater substrates. Hydrophilic antifouling coatings operate by capitalizing on the inhibitory effect of the surface hydration layer on the adsorption of organic macromolecules, which discourages subsequent attachment by fouling organisms [38]. Consequently, evaluating the anti protein-adsorption capacity of antifouling coatings is essential for assessing their fouling resistance. In this study, we employed two major plasma proteins, bovine serum albumin (BSA) and γ-globulin, to evaluate the anti-protein-adsorption performance of the coatings through static protein adsorption. The absorbance of BSA and γ-globulin solutions to different coating surfaces is presented in Figure 5a,b. Notably, for both protein solutions, the absorbance to the IPUDM-SB and HUDM-SB coatings was lower than the absorbance to the BG, IPUDM, and HUDM coatings. As depicted in Figure 5c, for the BSA solution, the ultraviolet absorbances at 280 nm were 0.51, 0.31, and 0.32 for the BG, HUDM, and IPUDM coatings, respectively. In the case of the γ-globulin solution, the corresponding ultraviolet absorbances for these coatings were 0.58, 0.36, and 0.3. Both IPUDM and HUDM coatings exhibited significantly lower ultraviolet absorbance at 280 nm for both types of proteins compared to the blank control BG. This result indicates that IPUDM and HUDM coatings have some level of resistance to protein adhesion. After the addition of sulfobetaine zwitterionic groups, the IPUDM-SB and HUDM-SB coatings showed a significant reduction in ultraviolet absorbance at 280 nm in both BSA and γ-globulin solutions. The ultraviolet absorbance for BSA solutions decreased to 0.11 and 0.35, respectively, for the two coatings, while the ultraviolet absorbance for γ-globulin solutions decreased to 0.14 and 0.24. These findings indicate that both coatings exhibit excellent fouling resistance, primarily due to the highly hydrophilic nature of the zwitterionic groups on the coating surfaces. These groups, through electrostatically induced hydration interactions, form a hydration layer, hindering the interaction between proteins and the surface. The zwitterionic groups thus provide the coatings with outstanding anti-protein-adhesion properties [39].

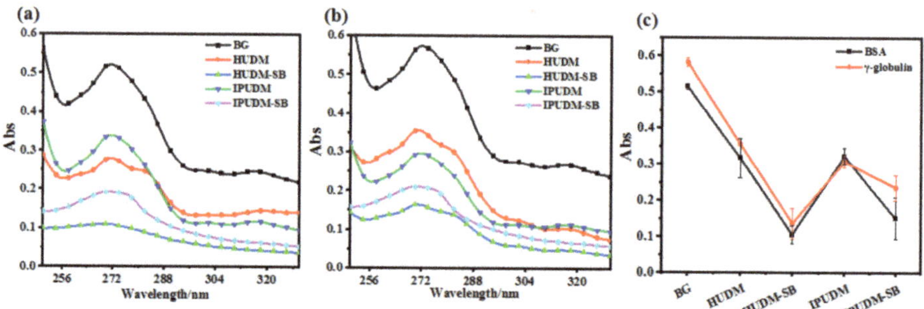

Figure 5. The absorbance at 280 nm of protein solutions attached to the surfaces of BG, HUDM, HUDM-SB, IPUDM, and IPUDM-SB coatings. (**a**) BSA; (**b**) γ-globulin; (**c**) Adsorption capacity of BSA and γ-globulin solutions on BG, HUDM, HUDM-SB, IPUDM, and IPUDM-SB coatings at 280 nm.

2.5. Antibacterial Performance of the Coatings

To investigate the antibacterial performance of IPUDM-SB and HUDM-SB coatings, four bacterial strains were employed for assessment, namely, Gram-negative *E. coli* (BW 25113), Gram-positive *S. aureus* (ATCC 25923), marine bacterium *V. alginolyticus* (ATCC 33787), and *Bacillus* sp. (MCCC 1B00342). The antibacterial effectiveness of the antifouling coatings was assessed by determining the number of bacterial colonies that developed on agar plates from a bacterial culture broth collected from the sample surface and comparing to the yield from the control sample BG. As shown in Figure 6a, visual images of agar plates from various samples clearly depict substantial numbers of bacterial colonies grown from culture broth collected from the unmodified IPUDM and HUDM coatings, indicating limited antibacterial properties. The antibacterial rates of IPUDM coatings against *E. coli*, *S. aureus*, *V. alginolyticus*, and *Bacillus* sp. were 19.5%, 28%, 36%, and 24.5%, respectively, while HUDM coatings displayed lower antibacterial effectiveness rates of 14%, 14.5%, 34%, and 12.5% against the same bacteria (Figure 6b). In contrast, IPUDM-SB coatings exhibited the highest antibacterial effectiveness against all four bacteria, with nearly no bacterial colonies on the agar plates and an antibacterial effectiveness rate of 99.9%. On the other hand, growth media collected from the HUDM-SB coatings yielded scattered bacterial colonies on agar plates for all four bacteria, with antibacterial effectiveness rates of 95.0%, 96.5%, 99.5%, and 97.5% (Figure 6b). The zwitterionic sulfobetaine-modified polyphenol coatings thus demonstrated remarkable antibacterial properties. This finding can be mainly attributed to the cationic quaternary ammonium salts on the surfaces of IPUDM-SB and HUDM-SB coatings, which actively engage negatively charged bacteria through electrostatic interactions, resulting in cell-membrane rupture, cytoplasm leakage, and bacterial death. The ions thus exert antibacterial effects [40,41].

2.6. Algal-Fouling Resistance of the Coatings

Marine environments harbor abundant marine microalgae that readily adhere to marine structure surfaces, forming conducive biofilms that provide favorable conditions for the subsequent attachment of larger marine fouling organisms. Consequently, in this study, two marine microalgae, *N. closterium* and *P. tricornutum*, served as models for marine fouling. We investigated the anti-biofouling capacity of zwitterionic-functionalized polyphenol coatings. As illustrated in Figure 7, BG, IPUDM, IPUDM-SB, HUDM, and HUDM-SB coatings were immersed in diluted marine microalgae cell solutions in an f/2 culture medium for co-culture. As shown in the images in Figure 7a–d, after 7 days of immersion, a substantial amount of greenish-yellow precipitate was observed at the bottom of the culture medium with the BG, IPUDM, and HUDM coatings. This result suggests extensive growth and proliferation of *N. closterium* and *P. tricornutum* in the culture media with these three samples. Furthermore, the concentrations of marine microalgae cells in the culture media with the IPUDM and HUDM coatings did not differ significantly from that in culture with the blank control BG, indicating that these two coatings did not inhibit algal cell proliferation, a finding that agrees with the results of the antibacterial assay. Before sulfobetaine was introduced, the IPUDM and HUDM coatings lacked effective antibacterial and anti-algal functional groups, and as a result, they did not exhibit resistance to fouling. However, after sulfobetaine was incorporated into the IPUDM and HUDM coatings, the IPUDM-SB and HUDM-SB coatings significantly inhibited the growth of both marine microalgae species. As shown in Figure 7e, the concentration of *N. closterium* cells in the IPUDM-SB coating culture medium was 4.45×10^5 cells/mL and the concentration of *P. tricornutum* cells was 23.75×10^5 cells/mL. In the culture media with HUDM-SB coatings, the concentration of *N. closterium* cells was 14.7×10^5 cells/mL, and the concentration of *P. tricornutum* cells was 6.31×10^5 cells/mL. Furthermore, after 7 days of immersion in water, IPUDM-SB and HUDM-SB coatings remained intact and did not exhibit significant signs of swelling, indicating their good stability underwater.

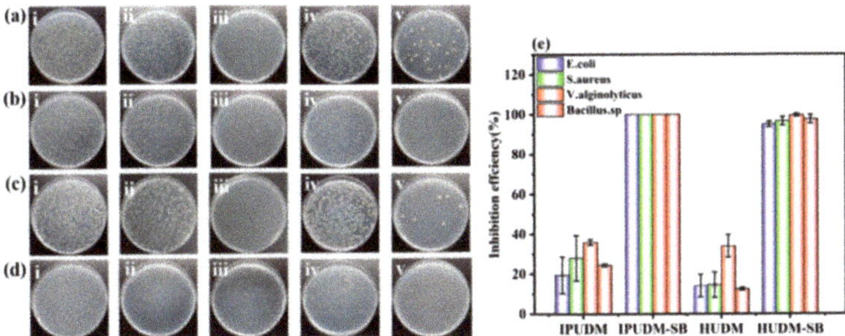

Figure 6. Digital photographs of antibacterial tests conducted with representative bacteria: (**a**) Gram-negative *E. coli*; (**b**) Gram-positive *S. aureus*; (**c**) Gram-negative *V. alginolyticus*, and (**d**) Gram-positive *Bacillus* sp. after 24 h of incubation on various coatings, namely (i) BG, (ii) IPUDM, (iii) IPUDM-SB, (iv) HUDM, and (v) HUDM-SB. Additionally, (**e**) the antibacterial efficiency of IPUDM, IPUDM-SB, HUDM, and HUDM-SB coatings was compared to BG in terms of their effects on *E. coli*, *S. aureus*, *V. alginolyticus*, and *Bacillus* sp., based on three parallel experiments.

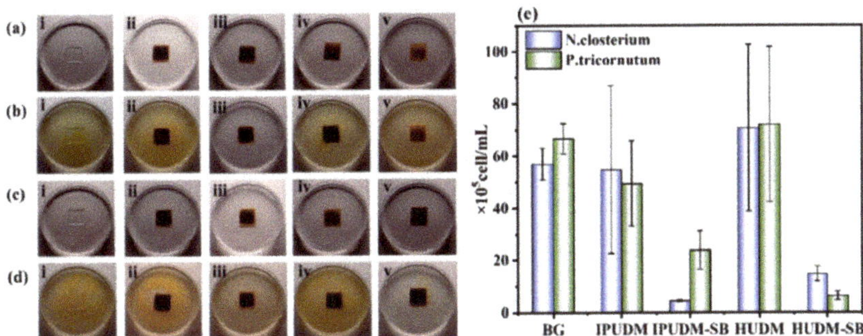

Figure 7. Photographs depicting algal inhibition: 1 day—(**a**) *N. closterium* and (**c**) *P. tricornutum*; 7 days—(**b**) *N. closterium* and (**d**) *P. tricornutum* in an f/2 medium with coatings (i) BG, (ii) IPUDM, (iii) IPUDM-SB, (iv) HUDM, and (v) HUDM-SB. Additionally, (**e**) shows the concentrations of *N. closterium* and *P. tricornutum* determined through cell counting with a hemocytometer after 7 days of cultivation, based on three parallel experiments.

For a more comprehensive analysis of the adhesion of marine algal cells on zwitterionic-functionalized polyphenol coatings, samples coated with these materials were immersed in diluted algal cell solutions for 7 days, then examined using fluorescence microscopy. Figure 8a,b depicts fluorescence microscopy images of *N. closterium* and *P. tricornutum* adhered to the BG, IPUDM, IPUDM-SB, HUDM, and HUDM-SB coatings. To quantify the coverage of algal cells on the coating surface, we measured the fluorescence intensity of the images using ImageJ (1.52a) software. The fluorescence microscopy images of the BG surface showed a consistent and bright fluorescence, indicating the uniform adhesion of a large number of algal cells to the surface, with coverages of 17.14% for *N. closterium* and 19.88% for *P. tricornutum* (Figure 8c). On the surfaces of the IPUDM and HUDM coatings, the coverage values of *N. closterium* were 1.95% and 3.36%, respectively, and those of *P. tricornutum* were 1.67% and 1.17%, respectively (Figure 8c). In contrast, the coverage values of *N. closterium* on the surfaces of the IPUDM-SB and HUDM-SB coatings were only 0.21% and 1.02%, while those of *P. tricornutum* were only 0.13% and 0.10% (Figure 8c). These results indicate that fewer algal cells adhered to the surfaces of the IPUDM-SB and HUDM-SB coatings, highlighting their superior anti-algal-adhesion properties. These findings can be attributed to several factors. First, the sulfobetaine groups on the surfaces of the IPUDM-SB and HUDM-SB coatings effectively inhibit the growth and proliferation of algal cells, reducing algal cell concentrations [42,43]. Second, as depicted in Figure 8d, the hydrophilicity of the IPUDM-SB and HUDM-SB coatings, combined with the dense hydration layer formed on their surfaces through electrostatic interactions, reduces the strength of adhesive forces between marine algal cells and the coatings [44]. It is thus easier to rinse the cells off with deionized water, resulting in improved resistance to algal cell adhesion.

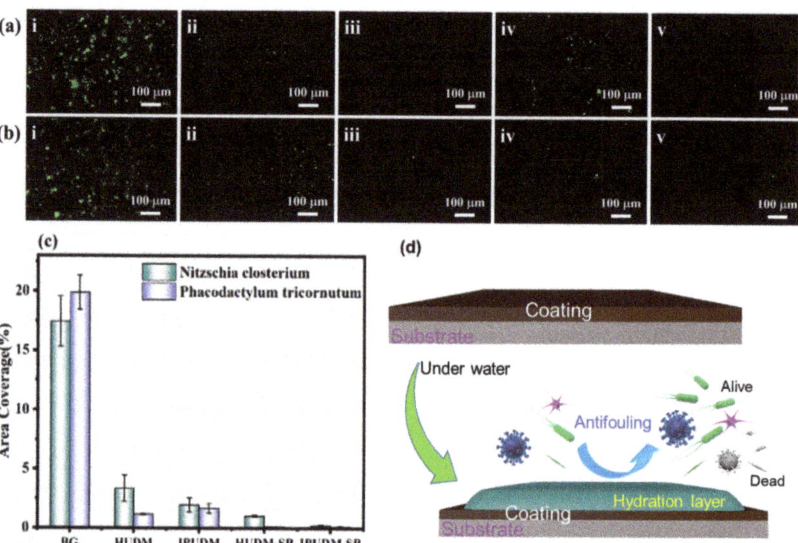

Figure 8. Fluorescent photographs of (**a**) *Nannochloropsis closterium* and (**b**) *Phaeodactylum tricornutum* adhesion after 7 days of cultivation on coatings (i) BG, (ii) IPUDM, (iii) IPUDM-SB, (iv) HUDM, and (v) HUDM-SB. The scale bars in the images represent 40 μm. (**c**) Statistical chart of algal density in examined fields using ImageJ software, based on five random fields at 40× magnification, covering 0.156 mm^2 per field. (**d**) Antifouling mechanism of the HUDM-SB and IPUDM-SB coatings.

3. Materials and Methods

3.1. Materials

Raw lacquer, purchased from Maoba Town, Hubei Province, China, was treated with ethanol to extract urushiol [45]. Hexamethylene diisocyanate (HMDI), isophorone diisocyanate (IPDI), dibutyltin dilaurate (DBTDL), trichloromethane, and N,N-dimethylethanolamine (DMEA) were purchased from Shanghai McLin Biochemical Co., Ltd. (Shanghai, China). Tetrahydrofuran, ethanol, 1,3-propanesultone, deionized water, BSA, and γ-globulin were obtained from Shanghai National Pharmaceutical Reagent Co., Ltd. (Shanghai, China). Phosphate-buffered saline (PBS, pH = 7.4) was procured from Shanghai Sangon Biotech Co., Ltd. (Shanghai, China). Artificial seawater (ASW) was prepared according to ASTM D1141–1998 (2013). *E. coli*, *V. alginolyticus*, *S. aureus*, and *Bacillus.* sp. were purchased from Beijing Baibo Wei Biotechnology Co., Ltd. (Beijing, China). *N. closterium* and *P. tricornutum* were obtained from Shanghai Guangyu Biotechnology Co., Ltd. (Shanghai, China).

3.2. Synthesis of HUDM-SB Monomers

Under a nitrogen atmosphere, 1.78 g of DMEA (0.20 mol), 3.36 g of HMDI (0.20 mol), and 20 mL of anhydrous trichloromethane were placed in a 250 mL three-necked round-bottom flask equipped with a thermometer, a reflux condenser, and a constant-pressure dropping funnel. Then, the mixture was stirred continuously at 40 °C for 2 h to facilitate the addition reaction between the alcohol hydroxyl and the NCO groups of DMEA and HMDI, respectively, forming a linear precursor (PHDE). Subsequently, 6.32 g of urushiol (0.20 mol) dissolved in 10 mL of anhydrous trichloromethane was slowly added to the aforementioned reaction system. The reaction system was then heated to 70 °C, and the reaction was allowed to proceed for an additional 12 h. The system was cooled to room temperature after the reaction was completed, and rotary evaporation was employed to remove the solvent and obtain the urushiol-based precursor (HUDM). Next, HUDM and 1,3-propanesultone were mixed in a 1:1 molar ratio in a 100 mL three-necked flask. The mixture was stirred continuously at 37 °C for 24 h, and rotary evaporation was used to remove the solvent. Finally, a black viscous liquid, the urushiol-based sulfobetaine monomer (HUDM-SB), was obtained by vacuum drying.

3.3. Synthesis of IPUDM-SB Monomers

In a nitrogen atmosphere, 1.78 g of DMEA (0.20 mol), 3.36 g of IPDI (0.20 mol), and 20 mL of anhydrous trichloromethane were placed in a 250 mL three-necked round-bottom flask equipped with a thermometer, a reflux condenser, and a constant-pressure dropping funnel. The mixture was stirred continuously at 40 °C for 2 h, facilitating the addition reaction between the alcohol hydroxyl and the NCO groups of DMEA and IPDI, respectively, resulting in the formation of a linear precursor (PIPDE). Subsequently, 6.32 g of urushiol (0.20 mol) dissolved in 10 mL of anhydrous trichloromethane was added slowly to the aforementioned reaction system. The reaction system was then heated to 70 °C, and the reaction was allowed to proceed for an additional 12 h. When the reaction was complete, the system was cooled to room temperature and rotary evaporation was employed to remove the solvent and obtain the urushiol-based precursor (IPUDM). Next, IPUDM and 1,3-propanesultone were mixed in a 1:1 molar ratio in a 100 mL three-necked flask. The mixture was stirred continuously at 37 °C for 24 h, and rotary evaporation was used to remove the solvent. Finally, a black viscous liquid, the urushiol-based sulfobetaine monomer (IPUDM-SB), was obtained after vacuum drying.

3.4. Synthesis of HUDM-SB and IPUDM-SB Coatings

First, 2.5 cm × 2.5 cm slides of bare glass (BG) underwent sequential cleaning through ultrasonication in acetone, ethanol, and deionized water for 10 min, which was followed by N_2 drying. A predetermined quantity of HUDM-SB or IPUDM-SB was dissolved in tetrahydrofuran (40 wt%), and curing agent DBTDL (0.05 wt%) was incorporated into the solution. After thorough stirring, the solution was drop-cast onto the BG slides and left at room temperature until the solvent evaporated completely, resulting in the formation of cured coatings. HUDM and IPUDM coatings were prepared as control samples using the same method.

3.5. Characterizations

The chemical structures of the HUDM, IPUDM, HUDM-SB, and IPUDM-SB monomers were characterized using Fourier-transform infrared (FTIR) spectroscopy (Thermo Fisher Nicolet 5700, Waltham, MA, USA). The spectra were recorded in the 4000–400 cm^{-1} range with 4 cm^{-1} resolution and 32 scans. The composition of the surface elements and contents of the HUDM, IPUDM, HUDM-SB, and IPUDM-SB coatings were characterized using X-ray photoelectron spectroscopy (XPS, Thermo Electron Fisher VG MultiLab 2000, Waltham, MA, USA) with monochromatic Al Kα radiation (1254.0 eV). The binding energies were calibrated using the C1s peak at 284.8 eV. The hydrophilicity of the surfaces of the different coatings was assessed using static water contact angle measurements (DSA25, Kruss, Hamburg, Germany). Deionized water (2 μL) was added at room temperature to conduct the measurements, with five measurements taken for each sample to obtain average values and standard deviations. The glass transition temperatures of the coatings were determined using differential scanning calorimetry (DSC, Mettler-Toledo 3, Greifensee, Switzerland). The measurements were conducted under a nitrogen atmosphere with a 50 mL/min flow rate and a 10 °C/min heating rate from −30 to 160 °C. The thermal stability of the coatings was analyzed by thermogravimetric analysis (TGA, Mettler-Toledo 3, Greifensee, Switzerland). The measurements were conducted under a nitrogen atmosphere with a 50 mL/min flow rate and a 10 °C/min heating rate from 30 to 600 °C. The coating adhesion of droplets was measured using a surface-tension analyzer (Kruss K100MK2, Hamburg, Germany). Deionized water (6 μL) was used at room temperature to conduct the measurements with an immersion depth and time of 0.1 mm and 0.01 min, respectively, and five measurements were conducted for each sample to obtain average values and standard deviations. The pencil-scratch method was used to test the hardness of the coatings according to the GB/T 6739–2006/ ISO 15184–2012 standards [46]. The cross-cut method was used to assess the adhesion of the coatings according to the GB/T 1720–1979 standard [47].

3.6. Testing of Surface-Protein Adsorption

Protein-adhesion experiments were conducted according to the established protocols described in the references. The Micro BCATM protein assay was utilized to measure the adhesion of samples to BSA. Initially, the coating samples, including BG, HUDM, IPUDM, HUDM-SB, and IPUDM-SB (2.5 cm × 2.5 cm), were sterilized under ultraviolet (UV) light for 30 min. Subsequently, the samples were equilibrated in sterilized PBS for 2 h. Next, the samples were immersed in a protein solution with a 2 mg/mL concentration. Subsequently, the samples were incubated at 37 °C for 4 h and gently rinsed three times with PBS. They were then placed in 4 mL of PBS containing 2 wt% sodium dodecyl sulfate (SDS) and subjected to 30 min of ultrasonication in the SDS solution to remove any BSA that had not adhered to the surface of the coatings. The absorbance at 280 nm was measured using ultraviolet–visible (UV–vis) spectroscopy (TU-1810, Beijing Purkinje General Instrument Co., Beijing, China) to determine the concentration of BSA that had adhered to the surface of the coatings. The resistance of each coating to protein adsorption by γ-globulin was evaluated by the same methodology.

3.7. Assessment of Antibacterial Activity

The antibacterial activity of the coatings was evaluated against four bacterial strains: Gram-negative *E. coli* (BW 25113) and *V. alginolyticus* (ATCC 33787), and Gram-positive *S. aureus* (ATCC 25923) and *Bacillus*. sp. (MCCC 1B00342). Strains of *E. coli* and *S. aureus* were preserved in a 1:1 solution of Luria–Bertani (LB) broth and 40% (*v*/*v*) glycerol and stored at −80 °C. Strains of marine bacterium *V. alginolyticus* and *Bacillus*. sp. were preserved in a 1:1 solution of 2216E medium and 40% (*v/v*) glycerol and stored at −80 °C. The bacterial strains were cultivated in fresh LB broth (2216E medium for marine bacterium) at 37 °C (30 and 28 °C for *V. alginolyticus* and *Bacillus*. sp., respectively) before the antibacterial experiments, with shaking at 170 rpm for 20 h until O.D.$_{600nm}$ reached 1.8–2.0. Subsequently, BG slides and HUDM, IPUDM, HUDM-SB, and IPUDM-SB plates (2.5 cm × 2.5 cm) were wiped with anhydrous ethanol, disinfected using 30 min of UV irradiation, and placed in plastic Petri dishes. The bacteria were diluted to approximately 10^5–10^6 CFU/mL with fresh LB broth (2216E medium for marine bacterium) to prepare the bacterial suspensions, and 100 μL of the diluted bacterial suspension was inoculated onto the BG slides and onto the HUDM, IPUDM, HUDM-SB, and IPUDM-SB coatings. The samples were covered with plastic wrap and incubated at 37 °C for 24 h. Subsequently, the samples were gently rinsed with 10 mL of sterile PBS to remove the non-adherent bacteria. The antibacterial rate (*A.R.*) of the HUDM, IPUDM, HUDM-SB, and IPUDM-SB coatings was calculated based on the count of bacterial colonies in the PBS rinse solution on agar plates using the following formula:

$$A.R. = \frac{N_{contriol} - N_{sample}}{N_{control}} \times 100\%$$

where $N_{control}$ represents the colony count (in CFU/mL) on BG plates and N_{sample} represents the colony count (in CFU/mL) on the HUDM, IPUDM, HUDM-SB, and IPUDM-SB coatings. Each sample was measured three times to obtain the average and the standard deviation.

3.8. Algal Biofouling Assessment

Algal growth and attachment experiments were conducted to evaluate the effects of HUDM-SB and IPUDM-SB coatings on algae. Algal cells, specifically *N. closterium* and *P. tricornutum*, were cultured in an f/2 medium using ASW with a 22 ± 2 °C growth temperature and a 12 h:12 h light:dark cycle. The algal cells were diluted to a concentration of 10^5–10^6 cells/mL with fresh culture medium for use in algal cell reproduction and adhesion experiments after seven days of cultivation. BG slides and HUDM, IPUDM, HUDM-SB, and IPUDM-SB coatings were sterilized under UV light for 30 min and immersed in 30 mL of culture medium containing algal cells. The concentration of algal cells was determined using a hemocytometer after seven days of immersion, and the images were recorded to monitor algal growth. The BG slides and the HUDM, IPUDM, HUDM-SB, and IPUDM-SB coatings were removed from the experimental culture medium following the cultivation period and rinsed with 20 mL of sterile PBS solution to remove all non-adherent algal cells. Subsequently, the algae adhered onto the surfaces of BG slides, HUDM, IPUDM, HUDM-SB, and IPUDM-SB coatings were examined using an Epi-illumination fluorescence microscope (Eclipse Ci-L plus, Nikon, Tokyo, Japan) with a DAPI fluorescence filter kit. The images of five random fields (20× magnification, 0.156 mm^2/per field) under a dark field were recorded for each sample. ImageJ software was employed to analyze the fluorescence microscopy images and determine the algal coverage on the surfaces of the BG slides and the HUDM, IPUDM, HUDM-SB, and IPUDM-SB coatings. Three parallel experiments were conducted to calculate the standard deviation.

4. Conclusions

In summary, environmentally friendly sulfobetaine-functionalized polyurushiol coatings, namely HUDM-SB and IPUDM-SB, were successfully synthesized. These coatings utilize sustainable natural urushiol as the primary material and two types of diisocyanates as crosslinkers. They incorporate zwitterionic sulfobetaine, which has remarkable water affinity, as their hydrophilic component. These zwitterions form a dense hydration layer underwater, effectively deterring the adhesion of nonspecific proteins, marine microalgae, and other fouling organisms. The introduction of sulfobetaine significantly enhances the hydrophilicity of polyurushiol coatings, resulting in water contact angles of 56.2° and 57.0° for the HUDM-SB and IPUDM-SB coatings, respectively. These results demonstrate the coatings' excellent resistance to protein adhesion. Furthermore, the sulfobetaine moieties on the coating surfaces actively engage with negatively charged bacteria through electrostatic interactions, rupturing their cell membranes and resulting in superior antibacterial properties. The IPUDM-SB coatings demonstrate remarkable antibacterial activity, with 99.9% prevention of the growth of prevalent Gram-negative bacteria (*E. coli* and *V. alginolyticus*) and Gram-positive bacteria (*S. aureus* and *Bacillus* sp.). Similarly, the HUDM-SB coatings also exhibit significant antibacterial activity, achieving rates exceeding 95.0% against these four bacterial species. Moreover, the sulfobetaine moieties on the coating surfaces effectively inhibit the growth and proliferation of algal cells. Additionally, they create a dense hydration layer through electrostatic interactions, reducing adhesive forces between marine microalgae and the coating surfaces. This layer effectively prevents biofouling by marine microalgae without the use of fouling-release agents. These coatings offer a green and environmentally friendly antifouling strategy, addressing economic losses caused by marine biofouling and environmental pollution arising from the misuse of fouling-release agents.

Author Contributions: Conceptualization: F.L. and K.X.; methodology: K.X., C.S., H.X. and W.L.; validation: W.L., J.C., Z.Y. and F.L.; formal analysis: H.X., G.Z., X.Z. and Y.X.; investigation: Y.X. and F.L.; writing original draft preparation: F.L.; writing review and editing: J.C. and F.L.; supervision: J.C. and Y.X.; project administration; F.L., Y.X. and J.C.; funding acquisition: Y.X., X.Z. and F.L. All authors have read and agreed to the published version of the manuscript.

Funding: We acknowledge the generous financial support of the National Natural Science Foundation of China (Grant No 32301539, 21978050), Fujian Province Science and Technology Project: School-Enterprise Cooperation in Science and Engineering (Grant No 2020H6029), Marine Economic Development Special Fund Project of Fujian Province (Grant No FUHJF-L-2022-12), and Natural Science Foundation of Fujian Province (Grant No. 2023J011574, 2019J01388, 2022J05234), Fujian College Students' Innovative Entrepreneurial Training Plan Program (Grant No. s202210395051).

Institutional Review Board Statement: Not applicable.

Informed Consent Statement: Not applicable.

Data Availability Statement: The data presented in this study are available on request from the corresponding author.

Conflicts of Interest: The authors declare no conflict of interest.

References

1. Jin, H.; Tian, L.; Zhao, J.; Ren, L. Bioinspired marine antifouling coatings: Status, prospects, and future. *Prog. Mater. Sci.* **2022**, *124*, 100889. [CrossRef]
2. Pourhashem, S.; Seif, A.; Saba, F.; Nezhad, E.G.; Ji, X.; Zhou, Z.; Zhai, X.; Mirzaee, M.; Duan, J.; Rashidi, A. Antifouling nanocomposite polymer coatings for marine applications: A review on experiments, mechanisms, and theoretical studies. *J. Mater. Sci. Technol.* **2022**, *118*, 73–113. [CrossRef]
3. Romeu, M.J.; Mergulhão, F. Development of Antifouling Strategies for Marine Applications. *Microorganisms* **2023**, *11*, 1568. [CrossRef] [PubMed]
4. Ma, C.; Wang, W.; Li, W.; Sun, T.; Feng, H.; Lv, G.; Chen, S. Full solar spectrum-driven Cu_2O/PDINH heterostructure with enhanced photocatalytic antibacterial activity and mechanism insight. *J. Hazard. Mater.* **2023**, *448*, 130851. [CrossRef] [PubMed]
5. Ding, T.; Xu, L. A Cu_2O-based marine antifouling coating with controlled release of copper ion mediated by amphiphilic PLMA-b-PDMAEMA copolymers. *Prog. Org. Coat.* **2022**, *170*, 107003. [CrossRef]
6. Soon, Z.Y.; Jung, J.; Jang, M.; Kang, J.; Jang, M.; Lee, J.; Kim, M. Zinc pyrithione (ZnPT) as an antifouling biocide in the marine environment—A literature review of its toxicity, environmental fates, and analytical methods. *Water Air Soil Pollut.* **2019**, *230*, 310. [CrossRef]
7. Lee, S.; Haque, M.N.; Lee, D.; Rhee, J. Comparison of the effects of sublethal concentrations of biofoulants, copper pyrithione and zinc pyrithione on a marine mysid-A multigenerational study. *Comp. Biochem. Physiol. Part C Toxicol. Pharmacol.* **2023**, *271*, 109694. [CrossRef]
8. Almond, K.M.; Trombetta, L.D. The effects of copper pyrithione, an antifouling agent, on developing zebrafish embryos. *Ecotoxicology* **2016**, *25*, 389–398. [CrossRef]
9. Li, Q.; Huang, M.; Li, F.; Ling, Z.; Meng, Y.; Chen, F.; Ji, Z.; Wang, S. Biomimetic stable cellulose based superhydrophobic Janus paper sheets engineered with industrial lignin residues/nano-silica for efficient oil-water separation. *Ind. Crops Prod.* **2024**, *207*, 117774. [CrossRef]

10. Jiang, X.; Li, Q.; Li, X.; Meng, Y.; Ling, Z.; Ji, Z.; Chen, F. Preparation and Characterization of Degradable Cellulose—Based Paper with Superhydrophobic, Antibacterial, and Barrier Properties for Food Packaging. *Int. J. Mol. Sci.* **2022**, *23*, 11158. [CrossRef]

11. Chen, J.; Jian, R.; Yang, K.; Bai, W.; Huang, C.; Lin, Y.; Zheng, B.; Wei, F.; Lin, Q.; Xu, Y. Urushiol-based benzoxazine copper polymer with low surface energy, strong substrate adhesion and antibacterial for marine antifouling application. *J. Clean. Prod.* **2021**, *318*, 128527. [CrossRef]

12. Sathishkumar, G.; Gopinath, K.; Zhang, K.; Kang, E.T.; Xu, L.; Yu, Y. Recent progress in tannic acid-driven antibacterial/antifouling surface coating strategies. *J. Mater. Chem. B* **2022**, *10*, 2296–2315. [CrossRef] [PubMed]

13. Chen, J.; Zhao, J.; Lin, F.; Zheng, X.; Jian, R.; Lin, Y.; Wei, F.; Lin, Q.; Bai, W.; Xu, Y. Polymerized tung oil toughened urushiol-based benzoxazine copper polymer coatings with excellent antifouling performances. *Prog. Org. Coat.* **2023**, *177*, 107411. [CrossRef]

14. Zhang, C.; Qi, Y.; Zhang, S.; Xiong, G.; Wang, K.; Zhang, Z. Anti-marine biofouling adhesion performance and mechanism of PDMS fouling-release coating containing PS-PEG hydrogel. *Mar. Pollut. Bull.* **2023**, *194*, 115345. [CrossRef]

15. Wang, F.; Zhang, H.; Yu, B.; Wang, S.; Shen, Y.; Cong, H. Review of the research on anti-protein fouling coatings materials. *Prog. Org. Coat.* **2020**, *147*, 105860. [CrossRef]

16. Li, K.; Qi, Y.; Zhou, Y.; Sun, X.; Zhang, Z. Microstructure and properties of poly (ethylene glycol)-segmented polyurethane antifouling coatings after immersion in seawater. *Polymers* **2021**, *13*, 573. [CrossRef] [PubMed]

17. Sun, D.; Li, P.; Li, X.; Wang, X. Protein-resistant surface based on zwitterion-functionalized nanoparticles for marine antifouling applications. *New J. Chem.* **2020**, *44*, 2059–2069. [CrossRef]

18. Ye, Z.; Zhang, P.; Zhang, J.; Deng, L.; Zhang, J.; Lin, C.; Guo, R.; Dong, A. Novel dual-functional coating with underwater self-healing and anti-protein-fouling properties by combining two kinds of microcapsules and a zwitterionic copolymer. *Prog. Org. Coat.* **2019**, *127*, 211–221. [CrossRef]

19. Zheng, L.; Sundaram, H.S.; Wei, Z.; Li, C.; Yuan, Z. Applications of zwitterionic polymers. *React. Funct. Polym.* **2017**, *118*, 51–61. [CrossRef]

20. Chen, P.; Lang, J.; Zhou, Y.; Khlyustova, A.; Zhang, Z.; Ma, X.; Liu, S.; Cheng, Y.; Yang, R. An imidazolium-based zwitterionic polymer for antiviral and antibacterial dual functional coatings. *Sci. Adv.* **2022**, *8*, abl8812. [CrossRef]

21. Zhang, J.; Wu, M.; Peng, P.; Liu, J.; Lu, J.; Qian, S.; Feng, J. "Self-Defensive" antifouling zwitterionic hydrogel coatings on polymeric substrates. *ACS Appl. Mater. Interfaces* **2022**, *14*, 56097–56109. [CrossRef] [PubMed]

22. Zhang, H.; Li, Y.; Tian, S.; Qi, X.; Yang, J.; Li, Q.; Lin, C.; Zhang, J.; Zhang, L. A switchable zwitterionic ester and capsaicin copolymer for multifunctional marine antibiofouling coating. *Chem. Eng. J.* **2022**, *436*, 135072. [CrossRef]

23. Guan, Y.; Li, S.-L.; Fu, Z.; Qin, Y.; Wang, J.; Gong, G.; Hu, Y. Preparation of antifouling TFC RO membranes by facile grafting zwitterionic polymer PEI-CA. *Desalination* **2022**, *539*, 115972. [CrossRef]

24. Li, D.; Li, K.; Fang, J. Research Progress on Modification and Application of Raw Lacquer. *ChemistrySelect* **2022**, *7*, e202200943. [CrossRef]

25. Chen, Y.; Zhang, G.; Zhang, G.; Ma, C. Rapid curing and self-stratifying lacquer coating with antifouling and anticorrosive properties. *Chem. Eng. J.* **2021**, *421*, 129755. [CrossRef]

26. Chen, S.; Wang, L.; Lin, X.; Ni, P.; Liu, H.; Li, S. Catechol derivative urushiol's reactivity and applications beyond traditional coating. *Ind. Crops Prod.* **2023**, *197*, 116598. [CrossRef]

27. Ma, J.; Lee, G.H.; Kim, J.H.; Kim, S.W.; Jo, S.; Kim, C. A transparent self-healing polyurethane–isophorone-diisocyanate elastomer based on hydrogen-bonding interactions. *ACS Appl. Polym. Mater.* **2022**, *4*, 2497–2505. [CrossRef]

28. Oh, J.; Kim, Y.K.; Hwang, S.; Kim, H.; Jung, J.; Jeon, C.; Kim, J.; Lim, S.K. Synthesis of thermoplastic polyurethanes containing bio-based polyester polyol and their fiber property. *Polymers* **2022**, *14*, 2033. [CrossRef]

29. Gaddam, S.K.; Arukula, R. Renewable soft segment-induced anionic waterborne polyurethane dispersions with enriched bio-content. *J. Polym. Res.* **2022**, *29*, 59. [CrossRef]

30. Cheng, Q.; Jia, X.; Cheng, P.; Zhou, P.; Hu, W.; Cheng, C.; Hu, H.; Xia, M.; Liu, K.; Wang, D. Improvement of the filtration and antifouling performance of a nanofibrous sterile membrane by a one-step grafting zwitterionic compound. *New J. Chem.* **2022**, *46*, 15423–15433. [CrossRef]

31. Yuan, Y.; Tan, W.; Zhang, J.; Li, Q.; Guo, Z. Water-soluble amino functionalized chitosan: Preparation, characterization, antioxidant and antibacterial activities. *Int. J. Biol. Macromol.* **2022**, *217*, 969–978. [CrossRef] [PubMed]

32. Zhao, J.; Chen, J.; Zheng, X.; Lin, Q.; Zheng, G.; Xu, Y.; Lin, F. Urushiol-Based Benzoxazine Containing Sulfobetaine Groups for Sustainable Marine Antifouling Applications. *Polymers* **2023**, *15*, 2383. [CrossRef]

33. Morang, S.; Karak, N. Nanocomposites of waterborne polyurethanes. In *Eco-Friendly Waterborne Polyurethanes*; CRC Press: Boca Raton, FL, USA, 2022; pp. 83–100.

34. Bi, J.; Yan, Z.; Hao, L.; Elnaggar, A.Y.; El-Bahy, S.M.; Zhang, F.; Azab, I.H.E.; Shao, Q.; Mersal, G.A.; Wang, J. Improving water resistance and mechanical properties of waterborne acrylic resin modified by octafluoropentyl methacrylate. *J. Mater. Sci.* **2023**, *58*, 1452–1464. [CrossRef]

35. Chou, Y.-N.; Ou, M. Zwitterionic Surface Modification of Aldehydated Sulfobetaine Copolymers for the Formation of Bioinert Interfaces. *ACS Appl. Polym. Mater.* **2023**, *5*, 5411–5428. [CrossRef]

36. Liu, Z.; Xiao, Y.; Ma, X.; Geng, X.; Ye, L.; Zhang, A.; Feng, Z. Preparation and characterisation of zwitterionic sulfobetaine containing siloxane-based biostable polyurethanes. *Mater. Adv.* **2022**, *3*, 4608–4621. [CrossRef]

37. Chen, Z. Surface hydration and antifouling activity of zwitterionic polymers. *Langmuir* **2022**, *38*, 4483–4489. [CrossRef]

38. Wang, T.; Zhang, J.; Cai, Y.; Xu, L.; Yi, L. Protein-resistant amphiphilic copolymers containing fluorosiloxane side chains with controllable length. *ACS Appl. Polym. Mater.* **2022**, *4*, 7903–7910. [CrossRef]

39. Sarker, P.; Lu, T.; Liu, D.; Wu, G.; Chen, H.; Sajib, M.S.J.; Jiang, S.; Chen, Z.; Wei, T. Hydration Behaviors of Nonfouling Zwitterionic Materials. *Chem. Sci.* **2023**, *14*, 7500–7511. [CrossRef]

40. Chen, C.; Li, Z.; Li, X.; Kuang, C.; Liu, X.; Song, Z.; Liu, H.; Shan, Y. Dual-functional antimicrobial coating based on the combination of zwitterionic and quaternary ammonium cation from rosin acid. *Compos. Part B* **2022**, *232*, 109623. [CrossRef]

41. Peng, J.; Li, K.; Du, Y.; Yi, F.; Wu, L.; Liu, G. A robust mixed-charge zwitterionic polyurethane coating integrated with antibacterial and anticoagulant functions for interventional blood-contacting devices. *J. Mater. Chem. B* **2023**, *11*, 8020–8032. [CrossRef]

42. Malouch, D.; Berchel, M.; Dreanno, C.; Stachowski-Haberkorn, S.; Chalopin, M.; Godfrin, Y.; Jaffrès, P. Evaluation of lipophosphoramidates-based amphiphilic compounds on the formation of biofilms of marine bacteria. *Biofouling* **2023**, *39*, 591–605. [CrossRef] [PubMed]

43. Zou, W.; Gu, J.; Li, J.; Wang, Y.; Chen, S. Tailorable antibacterial and cytotoxic chitosan derivatives by introducing quaternary ammonium salt and sulfobetaine. *Int. J. Biol. Macromol.* **2022**, *218*, 992–1001. [CrossRef] [PubMed]

44. Qu, K.; Yuan, Z.; Wang, Y.; Song, Z.; Gong, X.; Zhao, Y.; Mu, Q.; Zhan, Q.; Xu, W.; Wang, L. Structures, properties, and applications of zwitterionic polymers. *ChemPhysMater* **2022**, *1*, 294–309. [CrossRef]

45. Xu, H.; Lu, Z.; Zhang, G. Synthesis and properties of thermosetting resin based on urushiol. *RSC Adv.* **2012**, *2*, 2768–2772. [CrossRef]

46. *GB/T 6739-2006/ISO 15184–2012*; Paints and Varnishes-Determination of Film Hardness by Pencil Test. ISO: Geneva, Switzerland, 2012.

47. *GB/T 1720–1979*; Method of Test for Adhesion of Paint Films. China Petroleum and Chemical Industry Federation: Beijing, China, 1979.

Article

The Neglected Role of Asphaltene in the Synthesis of Mesophase Pitch

Mingzhi Wang, Yulin Li, Haoyu Wang, Junjie Tao, Mingzhe Li, Yuzhu Shi and Xiaolong Zhou *

International Joint Research Center of Green Energy Chemical Engineering, East China University of Science and Technology, Shanghai 200237, China; y15220008@mail.ecust.edu.cn (M.W.); a2663093157@163.com (Y.L.); a893748291@163.com (H.W.); 21010763@mail.ecust.edu.cn (J.T.); 13002928623@163.com (M.L.); Y15230015@mail.ecust.edu.cn (Y.S.)
* Correspondence: xiaolong@ecust.edu.cn

Abstract: This study investigates the synthesis of mesophase pitch using low-cost fluid catalytic cracking (FCC) slurry and waste fluid asphaltene (WFA) as raw materials through the co-carbonization method. The resulting mesophase pitch product and its formation mechanism were thoroughly analyzed. Various characterization techniques, including polarizing microscopy, softening point measurement, Fourier-transform infrared spectroscopy (FTIR), and thermogravimetric analysis (TGA), were employed to characterize and analyze the properties and structure of the mesophase pitch. The experimental results demonstrate that the optimal optical texture of the mesophase product is achieved under specific reaction conditions, including a temperature of 420 °C, pressure of 1 MPa, reaction time of 6 h, and the addition of 2% asphaltene. It was observed that a small amount of asphaltene contributes to the formation of mesophase pitch spheres, facilitating the development of the mesophase. However, excessive content of asphaltene may cover the surface of the mesophase spheres, impeding the contact between them and consequently compromising the optical texture of the mesophase pitch product. Furthermore, the inclusion of asphaltene promotes polymerization reactions in the system, leading to an increase in the average molecular weight of the mesophase pitch. Notably, when the amount of asphaltene added is 2%, the mesophase pitch demonstrates the lowest I_D/I_G value, indicating superior molecular orientation and larger graphite-like microcrystals. Additionally, researchers found that at this asphaltene concentration, the mesophase pitch exhibits the highest degree of order, as evidenced by the maximum diffraction angle (2θ) and stacking height (Lc) values, and the minimum d_{002} value. Moreover, the addition of asphaltene enhances the yield and aromaticity of the mesophase pitch and significantly improves the thermal stability of the resulting product.

Keywords: mesophase pitch; FCC slurry; molecular orientation; stability; temperature; pressure; reaction time; graphite-like microcrystals; asphaltene; softening point

Citation: Wang, M.; Li, Y.; Wang, H.; Tao, J.; Li, M.; Shi, Y.; Zhou, X. The Neglected Role of Asphaltene in the Synthesis of Mesophase Pitch. *Molecules* **2024**, *29*, 1500. https://doi.org/10.3390/molecules29071500

Academic Editors: José A.P. Coelho and Roumiana P. Stateva

Received: 20 February 2024
Revised: 20 March 2024
Accepted: 21 March 2024
Published: 27 March 2024

1. Introduction

In petroleum mixtures, the heavy fraction contains a significant amount of macro-molecular non-hydrocarbon compounds, which are commonly classified as resins and asphaltenes. However, there is currently no internationally standardized definition or precise boundary for distinguishing between resins and asphaltenes [1–3]. The small molecules of n-alkanes that are insoluble in non-polar solvents but soluble in benzene are generally referred to as asphaltenes, representing the most polar and highest molecular weight non-hydrocarbon components in petroleum. The choice of organic solvent for asphaltene separation varies, and therefore, when presenting asphaltene content data, it is necessary to specify the solvent used, such as n-pentane asphaltene or n-heptane asphaltene. The amount of precipitated asphaltenes decreases with an increase in the relative molecular weight and solubility of n-alkanes in the asphaltene fraction. Resins, characterized by their relatively high polarity and molecular weight, are polydisperse macromolecular non-hydrocarbon compounds, ranking second only to asphaltenes in terms of polarity and molecular weight. However, there is no clear demarcation between resins and asphaltenes. Consequently, the reported data regarding resin content in petroleum may exhibit significant variations depending on the analysis methods employed. Chinese researchers have defined the components obtained by separating saturated and aromatic components from the n-heptane-soluble fraction of residual oil using alumina liquid chromatography as resins. In terms of appearance, resins appear as dark brown substances, typically existing as amorphous solids or viscous liquids. They melt at high temperatures and exhibit a relative density of approximately 1.0. Asphaltene, on the other hand, is a dark brown or black amorphous solid that does not melt when subjected to high temperatures. It is relatively brittle and prone to fracturing into pieces. The relative density of asphaltenes is also around 1.0, slightly higher than that of resins [4–7].

Due to the abundance of aromatic structures in asphaltene and resin, they exhibit relatively low hydrogen-to-carbon ratios, with asphaltene having a lower ratio compared to resin. The hydrogen-to-carbon ratio of resin is generally around 1.4, while for asphaltene, it ranges approximately between 1.1 and 1.3. Determining the average relative molecular weight of resin and asphaltene is a challenging task due to the presence of numerous heteroatoms in their molecular structures. The association between different molecules leads to the formation of supramolecular structures at various levels. Consequently, data obtained using different measurement methods can vary significantly. Therefore, it is crucial to specify the conditions and methods used when providing data on the average relative molecular weight of resin and asphaltene. Only data obtained using the same method under equivalent conditions can be considered comparable. The vapor pressure osmometry (VPO) method, which employs vapor pressure permeation, is commonly used to measure the average relative molecular weight [8–10]. The basic structure of resin and asphaltene molecules in petroleum consists of a dense aromatic ring system as the core, with multiple fused aromatic rings, surrounded by several cycloalkane rings. These aromatic and cycloalkane rings are attached to various length-variant normal or isomeric side chains. The molecular branches or ring systems may contain elements such as sulfur (S), nitrogen (N), and oxygen (O) and trace amounts of metal elements, as shown in Figure 1a [11–13]. This core structure, composed of a fused aromatic ring system, is the fundamental unit of resin and asphaltene molecules, also referred to as the unit structure or unit sheet. A resin or asphaltene molecule is composed of several such unit structures connected by methylene bridges of varying lengths, with oxygen or sulfur present between these bridge structures. Compared to resin, asphaltene has a higher average relative molecular weight and a higher level of aromatization. In terms of the number of rings per unit structure, asphaltene significantly exceeds resin. The resin unit structure contains approximately five aromatic rings, while the asphaltene unit structure contains around seven to ten aromatic rings. The aromatic rings in asphaltene primarily undergo peri-condensation, while in resin, both peri-condensation and kata-condensation occur between aromatic rings [14–16].

Graphite is a well-known hexagonal crystal structure composed of pure carbon. It consists of numerous overlapping layers of carbon atoms arranged in a network formation. Within each layer, the carbon atoms form regular hexagons, and the interlayer spacing is approximately 0.335 nm. The X-ray diffraction spectrum of graphite exhibits a sharp (002) peak at around 26°, indicating the ordered arrangement of carbon atom layers. In the early 1960s, T.F. Yen discovered that asphaltene, separated from petroleum, also exhibits a (002) peak at around 26° in its X-ray diffraction spectra. This suggests that asphaltene possesses a similar ordered structure to graphite. In other words, within asphaltene molecules, there are partially ordered crystalline structures formed by unit sheets with fused aromatic rings as the core. This ordering arises due to the overlap and synergistic effect of the π electron clouds of aromatic rings within and between the molecules. However, the presence of non-planar cycloalkanes and alkyl chains connected to these aromatic ring systems in asphaltene results in slightly larger interlayer spacing compared to graphite, approximately

0.36 nm. Figure 1b represents the crystalline structure particles of asphaltene [17–20]. Scientists also utilize the property of petroleum dispersion systems to determine the content of asphaltene by using a significant amount of low molecular weight n-alkanes to dilute crude oil or residue. By reducing the aromaticity and viscosity of the dispersion medium, the solvation layer of the gum in the micelles is disrupted, leading to the formation of larger aggregates of asphaltene, which then separate and precipitate into a distinct phase. The basic structure of asphaltene molecules is centered on a polycyclic aromatic ring system composed of multiple aromatic rings, surrounded by multiple cyclo-alkane rings, aromatic rings, and alkyl side chains. T.F. Yen proposed the model shown in Figure 1c to represent this structure [20–23]. However, in the production process of mesophase pitch, asphaltene is considered a by-product, and many researchers aim to minimize its content. However, if the content of asphaltene is too low, it is detrimental to the formation of mesophase pitch; but if the content of asphaltene is too high, it will destroy the optical texture of mesophase pitch, which is not conducive to being used as a precursor for producing carbon fibers; therefore, it is important to control the amount of asphaltene in the product. In our research, we plan to investigate the influence of asphaltene on the formation of mesophase pitch. Initially, our research group used FCC slurry to synthesize mesophase pitch [24], exploring optimal conditions such as temperature, pressure, and reaction time. Subsequently, asphaltene was employed as a co-carbonizing agent, and by varying the amount of asphaltene added, we continuously screened the reaction conditions to identify the production process of mesophase pitch with the best reaction efficiency. In this study, we incorporate WFA as an additive to analyze and discuss its impact on the structure, morphology, thermal properties, and optical texture of pyrolysis products. The objective of our research is to develop a co-carbonization agent that enhances the formation of high-quality mesophase pitch.

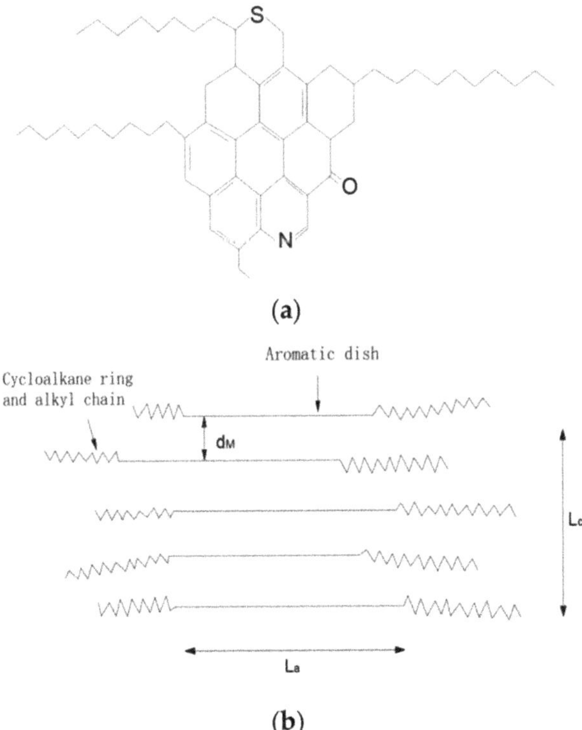

Figure 1. *Cont.*

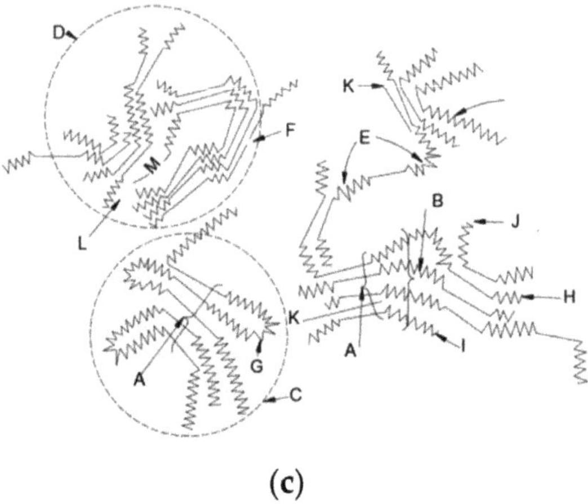

(c)

Figure 1. (**a**) Basic molecular structure of colloidal asphaltenaceous material; (**b**) asphaltene-like particles with a crystalline structure; (**c**) schematic diagram of the macroscopic structure of asphaltenes. The straight line indicates the aromatic ring system, and the zigzag line indicates the saturated structure (containing alkanes and naphthenes). A—grain; B—side chain bundle; C—a particle; D—a micelle; E—weak bond; F—one hole; G—intramolecular clusters; H—intermolecular cluster; I—gum; J—a single tablet; K—petroporphyrin; L—metal; M—insoluble component in micelle ; L_a—average diameter; L_c—stacking height; d_M—interlayer spacing.

2. Results and Discussion

2.1. Determination of Optimal Reaction Conditions

2.1.1. Determination of Reaction Temperature

Based on previous experimental results of the direct thermal condensation polymerization reaction of FCC-BL, the temperatures corresponding to the best product performance were found to be 400 °C and 420 °C. To validate the most suitable reaction temperature for the co-carbonization system, the research group opted to utilize asphaltene as the co-carbonizing agent with an addition amount of 1%, a system pressure of 1 MPa, and a reaction time of 6 h. The influence of different temperatures (390 °C, 400 °C, 410 °C, 420 °C, and 430 °C) on the co-carbonization method for preparing mesophase pitch was investigated, as depicted in Figure S3. In these micrographs, the corresponding part of the bright area is the anisotropic component, and the proportion of mesophase pitch products is determined by the ratio of anisotropic textures. From Figure S3, it is evident that different reaction temperatures have a substantial impact on the content and optical texture of the mesophase phase. At 390 °C and 400 °C, the content of the mesophase phase in the product is relatively low due to the reaction being a free radical reaction. At lower temperatures, the higher viscosity within the system hinders chain-initiated reactions, thereby significantly impeding the generation of mesophase phases. With increasing temperature, the proportion of the mesophase pitch shows an upward trend, accompanied by a transformation of the optical texture from a small flow structure to a large domain structure. At 420 °C, both the content of the mesophase pitch and the optical texture achieve optimal results. This is attributed to the elevated temperature intensifying the thermal reaction within the system, leading to abundant generation of free radicals from the decomposition of raw materials. This, in turn, facilitates further thermal cracking and condensation of the reaction. Simultaneously, the increased temperature reduces the viscosity of the reactants, promoting the condensation of aromatic hydrocarbons and the transfer of free radicals, resulting in the formation of a greater number of planar-fused ring macromolecules. These macromolecules arrange and stack freely, generating a significant quantity of mesophase pitch microspheres. These microspheres undergo fusion, growth, and rupture, ultimately forming a wide-ranging mesophase pitch asphalt. When the temperature exceeds 420 °C, the rates of thermal cracking and condensation reactions within the system increase significantly, leading to excessive carbonization of the intermediate products [25–27]. Consequently, the research group determined 420 °C as the optimal co-carbonization reaction temperature. An increase in temperature will increase the softening point of the product, but at the same time, it will reduce the yield of the product. Therefore, it is important to control the experimental temperature reasonably.

2.1.2. Determination of Reaction Time

Based on the previous experimental findings of the direct thermal condensation reaction of FCC-BL, the optimal reaction time was observed to be 6 h at a reaction temperature of 420 °C. In order to explore the most suitable reaction time, the research team decided to employ 1% asphaltene as a co-carbonizing agent, with a system pressure of 1 MPa and a reaction temperature of 420 °C. The influence of different reaction times (4 h, 5 h, 6 h, 7 h, and 8 h) on the production of mesophase pitch using the co-carbonization method was investigated. As illustrated in Figure S4, when the reaction was conducted for 4 h, the proportion of mesophase pitch in the resulting product was relatively low. However, after 5 h of reaction, there was a noticeable increase in the proportion of mesophase pitch. Upon reaching 6 h of reaction time, the proportion of mesophase pitch exceeded 90%, with the optical texture predominantly exhibiting a large domain structure. When the reaction time exceeds 6 h, the optical texture of the mesophase pitch in the product becomes disordered, and the domain structure significantly deteriorates. The reason is that the system contains a large number of aromatic structures and chain alkanes, which have high reactivity and greatly promote thermal cracking reactions. Consequently, the content of aromatic free radicals is considerably enhanced, thereby facilitating the abundant generation of planar-fused ring aromatic molecules. These larger molecules undergo directed arrangement and stacking, ultimately giving rise to the formation of mesophase microspheres. These microspheres interact, merge, grow, and eventually rupture, leading to the formation of sheet-like mesophase pitch, which exhibits a flow structure basin-like morphology under the influence of pressure. However, when the reaction time surpasses the critical threshold, excessive aggregation of intermediate molecules occurs, and the stacking between molecules tends to assume a disordered state, ultimately resulting in carbonization [28–30]. In light of the aforementioned observations, the research team established a reaction time of 6 h for their study.

2.1.3. Determination of Reaction Pressure

Based on prior experimental results of the direct thermal polycondensation reaction of FCC-BL, the system pressure yielding the best experimental outcome was approximately 1 or 2 MPa. To explore the most suitable pressure for the co-carbonization reaction, the research team employed asphaltene as a co-carbonization agent with a 1% addition rate, a reaction temperature of 420 °C, and a reaction time of 6 h. The impact of different reaction pressures (0, 0.5 MPa, 1 MPa, 1.5 MPa, and 2 MPa) on the production of mesophase pitch using the co-carbonization method was investigated. As shown in Figure S5, when the system pressure is low, a large number of light components will escape from the reaction system, causing a significant increase in the viscosity between the raw materials, greatly increasing the degree of polycondensation of the product molecules. The large product molecules have a serious impact on the optical texture, while the escaping light component substances inhibit the growth of mesophase pitch, making it tend to be disordered, which is manifested as a mosaic structure in the product. As the pressure gradually increases, the morphology of mesophase pitch gradually changes from a mosaic structure to a large-domain structure. When the pressure is 1.0 MPa, the optical texture of the product is the best, approaching 100%. However, when the pressure exceeds 1.5 MPa, the optical texture deteriorates significantly. This is due to the suppression of light component escape at pressures beyond the critical threshold, hindering the ordered accumulation of condensed ring macromolecules and the fusion and growth of mesophase pitch microspheres, ultimately disrupting the ordered structure of the mesophase pitch product. Consequently, the optical texture of the mesophase pitch product deteriorates. When the pressure reaches 2.0 MPa, excessive pressure impedes the coalescence of mesophase microspheres, resulting in incomplete development of mesophase pitch [31,32]. Thus, the research team determined a reaction pressure of 1.0 MPa.

The properties of the modified pitch were analyzed and are presented in Table 1. The hydrogen–carbon ratio of the product gradually decreases as the proportion of WFA increases, indicating a transformation of aliphatic structures to aromatic ring structures. The content of toluene insoluble substance (TI) does not show a discernible pattern of change, while the content of quinoline insoluble substance (QI) increases progressively. This can be attributed to the promotion of polymerization reactions within the system with increasing WFA, resulting in the formation of more quinoline-insoluble substances. The yield and softening point of the product exhibits significant improvements, which can be attributed to two factors: firstly, both the carbonization rate of the pitch generated by WFA and FCC thermal condensation increase, and secondly, the cooperative effect of WFA and FCC thermal condensation during the co-carbonization process leads to an overall increase in the softening point and yield of the product. Notably, the ash content remains unchanged throughout the entire co-carbonization process [33].

Table 1. Basic properties of different modified mesophase pitch.

Sample	Elemental Analysis (wt%)		H/C Ratio	Solubility (wt%)		Yield (%)	Softing Point	Ash (%)
	C	H		TI	QI			
MMP-0	89.77	6.86	0.92	23.8	1.3	76.3	243	0.20
MMP-1	90.86	6.27	0.83	30.7	3.2	79.5	267	0.23
MMP-2	91.63	5.59	0.73	29.9	3.7	82.7	289	0.22
MMP-3	92.47	5.23	0.68	35.2	8.6	84.3	307	0.26
MMP-5	94.45	4.68	0.59	40.3	17.8	89.2	344	0.25

2.2. Analysis of Mesophase's Crystal Structure

The characterization of mesophase pitch's microcrystalline size and molecular orientation is commonly performed using X-ray diffraction and Raman spectroscopy. Initially, the optical textures of five sample groups (MMP-0, MMP-1, MMP-2, MMP-3, and MMP-5) were examined. The specific reference standards for the optical texture of the product can be found in Table S2. Figure 2 illustrates the optical textures of the five types of mesophase pitch. From the diagram, it is evident that the presence of anisotropic components is relatively low when WFA is not added. However, the addition of WFA to the reaction system significantly influences the optical microstructure of mesophase pitch. Upon the initial addition of WFA to the reaction system, mesophase pitch gradually exhibits a large domain and extensively streamlined optical texture. When the WFA addition reaches 2%, the content of mesophase in the large domain streamlined structure reaches its maximum, approximately 95%. However, as the WFA content increases further, the content of the wide-area streamlined mesophase decreases gradually, and a distinct mosaic-like structure starts to appear on the surface. When the WFA addition reaches 5%, the optical texture of the mesophase pitch surface becomes severely damaged, with the presence of anisotropic components almost indiscernible. The surface transforms into a pronounced mosaic form. Based on this analysis, it can be inferred that the addition of WFA significantly influences the nucleation, growth, and aggregation of mesophase microspheres. Previous studies have indicated [34] that WFA acts as a nucleating agent to facilitate the formation of mesophase microspheres. When a small amount of WFA is added to the system, the number of nucleation sites in the system decreases, and the number of sites where the active molecules undergo π-π conjugation and adhere also decreases. Consequently, the system contains a higher proportion of highly mobile active molecules, promoting the growth and aggregation of mesophase microspheres. However, excessive WFA content leads to its accumulation on the surface of mesophase spheres, impeding their proximity and fusion, thereby creating a mosaic-like structure. Ultimately, this causes severe damage to the optical texture of mesophase pitch's surface.

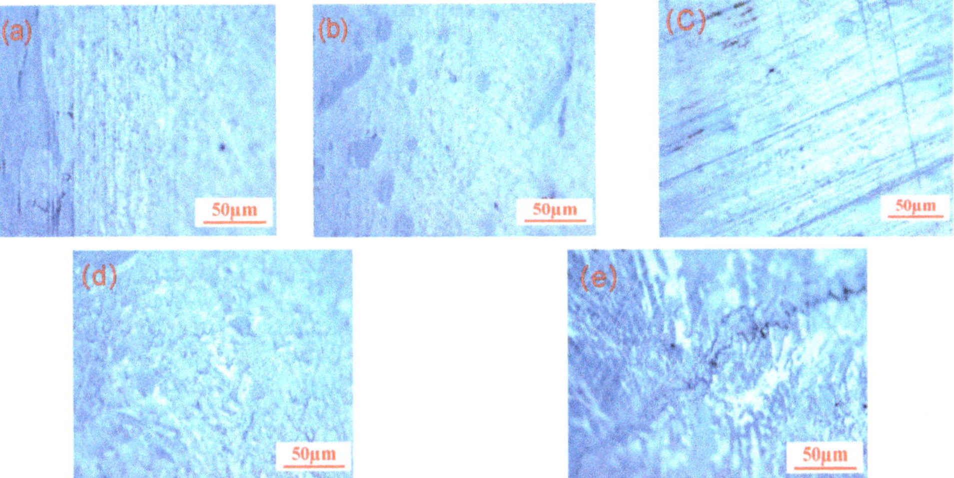

Figure 2. Optical texture picture of modified mesophase pitch with different WFA content. (**a**) 0%; (**b**) 1%; (**c**) 2%; (**d**) 3%; (**e**) 5%.

Figure 3 presents the XRD and Raman spectra of five types of mesophase pitch [35–39]. The microcrystalline size parameters of the mesophase pitch, calculated using the Bragg equation and Scheler equation in conjunction with these two characterization methods, are provided in Table 2. From Figure 3a, it is evident that all five mesophase pitches exhibit high-intensity carbon (002) diffraction peaks around $2\theta = 25°$. This indicates that these asphalts possess a significant degree of molecular orientation. The intensity order of the carbon (002) surface diffraction peaks for the five mesophase pitch is as follows: MMP-0 < MMP-1 < MMP-5 < MMP-3 < MMP-2, with the corresponding 2θ values following the order: MMP-0 < MMP-1 < MMP-5 < MMP-3 < MMP-2. Furthermore, by employing the Bragg equation and Scheler equation, the stacking height (Lc) and molecular spacing (d_{002}) of the molecular layers were calculated. As the 2θ value increases, the corresponding Lc value increases while the d_{002} value decreases, indicating a higher level of orderliness in the product. Table 2 reveals that with an increase in the proportion of WFA, the (002) diffraction peak initially shifts towards higher angles and then towards lower angles. This suggests that the structure of the mesophase pitch first tends towards orderliness and subsequently transitions to disorderliness. Notably, MMP-2 exhibits the highest 2θ value, indicating that when the WFA addition reaches 2%, the resulting mesophase pitch attains the highest degree of orderliness. To summarize, the content of WFA can alter the regularity of the carbon layer in pyrolysis products. What is more, the orderliness of mesophase pitch products primarily depends on their distinct molecular chemical compositions and molecular configurations.

The inclusion of WFA in the system introduces a significant number of aromatic rings and cycloalkanes. A small amount of WFA addition promotes the development of the mesophase pitch and increases its content. However, the excessive addition of WFA leads to an increased number of aromatic cores in polycyclic aromatic hydrocarbons, resulting in a higher quantity of aromatic nuclei. The structural variations among different aromatic nuclei reduce the conjugation of the mesophase pitch molecules, leading to inconsistent core orientations. Consequently, the planarity of the mesophase pitch molecules decreases, resulting in an increase in intermolecular interlayer spacing (d_{002}). On the other hand, despite the unfavorable effect of different aromatic core orientations on molecular stacking, the molecular structure of WFA contains a substantial number of short-chain alkanes. These alkanes contribute to enhancing the mobility of MMP molecules, thereby promoting an increase in the number of molecular stacking layers and the molecular stacking height. This observation aligns with recent research findings [37]. The presence of higher stacking layers and heights in the mesophase pitch products is also closely associated with the introduction of certain short-chain alkanes during the co-carbonization process.

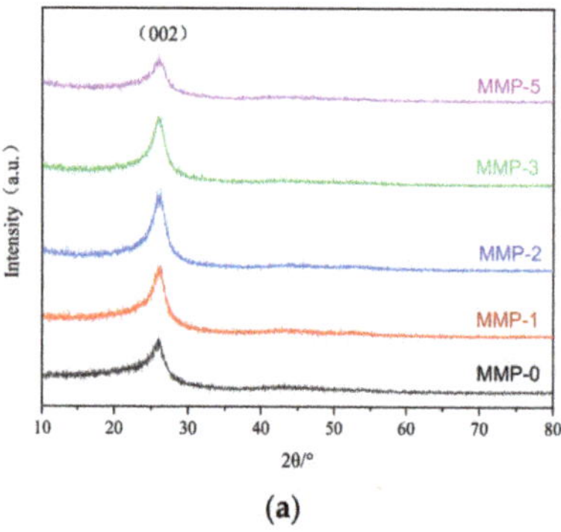

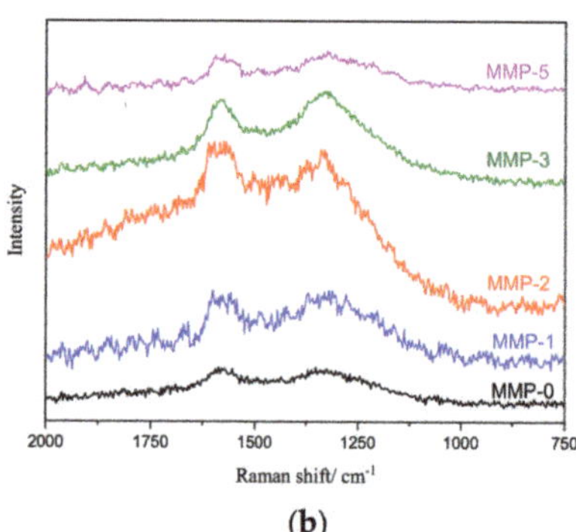

Figure 3. (**a**) XRD and (**b**) Raman patterns of the modified mesophase pitch.

Table 2. Microcrystalline parameters of different modified mesophase pitches.

Sample	XRD				Raman
	2θ/Degree	d_{002}/nm	L_c/nm	N	I_D/I_G
MMP-0	26.003	0.3424	1.910	6.578	1.03
MMP-1	26.216	0.3396	2.008	6.913	0.98
MMP-2	26.380	0.3376	2.228	7.599	0.92
MMP-3	26.277	0.3389	1.973	6.822	1.02
MMP-5	25.919	0.3434	1.881	6.478	1.06

The Raman spectrum of carbon materials exhibits two prominent peaks [37–39] within the primary spectral range of 1000~2000 cm^{-1}, namely the G peak and the D peak. The D peak, located around 1360 cm^{-1}, is attributed to lattice defects in graphite, the low symmetry of carbon structures, and the disordered arrangement of edges. An increase in structural disorder and a decrease in microcrystalline size contribute to the enhanced intensity of the D peak. The G peak is located near 1580 cm^{-1}. For mesophase pitch, the greater the intensity of the D peak, the higher its disorder, and the greater the intensity of the G peak, the higher its order. The relative intensity ratio of the D and G peaks, denoted as I_D/I_G, reflects the quality of molecular orientation and structure. A lower I_D/I_G value suggests better molecular orientation and larger graphite-like microcrystals. According to Table 3, as the proportion of WFA increases, the I_D/I_G values of the mesophase pitch products initially decrease and then increase. When the WFA addition reaches 2%, the I_D/I_G value is minimized, indicating that the ordered structure of the pyrolysis products first increases and then decreases. Notably, MMP-2 exhibits the highest degree of structural order, aligning with the findings from XRD analysis. In conclusion, the appropriate addition of WFA promotes an increase in the orderliness of the mesophase pitch in the product. However, excessive amounts of WFA may lead to an elevation in the disorder of the mesophase pitch.

Table 3. Proton distribution of the different modified mesophase pitches determined by ^{1}H-NMR and ^{13}C-NMR analysis.

Sample	^{1}H-NMR (%)									^{13}C-NMR (%)		
	$H_{ar(mono)}$	$H_{ar(di)}$	$H_{a(poly)}$	H_{ar}	H_α	H_β	H_γ	H_{al}	H_{ar}/H_{al}	C_{sat}	C_{ar}	C_{ar}/C_{sat}
MMP-0	6.45	40.65	5.17	52.27	27.53	17.34	2.86	47.73	1.095	14.29	85.71	5.998
MMP-1	6.09	41.28	7.75	55.12	30.68	12.45	1.75	44.88	1.228	13.17	86.83	6.593
MMP-2	6.99	42.98	9.48	59.45	30.61	8.71	1.23	40.55	1.466	11.89	88.11	7.410
MMP-3	4.94	44.18	13.21	62.33	31.88	5.23	0.56	37.67	1.655	9.27	90.73	9.787
MMP-5	4.92	45.52	17.71	68.15	28.92	2.79	0.14	31.85	2.140	7.14	92.86	13.006

2.3. Molecular Structure Analysis of Mesophase Pitch

Fourier-transform infrared (FT-IR) spectroscopy is a commonly used method to provide functional group information in complex solid materials [40–42]; specific reference standards can be found in Table S3. In the FT-IR spectrum, specific peaks can be assigned to various functional groups present in the sample. For instance, the peak around 3040 cm^{-1} corresponds to the stretching vibration absorption of aromatic C-H bonds, while the peak near 2950 cm^{-1} is attributed to the methyl absorption peak of fatty hydrocarbons. Peaks within the range of 2930–2700 cm^{-1} represent the stretching vibration absorption of C-H bonds in fatty hydrocarbons. The absorption peak at 1600 cm^{-1} is associated with the stretching vibration of the aromatic skeleton, whereas the peak at 1450 cm^{-1} corresponds to the bending vibration absorption of methylene groups in cycloalkane structures and fatty branch chains. The absorption peak at 1380 cm^{-1} indicates the bending vibration of -CH$_3$ groups on the benzene ring, and the peaks at 880–750 cm^{-1} represent out-of-plane bending vibrations of aromatic C-H bonds. Notably, the peak at 880 cm^{-1} corresponds to an isolated out-of-plane bending vibration of aromatic C-H, while the peak at 750 cm^{-1} arises from the synergistic effect of four adjacent aromatic C-H bonds. The peaks within the range of 500–460 cm^{-1} gradually decrease, and the corresponding characteristic peaks in this range are caused by the outward bending vibration of the C-H plane of aromatic rings and substituted aromatic rings containing four adjacent hydrogens. From Figure 4a, it can be seen that before the addition of WFA, the content of aliphatic side chains in the mesophase pitch is slightly higher than that after the addition of WFA. Furthermore, with increasing amounts of WFA, the absorption peak of fatty branch chains in the mesophase pitch shows minimal variation, whereas the absorption peak of aromatic groups initially weakens and then strengthens. Upon analyzing the FT-IR spectrum, it could be observed that MMP consists of planar aromatic molecules with alkyl chain substituents on the aromatic ring. The ortho-substitution index (I_{os}), which indicates the relative size of aromatic molecules, can be calculated using Formula (2). The aromaticity (fa) of mesophase pitch can be determined using Formula 3, where A represents the absorption intensity of a specific peak. According to Figure 4b, as the proportion of WFA increases, the ortho-substitution index (I_{os}) of ortho-substituted aromatic rings initially increases and then decreases. Simultaneously, the fragrance of the product also increases with the addition of WFA. Based on these findings, it can be inferred that a small amount of WFA added to the reaction system can promote the dehydrogenation condensation reaction of small molecule aromatic rings, thereby facilitating the aromatization reaction. However, excessive amounts of WFA may restrict the accumulation of smaller planar aromatic molecules and inhibit the formation of larger planar bitter aromatic molecules.

The graphs displayed in Figure 5 illustrate the thermogravimetric (TG) and derivative thermogravimetric (DTG) curves of mesophase pitch. From the figure, it can be seen that the pyrolysis of the product is divided into two stages. The first stage is between 200–570 °C, and the five types of mesophase pitch exhibit weight loss starting at approximately 200 °C. This reduction in sample mass primarily stems from the evaporation of light components and volatile substances. As the temperature reaches 380 °C, the sample experiences a continual decrease in mass due to molecular dissociation and dehydrogenation. The second stage is at 570–800 °C. Low molecular weight gases (various alkanes) are mainly produced through side chain cracking reactions, polymerization reactions, and aromatization reactions, and the weight loss curves of the five mesophase pitches begin to stabilize. In the absence of WFA in the reaction system, the carbon yield of mesophase pitch is minimal. However, with an increasing proportion of WFA, the carbon yield of mesophase pitch gradually rises. This observation suggests that WFA can enhance the thermal stability of mesophase pitch. The reason for this phenomenon is that the addition of WFA can effectively promote the thermal polymerization reaction. Notably, when the WFA ratio reaches 5%, the carbon yield can reach a substantial value of up to 93%. These findings provide evidence that an appropriate amount of WFA can stimulate thermal polymerization reactions [43,44]. For applications requiring improved thermal stability of the reaction system, it is advisable to explore suitable co-carbonizing agents that facilitate the transformation of small molecules into planar molecular structures during thermal decomposition.

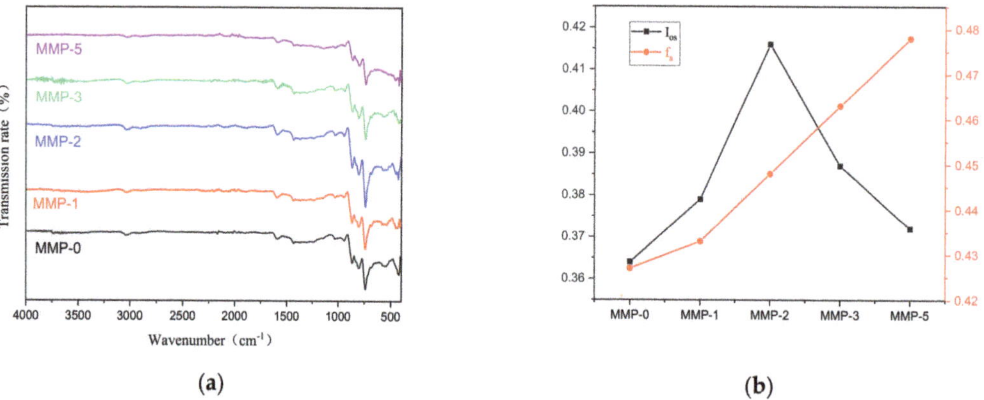

Figure 4. (a) FTIR spectra and (b) FTIR structural index parameters of different modified mesophase pitches.

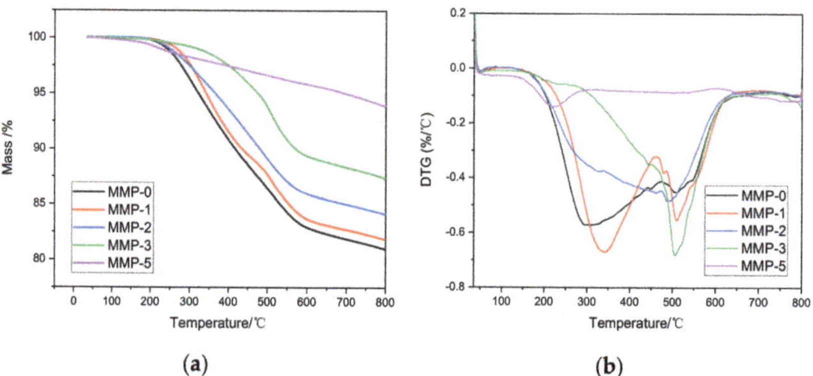

Figure 5. (a) TG and (b) DTA results of the different modified mesophase pitches.

The MALDI-MS spectrum presented in Figure 6 exhibits a broad distribution of molecular weights for mesophase pitch, ranging from 100 to 3000 m/z. The average molecular weights of asphaltene, MMP-0, MMP-1, MMP-2, MMP-3, and MMP-5 are determined to be 343, 392, 446, 540, 587, and 612, respectively. The fluctuations in average molecular weight suggest that the composition of the five mesophase pitches comprises polycyclic aromatic hydrocarbon molecules with different numbers of units, typically characterized by 2–4 hexagonal rings. Polycyclic aromatic hydrocarbon molecules can be categorized into monomers ($m/z \approx$ 200–400), dimers ($m/z \approx$ 400–700), trimers ($m/z \approx$ 700–1000), and tetramers ($m/z \approx$ 1000–1200) based on the number of aromatic units. In the absence of WFA in the system, mesophase pitch predominantly consists of monomers, resulting in a lower average molecular weight. This phenomenon can be attributed to the dealkylation reaction during thermal polymerization, leading to a reduction in molecular weight. Additionally, the dealkylation reaction generates a significant number of free radicals, causing a rapid increase in system viscosity, which hinders the progress of the polymerization reaction. However, with the introduction of WFA, the average molecular weight of mesophase pitch gradually increases. This effect arises from the synergistic interaction between the two raw materials during co-carbonization. Specifically, the abundant cycloalkane structure in the molecular composition of WFA reduces the dealkylation reaction rate during FCC-BL pyrolysis. Moreover, it collaborates with the rich concentration of short fatty hydrocarbon components in FCC-BL, collectively reducing the viscosity of the reaction system. Consequently, this facilitates the polymerization reaction of the raw material molecules during the thermal treatment process. Based on the aforementioned details, it can be deduced that the components generated from WFA pyrolysis promote the formation of large polycyclic aromatic molecules through the cross-linking of aromatic units [45,46]. As a result, the molecular weight of these anisotropic components progressively increases with the augmentation of WFA addition.

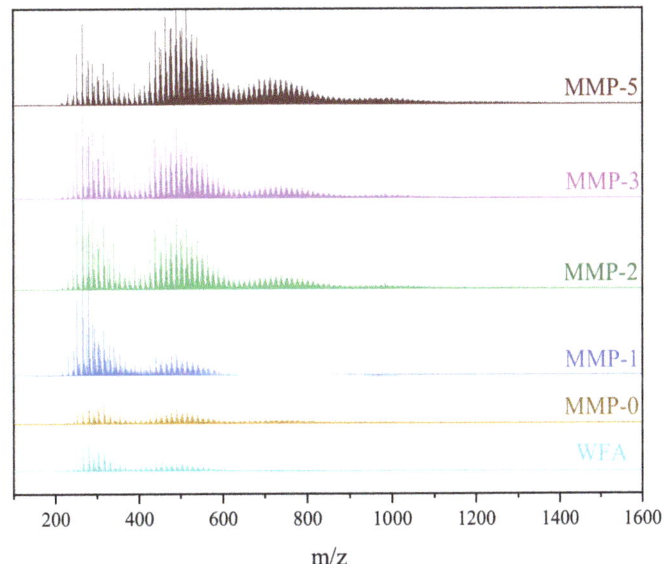

Figure 6. MALDI-TOF MS spectra of the different modified mesophase pitches.

2.4. ^{1}H-NMR Analysis and ^{13}C-NMR Analysis

^{1}H-NMR analysis and ^{13}C-NMR analysis are indispensable techniques employed for determining the relative proportions of hydrogen and carbon in the aliphatic and aromatic constituents of diverse mesophase samples [47,48]. Specific reference standards can be found in Table S4. Table 3 provides a comprehensive summary of the hydrogen and carbon atom contents in intermediate-phase pitch samples prepared using varying amounts of WFA as a co-carbonizing agent. The aliphatic protons encompass H_α, H_β, and H_r while the aromatic protons encompass monoaromatic, biaromatic, and polyaromatic protons. The chemical shifts of aromatic protons in ^{1}H-NMR spectroscopy lie within the range of 6.0 to 9.0 ppm. Among these, monoaromatic protons ($H_{ar(mono)}$) exhibit chemical shifts in the range of 6.0 to 7.1 ppm, biaromatic protons ($H_{ar(di)}$) exhibit chemical shifts in the range of 7.1 to 8.2 ppm, and polyaromatic protons ($H_{ar(poly)}$) exhibit chemical shifts in the range of 8.2 to 9.0 ppm. The chemical shifts of aliphatic protons (H_{al}) predominantly concentrate in the range of 0.5 to 4.5 ppm. Specifically, the chemical shift of H_α lies within the range of 2.1 to 4.5 ppm, that of H_β lies within 1.1 to 2.1 ppm, and that of H_r lies within 0.5 to 1.1 ppm. In ^{13}C-NMR spectroscopy, the chemical shift of saturated carbon (C_{sat}) is observed at 5 to 50 ppm, while aromatic carbon (C_{ar}) exhibits a chemical shift of 100 to 160 ppm.

The data of the ^{1}H-NMR spectrum reveal that the addition of WFA leads to an increase in the ratio of H_{ar} and a decrease in the ratio of H_{al}. This observation suggests that the co-carbonization reaction facilitates the conversion of linear aliphatic chains into bridged structures. The predominant reaction in this co-carbonization process is the cyclization reaction, which involves a dehydrogenation polymerization step that generates hydrogenated aromatic hydrocarbon structures. These structures subsequently undergo rapid conversion into aromatic hydrocarbons. In the reaction system, if the content of H_α is higher than that of H_β and H_r, this indicates that the majority of alkyl groups are connected to the aromatic ring in the form of methylene. The presence of alkyl groups promotes catalytic cracking and the formation of more aromatic hydrocarbons, thereby enhancing the performance of the mesophase pitch. With increasing WFA content, the proportion of H_β and H_r decreases, primarily due to the occurrence of cracking reactions at the β and r positions of the fatty chain under high-temperature conditions. Similarly, an increase in the hydrogen aromaticity index signifies the higher content of aromatic hydrogen in the prepared mesophase structure. The ^{13}C-NMR data also demonstrate that the mesophase pitch prepared with WFA exhibits higher content of aromatic carbon compared to the pitch prepared without WFA.

The data presented in Table 3 demonstrate that the carbon aromaticity (C_{ar}/C_{sat}) is higher compared to the hydrogen aromaticity (H_{ar}/H_{al}), suggesting that the formation of condensed aromatic structures involves dehydrogenation, condensation, and polymerization reactions. In conclusion, the condensation and polymerization reactions occurring during the co-carbonization process entail the removal of hydrogen, leading to the generation of high molecular weight aromatic polymers.

2.5. Morphology and Structure of Final Co-carbonization Products at High Temperature (SEM)

The SEM images in Figure 7 depict the morphology and structure of the pyrolysis products obtained from FCC-BL with varying amounts of added WFA. Figure 7a reveals that in the absence of WFA, the surface of the product mesophase pitch exhibits numerous granular impurities. At this time, the granular impurities mainly include unmelted intermediate phases and quinoline-insoluble substances. However, in Figure 7b, the addition of 1% WFA leads to a significant reduction in the number of granular impurities and the development of a small flow structure is noticeably improved compared to the system without WFA. Figure 7c demonstrates that with a WFA addition of 2%, almost all the granular impurities on the pyrolysis product surface disappear, resulting in a complete layered structure. On the other hand, Figure 7d shows that at WFA content of 3%, substantial accumulation of impurities occurs on the pyrolysis product surface and the main component of impurities at this time is quinoline insoluble substances, leading to a significant impact on the layered structure. Finally, in Figure 7e, it is evident that when the WFA content reaches 5%, the pyrolysis product surface is nearly entirely covered by impurities, resulting in the almost complete disappearance or coverage of the layered structure [49–52].

In conclusion, the observed variations in microstructure are strongly correlated with the pyrolysis and carbonization processes. The addition of WFA facilitates the nucleation and growth of the mesophase pitch, thereby enhancing aromaticity. However, when the WFA content surpasses a critical threshold, it has the potential to aggregate on the surface of pyrolysis products or mesophase microspheres, impeding their mutual approach and fusion. This phenomenon ultimately leads to the formation of a mosaic structure.

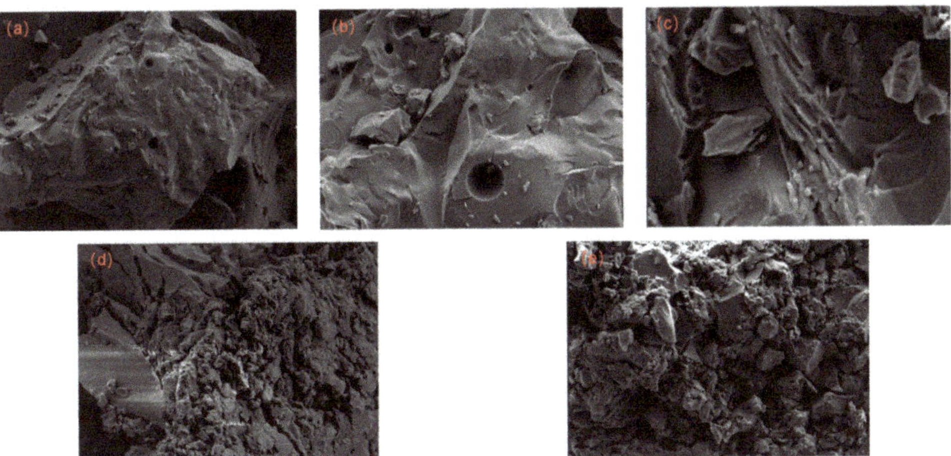

Figure 7. SEM images of different modified mesophase pitches with different WFA content. (**a**) 0%; (**b**) 1%; (**c**) 2%; (**d**) 3%; (**e**) 5%.

2.6. Experimental Mechanism

In general, the thermal decomposition of polycyclic aromatic hydrocarbons in the liquid phase can be divided into two stages: an initial thermal decomposition reaction and a subsequent thermal polymerization reaction. Peng et al. [53], in their study of n-butyl benzene under high-pressure conditions, found that styrene was the main product during the early stage of the reaction. The formation of styrene was attributed to the cleavage of the C_2-C_3 bond in the alkyl substituent, followed by rapid capture of hydrogen radicals to form styrene ethane, as depicted in Figure 8a. Savage investigated the pyrolysis mechanism of dodecyl pyrene and observed that the formation of the major product, ethyl pyrene, occurred through the removal of α-H free radicals at the C position by n-decane radicals, leading to the formation of α-C free radicals. Subsequently, the α-C free radicals underwent C_2-C_3 bond cleavage to yield vinyl pyrene. Based on their research, the dissociation energies were determined to be 375 kJ/mol, 428.4 kJ/mol, and 458.2 kJ/mol for benzyl, alkyl, and phenyl groups, respectively. This conclusion suggests that benzyl radicals, namely α-C free radicals, are more readily formed in aromatic compounds. Thus, it is widely accepted that the formation of α-C free radicals on alkyl side chains plays a significant role in the decomposition reactions of aromatic compounds containing long alkyl chains.

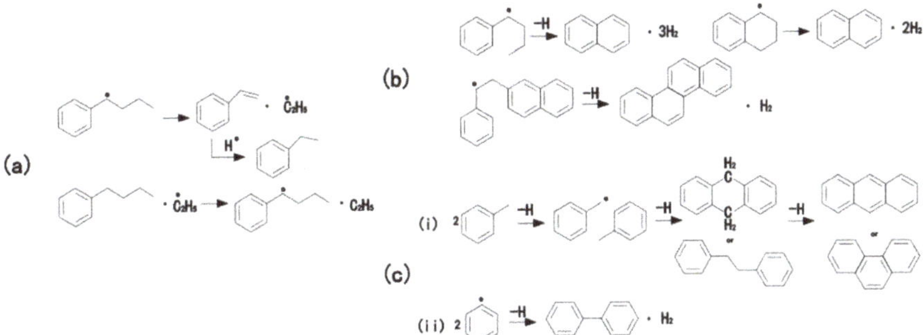

Figure 8. (**a**) The preliminary thermal decomposition reaction; (**b**) intramolecular condensation reaction; (**c**) condensation reaction between adjacent molecules. This includes: (**i**) Condensation reaction between adjacent molecules involving alkyl side chains. (**ii**)Thermal condensation reaction between adjacent molecules involving aromatic nuclei.

The main reaction of polycyclic aromatic hydrocarbons under medium temperature liquid phase cracking reaction conditions is condensation reaction. Based on the molecular structure of the reactants involved in the condensation reaction, condensation reactions can be roughly divided into the following types [54,55]: (1) intramolecular condensation reaction; for example, the conversion of n-butyl benzene to naphthalene or the formation of polycyclic aromatic hydrocarbons generally occurs on the -CH_2-CH_2- chain between two aromatic rings, as shown in Figure 8b. (2) Intermolecular condensation reactions; the intermolecular condensation reaction includes: the condensation reaction between adjacent molecules of alkyl side chains, as well as the thermal condensation reaction between adjacent molecules of aromatic nuclei, as shown in Figure 8c. The above different types of condensation reactions can be carried out in different ionization forms to obtain higher molecular weight polycyclic aromatic hydrocarbons.

The reaction mechanism employed in this experiment exhibits similarities to the Figure 8a mechanism in the initial stages, involving the generation of α-C free radicals on aromatic compounds within the system. In the later stages, the mechanism resembles the Figure 8b,c mechanisms, whereby the alkane moieties attached to the aromatic nuclei in aromatic compounds approach each other. During the polymerization process, these moieties, owing to their high electronegativity, engage in intramolecular or intermolecular cross-coupling reactions, leading to the formation of additional aromatic ring-containing molecules. This observation aligns with the conclusions drawn from the FT-IR characterization mentioned earlier.

The co-carbonizing agent asphaltene contains rich cyclic alkyl structures and long alkyl side chains, significantly reducing the viscosity of the reaction system, improving the flow-ability of the raw materials, increasing the probability of contact between raw material molecules, promoting the ordered stacking of planar aromatic macromolecules, and the subsequent fusion and growth of mesophase pitch microsphere, ultimately forming a large domain texture of mesophase pitch; In addition, asphaltene exhibits excellent reactivity during the reaction process, promoting the thermal reaction of the raw materials, leading to the generation of more free radicals in the reaction system, promoting the condensation of aromatic molecules, and facilitating the formation of mesophase pitch.

3. Experiment

3.1. Determination of Element Content

The carbon, hydrogen, nitrogen, and sulfur elements in the sample were quantified using a German elemental analyzer (Elementar vario EL III). The analyzer employs the difference subtraction method to calculate the oxygen element content.

3.2. Determination of Ash Content

We determined the ash content of the sample. We weighed the clean porcelain crucible and recorded it as m_0. Then, we placed 2.5 g of the sample into a crucible and weighed it, denoted as m_1. We rolled the quantitative filter paper into a conical shape, cut a section about 5 mm away from the tip, and then placed it into the crucible as the ignition core. After soaking the filter paper in the sample, we ignited and burned it until the sample stopped burning and there was no smoke. There was a shiny black substance remaining in the crucible, which is called residual carbon. We placed the crucible with the residue in a muffle furnace and burned it at 775 ± 25 °C for 1.5–2 h until the residue inside the crucible completely burned into white or grayish white powder. We cooled the crucible in air for 3 min before placing it in a dryer and weighing it to a constant weight, recorded as m_2. The ash content of the sample was calculated by Formula (1):

$$\text{Ash content} = \frac{m_2 - m_0}{m_1 - m_0} \times 100\% \tag{1}$$

3.3. Determination of Four Components

The four components of the oil slurry were determined using the IATROSCAN MK-6s TLC-FID thin-layer chromatograph from Shanghai branch of Japan's Yatelon Company.. A microinjector was used to point the sample on the chromatographic rod. The three developing agents were n-hexane, toluene, dichloromethane, and methanol (volume ratio 95:5), with solvent heights of 100 mm, 55 mm, and 25 mm, respectively. After unfolding the solution on the chromatographic rod, we placed the rod in a thin-layer chromatograph for detection and used the software provided by TLC-FID to analyze and calculate the content of aromatics, saturates, resins, and asphaltenes.

3.4. Infrared Spectroscopy Analysis (FT-IR)

The sample was subjected to infrared spectroscopy analysis using a Nicolet Magana IR500 infrared spectrometer (NICOLET Company, Wisconsin, USA). This technique provided information about the types of functional groups present in the test sample. The sample was compressed with KBr to form a pellet, and the spectral scanning range was set at 400–4000 cm^{-1}. The ortho-substitution index (I_{os}), which represents the relative size of aromatic molecules, was calculated according to Formula (2). The aromaticity (fa) of the mesophase pitches was determined using Formula (3):

$$I_{os} = \frac{Ab_{750}}{Ab_{750} + Ab_{815} + Ab_{880}} \tag{2}$$

$$fa = \frac{0.574Ab_{1600}}{Ab_{1600} + 0.16Ab_{1460} + 0.23Ab_{1330}} + 0.024 \tag{3}$$

Among them, Ab_{750}, Ab_{815}, Ab_{880}, Ab_{1330}, Ab_{1460}, and Ab_{1600} are the absorption intensities of the absorption peaks at 750, 815, 880, 1330, and 1460 cm^{-1}, respectively.

3.5. Nuclear Magnetic Resonance Hydrogen Spectroscopy (^{1}H-NMR) and Carbon Spectroscopy (^{13}C-NMR) Analysis

Nuclear magnetic resonance (NMR) analysis of asphaltene and its co-carbonization products is an important analytical method. In NMR analysis, the chemical shifts of carbon and hydrogen atoms in different environments and positions vary. NMR provides clear structural parameters of asphaltene, allowing for the deduction of its molecular structure. The B-L method, based on hydrogen NMR spectroscopy, is used to calculate the structural parameters of asphaltene. The instrument utilized in this section is the Ascend 600 nuclear magnetic resonance spectrometer from Beijing branch of Switzerland's Bruker Company. Deuterated chloroform (CDCl$_3$) served as the deuterated solvent. The sample was prepared by dissolving 100 mg of the sample in 2.0 mL of deuterated chloroform. The hydrogen spectrum was scanned 128 times, while the carbon spectrum was scanned 512 times. An AVANCE III 400 MHz nuclear magnetic resonance spectrometer(Bruker Company, Switzerland) was employed for NMR analysis of the sample, providing information on the sample's structural composition. Deuterated chloroform was used as the solvent, and the B-L method was employed for structural analysis.

3.6. Determination of Softening Point

The determination of the softening point follows the global method outlined in the GB/T4507-2014 standard. Methyl silicone oil was chosen as the heating medium. The sample was initially ground and then heated until it reached a flowable state. Simultaneously, the copper ring was preheated and placed on a flat metal plate. The sample was then poured into two copper rings, slightly above the plane of the copper ring. After solidification, any excess sample was scraped off from the top of the copper ring to ensure a smooth surface. The copper ring was positioned for testing, and each sample underwent two parallel experiments. The final experimental results were averaged. If the sample's softening point exceeds 180 °C, the DMA Q800 dynamic mechanical thermal analyzer can be used for measurement.

3.7. Determination of Quinoline Insoluble Content and Toluene Insoluble Substance

The sample was pulverized and subjected to a drying process at 60 °C for 12 h. A quantity of 1 g of the sample was carefully weighed and enclosed in filter paper, followed by placement in a Soxhlet extractor. Initially, extraction was performed using toluene as the solvent at a temperature of 130 °C. Extraction continued with quinoline as the solvent at 110 °C until the reflux solution within the extractor became colorless, indicating completion of the extraction process. Subsequently, the solvent was eliminated from the extracted solution through distillation. The resulting mixture, together with the filter paper, was introduced into a vacuum drying oven and subjected to a drying period of 2 h at 100 °C. Each component was then accurately weighed, leading to the determination of the final content. The processing method for measuring and calculating toluene insoluble matter is as follows: Following the aforementioned treatment method, place the mesophase pitch sample in a filter paper cylinder, and then place the filter paper cylinder in a Soxhlet extraction apparatus and extract with toluene. After extraction is complete, the remaining pitch in the filter paper cylinder is the toluene-insoluble component, and parallel samples were measured for each group of samples, and the average value was taken.

3.8. Polarization Microscope Characterization

The optical texture of the sample was examined using an LV100N POL polarizing microscope, manufactured by the Nikon Corporation of Tokyo, Japan. The process involved mixing epoxy resin A adhesive and B adhesive in a specific ratio, which was then poured into a mold and combined with the sample before being subjected to heat treatment for solidification. The polishing procedure was conducted using an MP-2B fully automated polishing machine, produced by Fujian Testing Instrument and Equipment (Xiamen, China) Co., Ltd. Initially, polishing was performed using 300–5000 mesh sandpaper, followed by a polishing step with 10,000 mesh polishing paste to achieve a mirror-like finish on the specimen. Subsequently, the optical texture of the sample was examined under the polarization microscope at various magnifications. Among them, the area corresponding to the reflection and brightness of the sample under microscope observation is the anisotropic textures, which is the mesophase pitch; the dark areas are disordered isotropic asphalt. We imported the captured optical morphology image into ImageJ software and calculated the number of mesophase pitch pixels and the approximate proportion of the total number of pixels in the image. The result obtained is the mesophase pitch content corresponding to this section, which is then converted into volume content to obtain the mesophase pitch content of the asphalt sample corresponding to this stage.

3.9. X-ray Diffraction Analysis (XRD)

The crystal structure of the sample was determined using a D/max2550VB/PC X-ray diffractometer. The testing process employed a Cu target with a wavelength (λ) of 0.15406 nm and covered an angular range of 5–75°. The microcrystalline parameters, such as those shown in Formulas (2)–(4), were calculated utilizing the Scheler equation and the Bragg equation:

$$d_{002} = \frac{\lambda}{2sin\theta_{002}} \tag{4}$$

$$L_c = \frac{K\lambda}{\beta_{002}cos\theta_{002}} \tag{5}$$

$$N = \frac{L_c}{d_{002}} + 1 \tag{6}$$

Herein, θ_{002} represents the diffraction angle of the (002) crystal plane, while d_{002} denotes the interlayer spacing of graphite, measured in nm. The parameter K is a constant associated with the shape of the crystalline grains, typically set as K = 0.94. Lc refers to the average thickness of the crystal's stacked layers, measured in nm. β_{002} represents the half-peak width of the (002) peak, measured in radians. Lastly, N signifies the average number of layers in the graphite microcrystalline structure.

3.10. Time-of-Flight Mass Spectrometry (TOF-MS)

Asphaltene exhibits a wide distribution of molecular weights. Analyzing the average molecular weight and distribution of asphaltene and its co-carbonization products allows for the determination of the peak molecular weight present in asphaltene, thereby aiding in the elucidation of its average molecular structure. The instrument utilized in this study is the 4800 plus time-of-flight mass spectrometer, manufactured by ABS Company in Singapore. This instrument was employed to analyze samples of asphaltene and modified mesophase pitch to investigate their molecular characteristics.

3.11. Thermogravimetric Analysis (TG) and Differential Thermal Analysis (DTG)

The thermal weight loss behavior of the sample was investigated using a TGA 8000 thermogravimetric analyzer manufactured from Shanghai branch of Perkin-Elmer Company in the USA. The experimental measurements were performed under a controlled nitrogen atmosphere, with the temperature ramped up from ambient conditions to 800 °C at a constant heating rate of 10 °C/min. By meticulously recording the alterations in sample mass as a function of temperature, both the weight loss curve and weight loss rate curve of the sample were obtained.

3.12. Scanning Electron Microscope (SEM) Characterization

The surface topography of modified mesophase pitch samples (MMP-0, MMP-1, MMP-2, MMP-3, and MMP-5) was examined using the FEI Quanta 200FEG field emission environmental scanning electron microscope manufactured by Philips. This instrument, with a resolution of 6.0 nm, an acceleration voltage range of 20–25 KV, and adjustable magnification from 100,000 to 200,000 times, allowed for detailed observation of the sample's surface morphology. Additionally, the JEOL S4800 scanning electron microscope was utilized to explore the microstructure of the mentioned materials.

3.13. Raman Spectroscopy Analysis

The microcrystalline content and arrangement of carbon layers in the sample were characterized using the DXR3xi Raman Imaging Microscope of The East China University of Science and Technology. The main parameters of the instrument are as follows: laser: 785 nm; laser excitation: 50–3500 cm^{-1}; 10× objective; and exposure time: 1.0 s. The determination of microcrystalline content and carbon layer arrangement primarily relies on the ratio between the D-band, located near 1350 cm^{-1}, and the G-band, positioned at 1625 cm^{-1}. I_D/I_G is the ratio of the intensities of the D and G peaks in the Raman spectrum.

4. Materials and Methods

4.1. Materials

The FCC slurry utilized in this study was obtained from the Baling Petrochemical Plant [24] (in the following text, the FCC slurry will be abbreviated as "FCC BL"). Prior to experimentation, the FCC slurry underwent a deasphaltene treatment process, resulting in a treated FCC-BL sample with an asphaltene content of less than 0.5%. The performance characteristics of the FCC-BL sample are depicted in Table 4. The research team conducted direct thermal condensation treatment on the FCC slurry and analyzed the properties of the mesophase pitch products. The findings revealed that the optimal conditions for producing mesophase pitch from FCC-BL were a temperature range of 400 °C to 420 °C, a pressure range of 1 to 2 MPa, and a reaction time between 4 to 8 h. Some other physical properties of FCC-BL are referred to in Table S1.

Table 4. Basic properties of the experimental materials.

Material	Elemental Analysis (wt%)					H/C Ratio	Solubility (wt%)		Softing Point	Ash (wt%)
	C	H	N	O	S		TI	QI		
FCC-BL	90.55	7.66	1.02	0.32	0.45	1.02	<0.1	<0.1	20	0.02
WFA	86.20	8.46	1.21	1.80	2.33	1.18	22.5	1.6	145	0.53

TI: Toluene insoluble substance; QI: quinoline insoluble substance.

4.2. Obtaining Co-Carbonizing Agents

A supercritical fluid extraction and fractionation experimental device was employed to separate asphaltene, as illustrated in Figure S1. The experimental setup involved utilizing 500 g of Tahe crude oil as the raw material and n-hexane as the solvent. The extraction and fractionation tower was operated with controlled temperatures, where the top temperature was maintained at approximately 240 °C, with the upper part of the tower at 230 °C, the lower part at 220 °C, and the bottom at 210 °C. Each experimental run involved injecting 500 g to 1000 g of the sample, with a pressure range of 2.0 to 10.0 MPa and a linear pressure increase rate ranging from 0.25 to 2.50 MPa/min. The solvent circulation rate was set between 50 and 150 mL/min. The entire process was conducted under constant pressure conditions, and the final product, asphaltene, was obtained at the specified nozzle. The properties of the obtained asphaltene are summarized in Table 4.

4.3. Preparation of Modified Pitch

Upon discharge from the experimental apparatus, the asphaltene appears as a dense black solid. Initially, the asphaltene was subjected to a drying process at room temperature for a minimum of 6 h. Subsequently, the asphaltene was pulverized into a powdered form. Different proportions of asphaltene (0%, 1%, 2%, 3%, and 5%) were then mixed with 50 g of FCC slurry. The mixture was heated to various temperatures (ranging from 400 to 420 °C) and exposed to different reaction times (4 to 8 h) and reaction pressures (ranging from 0 to 2 MPa). The rotational speed of the reaction was maintained between 300 and 500 rad/min. Throughout the reaction, it was necessary to periodically vent the generated gases resulting from the heating process to maintain a constant gas pressure. Once the reaction was complete, the temperature inside the reaction vessel was allowed to naturally cool to room temperature. Subsequently, the co-carbonization reaction products were extracted and subjected to characterization analysis. A picture of the reaction vessel is shown in Figure S2. These samples are designated as MMP-X, where X represents the percentage of asphaltene used and MMP stands for modified mesophase pitch. The reaction mechanism of this experiment is shown in Figure 9 [56,57].

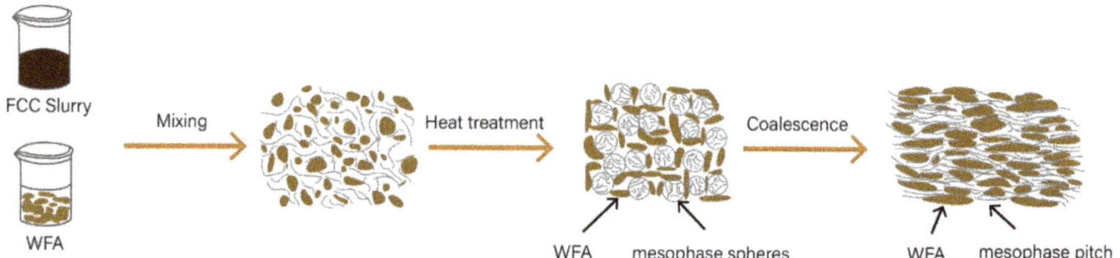

Figure 9. The schematic drawing of the influential mechanism of WFA and FCC-BL in pyrolysis and carbonization.

5. Conclusions

This study presents a methodology for effectively reutilizing asphaltene and producing high-quality mesophase pitch. The research group employed asphaltene as a modifying agent and utilized co-carbonization techniques to synthesize mesophase pitch. The addition of asphaltene had a discernible impact on the nucleation of the mesophase, resulting in an increase in the formation of mesophase small spheres. Moreover, the introduction of asphaltene during the carbonization process led to an augmentation of alkyl and aromatic hydrocarbons, thereby facilitating the formation of mesophase and enhancing both the quantity and quality of the resulting asphalt. Under specific conditions including a temperature of 420 °C, a pressure of 1 MPa, and a reaction time of 6 h, a mesophase pitch with a satisfactory softening point, broad and extensive texture, high product yield, and well-defined carbon layer structure could be obtained when asphaltene was added at a concentration of 2%. The addition of a small amount of asphaltene was found to facilitate mesophase transformation and improve the degree of optical anisotropy. The presence of a higher number of naphthenic structures in asphaltene exerted a more pronounced hydrogen transfer effect, promoting the generation of mesophase small spheres and the formation of wide-area mesophase pitch. Nevertheless, when the asphaltene content surpassed a certain threshold, it tended to cover the surface of mesophase small spheres, impeding their contact and compromising the optical texture of the resultant mesophase pitch products. Additionally, asphaltene contributed to the enhancement of polymerization reactions within the system, leading to an increased average molecular weight of the mesophase pitch. The addition of 2% asphaltene resulted in the minimum I_D/I_G value for mesophase pitch, indicating superior molecular orientation and larger graphite-like microcrystal size. Furthermore, it was observed that the maximum θ and Lc values and the minimum d_{002} value were achieved when 2% asphaltene was added, suggesting the highest degree of order in the mesophase pitch. The inclusion of asphaltene also elevated the yield and aromaticity of the mesophase pitch, while significantly improving its thermal stability.

Supplementary Materials: The following supporting information can be downloaded at: https://www.mdpi.com/article/10.3390/molecules29071500/s1. Figure S1. Schematic diagram of asphaltene separation equipment; Figure S2. Schematic diagram of carbonization; Figure S3. Optical structure of modified mesophase pitch under different reaction temperature; Figure S4. Optical structure of modified mesophase pitch under different reaction time; Figure S5. Optical structure of modified mesophase pitch under different reaction pressures; Table S1. Composition and properties of FCC-BL; Table S2. Classification of the optical microstructure of mesophase pitch; Table S3. The FT-IR characteristic absorption peaks of common structural; Table S4. The proton chemical shift in the 1H-NMR.

Author Contributions: Conceptualization, X.Z.; methodology, X.Z.; software, M.W.; validation, X.Z.; formal analysis, X.Z. and M.W.; investigation, M.W., H.W., M.L., Y.L., J.T., Y.S.; resources, X.Z.; data curation, M.W.; writing—original draft preparation, M.W.; writing—review and editing, M.W.; visualization, X.Z.; supervision, X.Z.; project administration, X.Z.; funding acquisition, X.Z. All authors have read and agreed to the published version of the manuscript.

Funding: The Fundamental Research Funds for the Central Universities.

Institutional Review Board Statement: Not applicable.

Informed Consent Statement: Not applicable.

Data Availability Statement: Data are contained within the article and Supplementary Materials.

Conflicts of Interest: The authors declare no conflicts of interest.

References

1. Talebi, A.; Shafiei, M.; Kazemzadeh, Y.; Escrochi, M.; Riazi, M. Asphaltene prevention and treatment by using nanomaterial: A comprehensive review. *J. Mol. Liq.* **2023**, *382*, 121891. [CrossRef]
2. Moud, A.A. Asphaltene induced changes in rheological properties: A review. *Fuel J. Fuel Sci.* **2022**, *316*, 123372. [CrossRef]
3. Mohammed, I.; Mahmoud, M.; Al Shehri, D.; El-Husseiny, A.; Alade, O. Asphaltene precipitation and deposition: A critical review. *J. Pet. Sci. Eng.* **2020**, *197*, 107956. [CrossRef]
4. Yonebayashi, H. Asphaltene Flow Assurance Risks: How Are Pitfalls Brought into the Open? *J. Jpn. Pet. Inst.* **2021**, *64*, 51–66. [CrossRef]

5. Nguyen, M.T.; Nguyen, D.L.T.; Xia, C.; Nguyen, T.B.; Shokouhimehr, M.; Sana, S.S.; Grace, A.N.; Aghbashlo, M.; Tabatabaei, M.; Sonne, C.; et al. Recent advances in asphaltene transformation in heavy oil hydroprocessing: Progress, challenges, and future perspectives. *Fuel Process. Technol.* **2021**, *213*, 106681. [CrossRef]

6. Zheng, F.; Shi, Q.; Vallverdu, G.S.; Giusti, P.; Bouyssiere, B. Fractionation and Characterization of Petroleum Asphaltene: Focus on Metalopetroleomics. *Processes* **2020**, *8*, 1504. [CrossRef]

7. Tazikeh, S.; Shafiei, A.; Yerkenov, T.; Abenov, A.; Seitmaganbetov, N.; Atabaev, T.S. A systematic and critical review of asphaltene adsorption from macroscopic to microscopic scale: Theoretical, experimental, statistical, intelligent, and molecular dynamics simulation approaches. *Fuel* **2022**, *329*, 125379. [CrossRef]

8. Masoumeh, H.; Majid, A. Essential role of structure, architecture, and intermolecular interactions of asphaltene molecules on properties (self-association and surface activity). *Heliyon* **2022**, *8*, e12170.

9. Hassanzadeh, M.; Abdouss, M. A Comprehensive Review on the Significant Tools of Asphaltene Investigation. *Analysis and Characterization Techniques and Computational Methods. J. Pet. Sci. Eng.* **2022**, *208*, 109611. [CrossRef]

10. Ali, S.I.; Awan, Z.; Lalji, S.M. Laboratory evaluation experimental techniques of asphaltene precipitation and deposition controlling chemical additives. *Fuel* **2022**, *310*, 122194. [CrossRef]

11. Rashid, Z.; Wilfred, C.D.; Gnanasundaram, N.; Arunagiri, A.; Murugesan, T. A comprehensive review on the recent advances on the petroleum asphaltene aggregation. *J. Pet. Sci. Eng.* **2019**, *176*, 249–268. [CrossRef]

12. Enayat, S.; Tavakkoli, M.; Yen, A.; Misra, S.; Vargas, F.M. Review of the Current Laboratory Methods To Select Asphaltene Inhibitors. *Energy Fuels* **2020**, *34*, 15488–15501. [CrossRef]

13. Ahmed, M.A.; Abdul-Majeed, G.H.; Alhuraishawy, A.K. An Integrated Review on Asphaltene: Definition, Chemical Composition, Properties, and Methods for Determining Onset Precipitation. *SPE Prod. Oper.* **2023**, *38*, 215–242. [CrossRef]

14. Meng, J.; Sontti, S.G.; Zhang, X. Review of microscale dynamics of dilution-induced asphaltene precipitation under controlled mixing conditions. *arXiv* **2022**, arXiv:2211.02698. [CrossRef]

15. Al-Hosani, A.; Ravichandran, S.; Daraboina, N. Review of Asphaltene Deposition Modeling in Oil and Gas Production. *Energy Fuels* **2021**, *35*, 965–986. [CrossRef]

16. Gray, M.R.; Yarranton, H.W.; Chacón-Patiño, M.L.; Rodgers, R.P.; Bouyssiere, B.; Giusti, P. Distributed Properties of Asphaltene Nanoaggregates in Crude Oils: A Review. *Energy Fuels* **2021**, *35*, 18078–18103. [CrossRef]

17. Seitmaganbetov, N.; Rezaei, N.; Shafiei, A. Characterization of crude oils and asphaltenes using the PC-SAFT EoS: A systematic review. *Fuel* **2021**, *291*, 120180. [CrossRef]

18. Ahmadi, M.; Hou, Q.; Wang, Y.; Chen, Z. Interfacial and molecular interactions between fractions of heavy oil and surfactants in porous media: Comprehensive review. *Adv. Colloid Interface Sci.* **2020**, *283*, 102242. [CrossRef] [PubMed]

19. Zhang, J.; Li, C.; Shi, L.; Xia, X.; Yang, F.; Sun, G. The formation and aggregation of hydrate in W/O emulsion containing different compositions: A review. *Chem. Eng. J.* **2022**, *445*, 136800. [CrossRef]

20. Liang, W. *Petroleum Chemistry*, 2nd ed.; China University of Petroleum Press: Beijing, China, 2009; pp. 68–70.

21. Pagán, N.M.P.; Zhang, Z.; Nguyen, T.V.; Marciel, A.B.; Biswal, S.L. Physicochemical Characterization of Asphaltenes Using Microfluidic Analysis. *Chem. Rev.* **2022**, *122*, 7205–7235. [CrossRef]

22. Mazloom, M.S.; Hemmati-Sarapardeh, A.; Husein, M.M.; Behbahani, H.S.; Zendehboudi, S. Application of nanoparticles for asphaltenes adsorption and oxidation: A critical review of challenges and recent progress. *Fuel* **2020**, *279*, 117763. [CrossRef]

23. Zhang, L.; Chen, P.; Pan, S.; Liu, F.; Pauchard, V.; Pomerantz, A.E.; Banerjee, S.; Yao, N.; Mullins, O.C. Structure-Dynamic Function Relations of Asphaltenes. *Energy Fuels* **2021**, *35*, 13610–13632. [CrossRef]

24. Wang, M.; Yang, B.; Yu, T.; Yu, X.; Rizwan, M.; Yuan, X.; Nie, X.; Zhou, X. Research progress in the preparation of mesophase pitch from fluid catalytic cracking slurry. *RSC Adv.* **2023**, *13*, 18676–18689. [CrossRef] [PubMed]

25. Kumari, K.; Rani, S.; Kumar, P.; Prakash, S.; Dhakate, S.R.; Kumari, S. Study of mesophase pitch based carbon fibers: Structural changes as a function of anisotropic content. *J. Anal. Appl. Pyrolysis* **2023**, *171*, 105961. [CrossRef]

26. Wang, Y.; Li, M.; Zhao, Z.; Xu, G.; Ge, Y.; Wang, S.; Bai, J. Preliminary exploration of the mechanism governing the cell structure variation of mesophase coal pitch/carbon black composite carbon foam. *Diam. Relat. Mater.* **2023**, *136*, 110077. [CrossRef]

27. Zeng, C.; Zhang, M.; Fang, W.; Wang, C.; Yang, W.; Zhou, P.; Liao, M.; Su, Z.; Huang, D.; Huang, Q.; et al. Effects of high thermal conductivity chopped fibers on ablation behavior of pressureless sintered SiC–ZrC ceramics. *Ceram. Int.* **2023**, *49 Pt B*, 28844–28853. [CrossRef]

28. Yang, J.Y.; Kim, B.S.; Park, S.J.; Rhee, K.Y.; Seo, M.K. Preparation and characterization of mesophase formation of pyrolysis fuel oil-derived binder pitches for carbon composites. *Composites.* **2019**, *165*, 467–472. [CrossRef]

29. Liu, D.; Li, M.; Qu, F.; Yu, R.; Lou, B.; Wu, C.; Niu, J.; Chang, G. Investigation on Preparation of Mesophase Pitch by the Cocarbonization of Naphthenic Pitch and Polystyrene. *Energy Fuels* **2016**, *30*, 2066–2075. [CrossRef]

30. Ko, S.; Kang, D.; Jo, M.S.; Ha, S.J.; Jeon, Y.P. Anisotropic phase transition via high temperature thin-layer evaporation of a petroleum-based isotropic pitch. *J. Ind. Eng. Chem.* **2021**, *95*, 92–100. [CrossRef]

31. Wei, Y.; Chen, J.; Zhao, H.; Zang, X. Pressure-Strengthened Carbon Fibers from Mesophase Pitch Carbonization Processes. *J. Phys. Chem. Lett.* **2022**, *13*, 3283–3289. [CrossRef]

32. Huang, D.; Liu, Q.; Zhang, P.; Ye, C.; Han, F.; Liu, H.; Feng, Z.; Zhu, S.; Fan, Z.; Liu, J.; et al. Thermal response of the two-directional high-thermal-conductive carbon fiber reinforced aluminum composites with low interface damage by a vacuum hot pressure diffusion method. *J. Alloys Compd.* **2022**, *905*, 164195. [CrossRef]

33. Zhang, X.; Meng, Y.; Fan, B.; Ma, Z.; Song, H. Preparation of mesophase pitch from refined coal tar pitch using naphthalene-based mesophase pitch as nucleating agent. *Fuel* **2019**, *243*, 390–397. [CrossRef]
34. Zhang, X.; Ma, Z.; Meng, Y.; Xiao, M.; Fan, B.; Song, H.; Yin, Y. Effects of the addition of conductive graphene on the preparation of mesophase from refined coal tar pitch. *J. Anal. Appl. Pyrolysis* **2019**, *140*, 274–280. [CrossRef]
35. Cheng, Y.; Zhang, Q.; Fang, C.; Ouyang, Y.; Chen, J.; Yu, X.; Liu, D. Co-carbonization behaviors of petroleum pitch/waste SBS: Influence on morphology and structure of resultant cokes. *J. Anal. Appl. Pyrolysis* **2018**, *129*, 154–161. [CrossRef]
36. Guo, J.; Li, Z.; Li, B.; Chen, P.; Zhu, H.; Zhang, C.; Sun, B.; Dong, Z.; Li, X. Hydrogenation of coal tar pitch for improved mesophase pitch molecular orientation and carbon fiber processing. *J. Anal. Appl. Pyrolysis* **2023**, *174*, 106146. [CrossRef]
37. Hu, J.; Fang, C.; Zhou, S.; Cheng, Y.; Han, H. Microstructure characterization and thermal properties of the waste-styrene-butadiene-rubber (WSBR)-modified petroleum-based mesophase asphalt. *J. Mater. Sci. Technol.* **2019**, *35*, 852–857. [CrossRef]
38. Lim, C.; Kwak, C.H.; Ko, Y.; Lee, Y.S. Mesophase pitch production from fluorine-pretreated FCC decant oil. *Fuel* **2022**, *328*, 125244. [CrossRef]
39. Lim, C.; Ko, Y.; Kwak, C.H.; Kim, S.; Lee, Y.-S. Mesophase pitch production aided by the thermal decomposition of polyvinylidene fluoride. *Carbon Lett.* **2022**, *32*, 1329–1335. [CrossRef]
40. Fang, C.; Zhang, M.; Yu, R.; Liu, X. Effect of Preparation Temperature on the Aging Properties of Waste Polyethylene Modified Asphalt. *J. Mater. Sci. Technol.* **2015**, *31*, 320–324. [CrossRef]
41. Ramos-Fernández, J.; Martínez-Escandell, M.; Reinoso, F.R. Preparation of mesophase pitch doped with TiO$_2$ or TiC particles. *J. Anal. Appl. Pyrolysis* **2007**, *80*, 477–484. [CrossRef]
42. Cao, Q.; Guo, L.; Dong, Y.; Xie, X.; Jin, L. Autocatalytic modification of coal tar pitch using benzoyl chloride and its effect on the structure of char. *Fuel Process. Technol.* **2015**, *129*, 61–66. [CrossRef]
43. Kumar, S.; Srivastava, M. Catalyzing mesophase formation by transition metals. *J. Anal. Appl. Pyrolysis* **2015**, *112*, 192–200. [CrossRef]
44. Jin, Z.; Zuo, X.; Long, X.; Cui, Z.; Yuan, G.; Dong, Z.; Zhang, J.; Cong, Y.; Li, X. Accelerating the oxidative stabilization of pitch fibers and improving the physical performance of carbon fibers by modifying naphthalene-based mesophase pitch with C9 resin. *J. Anal. Appl. Pyrolysis* **2021**, *154*, 105009. [CrossRef]
45. Guo, J.; Lu, S.; Xie, J.; Chen, P.; Li, B.; Deng, Z.; Dong, Z.; Zhu, H.; Li, X. Preparation of mesophase pitch with domain textures by molecular regulation of ethylene tar pitch for boosting the performance of its carbon materials. *J. Anal. Appl. Pyrolysis* **2023**, *170*, 105932. [CrossRef]
46. Chai, L.; Lou, B.; Yu, R.; Wen, F.; Yuan, H.; Li, Z.; Zhang, Z.; Jun, L.; Liu, D. Study on structures and properties of isotropic pitches and carbon fibers from co-carbonization of aromatic-rich distillate oil and polyethylene glycol. *J. Anal. Appl. Pyrolysis* **2021**, *158*, 105260. [CrossRef]
47. Lee, S.; Eom, Y.; Kim, B.J.; Mochida, I.; Yoon, S.H.; Kim, B.C. The thermotropic liquid crystalline behavior of mesophase pitches with different chemical structures. *Carbon* **2015**, *81*, 694–701. [CrossRef]
48. Lin, X.; Sheng, Z.; He, J.; He, X.; Wang, C.; Gu, X.; Wang, Y. Preparation of isotropic spinnable pitch with high-spinnability by co-carbonization of coal tar pitch and bio-asphalt. *Fuel* **2021**, *295*, 120627. [CrossRef]
49. Gong, X.; Lou, B.; Yu, R.; Zhang, Z.; Guo, S.; Li, G.; Wu, B.; Liu, D. Carbonization of mesocarbon microbeads prepared from mesophase pitch with different anisotropic contents and their application in lithium-ion batteries. *Fuel Process. Technol.* **2021**, *217*, 106832. [CrossRef]
50. Cao, S.; Yang, J.; Li, J.; Shi, K.; Li, X. Preparation of oxygen-rich hierarchical porous carbon for supercapacitors through the co-carbonization of pitch and biomass. *Diam. Relat. Mater.* **2019**, *96*, 118–125. [CrossRef]
51. Loktev, A.S.; Arkhipova, V.A.; Bykov, M.A.; Sadovnikov, A.A.; Dedov, A.G. Cobalt–Samarium Oxide Composite as a Novel High-Performance Catalyst for Partial Oxidation and Dry Reforming of Methane into Synthesis Gas. *Pet. Chem.* **2023**, *63*, 317–326. [CrossRef]
52. Zhang, D.; Zhang, L.; Yu, Y.; Zhang, L.; Xu, Z.; Sun, X.; Zhao, S. Mesocarbon Microbead Production from Fluid Catalytic Cracking Slurry Oil: Improving Performance through Supercritical Fluid Extraction. *Energy Fuels* **2018**, *32*, 12477–12485. [CrossRef]
53. Peng, Y.; Schobert, H.H.; Song, C.; Hatcher, P.G. Thermal decomposition studies of jet fuel components. n-Butylbenzene and t-butylbenzene. *Am. Chem. Soc. Pet. Chem.* **1992**, *37*, 505–513.
54. Liao, G.; Shi, K.; Ye, C.; Fan, Z.; Xiang, Z.; Li, C.; Huang, D.; Han, F.; Liu, H.; Liu, J. Influence of resin on the formation and development of mesophase in fluid catalytic cracking (FCC) slurry oil. *J. Anal. Appl. Pyrolysis* **2023**, *172*, 105997. [CrossRef]
55. Kim, J.H.; Choi, Y.J.; Lee, S.E.; Im, J.S.; Lee, K.B.; Bai, B.C. Acceleration of petroleum based mesophase pitch formation by PET (polyethylene terephthalate) additive. *J. Ind. Eng. Chem.* **2021**, *93*, 476–481. [CrossRef]
56. Cheng, Y.; Yang, L.; Fang, C.; Guo, X. Co-carbonization behavior of petroleum pitch/graphene oxide: Influence on structure and mechanical property of resultant cokes. *J. Anal. Appl. Pyrolysis* **2016**, *122*, 387–394. [CrossRef]
57. Liu, S.; Xue, J.; Liu, X.; Chen, H.; Li, X. Pitch derived graphene oxides: Characterization and effect on pyrolysis and carbonization of coal tar pitch. *J. Anal. Appl. Pyrolysis.* **2020**, *145*, 104746. [CrossRef]

Review

Composition of Lignocellulose Hydrolysate in Different Biorefinery Strategies: Nutrients and Inhibitors

Yilan Wang [1,2], Yuedong Zhang [2,3,4], Qiu Cui [2,5], Yingang Feng [2,3,4,6,*] and Jinsong Xuan [1,*]

1 Department of Bioscience and Bioengineering, School of Chemistry and Biological Engineering, University of Science and Technology Beijing, 30 Xueyuan Road, Beijing 100083, China
2 CAS Key Laboratory of Biofuels, Shandong Provincial Key Laboratory of Synthetic Biology, Shandong Engineering Laboratory of Single Cell Oil, Qingdao Institute of Bioenergy and Bioprocess Technology, Chinese Academy of Sciences, 189 Songling Road, Qingdao 266101, China
3 Shandong Energy Institute, 189 Songling Road, Qingdao 266101, China
4 Qingdao New Energy Shandong Laboratory, 189 Songling Road, Qingdao 266101, China
5 State Key Laboratory of Microbial Technology, Shandong University, Qingdao 266237, China
6 University of Chinese Academy of Sciences, Beijing 100049, China
* Correspondence: fengyg@qibebt.ac.cn (Y.F.); jsxuan@sas.ustb.edu.cn (J.X.)

Abstract: The hydrolysis and biotransformation of lignocellulose, i.e., biorefinery, can provide human beings with biofuels, bio-based chemicals, and materials, and is an important technology to solve the fossil energy crisis and promote global sustainable development. Biorefinery involves steps such as pretreatment, saccharification, and fermentation, and researchers have developed a variety of biorefinery strategies to optimize the process and reduce process costs in recent years. Lignocellulosic hydrolysates are platforms that connect the saccharification process and downstream fermentation. The hydrolysate composition is closely related to biomass raw materials, the pretreatment process, and the choice of biorefining strategies, and provides not only nutrients but also possible inhibitors for downstream fermentation. In this review, we summarized the effects of each stage of lignocellulosic biorefinery on nutrients and possible inhibitors, analyzed the huge differences in nutrient retention and inhibitor generation among various biorefinery strategies, and emphasized that all steps in lignocellulose biorefinery need to be considered comprehensively to achieve maximum nutrient retention and optimal control of inhibitors at low cost, to provide a reference for the development of biomass energy and chemicals.

Keywords: lignocellulose; cellulase; biorefinery; hydrolysate; fermentable sugar; fermentation inhibitor; pretreatment; saccharification; cellulosome

Citation: Wang, Y.; Zhang, Y.; Cui, Q.; Feng, Y.; Xuan, J. Composition of Lignocellulose Hydrolysate in Different Biorefinery Strategies: Nutrients and Inhibitors. *Molecules* **2024**, *29*, 2275. https://doi.org/10.3390/molecules29102275

Academic Editors: José A.P. Coelho and Roumiana P. Stateva

Received: 26 March 2024
Revised: 6 May 2024
Accepted: 9 May 2024
Published: 11 May 2024

1. Introduction

Lignocellulosic biomass (LCB), as one of the most abundant renewable resources in the world, plays an increasingly important role in the circular economy and sustainable development, thus attracting great attention in various research areas. These studies are dedicated to developing techniques in the bioconversion process known as biorefinery which converts lignocellulosic substrates into products such as biofuels, bioplastics, bio-based chemicals, and bio-gases, to substitute non-renewable fossil-based fuels partially or completely [1–4]. The biorefinery processes of lignocellulosic biomass can be classified into two groups based on conversion approaches and intermediate products (platform molecules). One is the thermochemical conversion process involving pyrolysis which converts biomass into hydrogen and carbon monoxide (the syngas platform) and downstream chemical conversion or biological fermentation which synthesizes various downstream products [5]. The other is the biochemical or biological process that converts lignocellulosic biomass into sugars (the sugar platform) for subsequent conversion. Many studies and recent advances in thermochemical conversion processes have been reported and summarized [6–9]. In this review article, we will mainly focus on the strategies related to lignocellulosic biorefinery via the sugar platform.

Cellulose and hemicellulose are major polysaccharides in lignocellulose. In the process via the sugar platform, these carbohydrate polymers are enzymatically hydrolyzed and release fermentable sugars such as monosaccharides or oligosaccharides, generating lignocellulosic hydrolysate (sugar solution), which can be further utilized by microbes [10]. Lignocellulosic biomass is a tightly interwoven matrix, and pretreatment is a required step for breaking this highly complex structure with a recalcitrant nature by various chemical, physical, physicochemical, and biological methods [11]. Then, the saccharification of cellulose and hemicellulose polymers is achieved by enzymatic hydrolysis with cellulases and hemicellulases [12,13]. In addition to cellulose and hemicellulose, lignin is also one major component of lignocellulose, as a network phenolic polymer providing structural reinforcement and resilience [14]. Moreover, other substances such as small amounts of pectins, cutins, waxes, lipids, tannins, terpenes, alkaloids, and resins are also found in lignocellulosic biomass, and they vary significantly with the species [15,16]. These structural constitutes can be retained in the lignocellulosic hydrolysate with varying degrees in different biorefinery strategies and processes. Most pretreatment processes are accompanied by complex physical and chemical changes, and all lignocellulosic compositions may undergo chemical reactions and produce new compounds that are potentially present in the lignocellulosic hydrolysate [17–19]. Therefore, the chemical composition of lignocellulosic hydrolysate is often complex and heterogeneous (Figure 1).

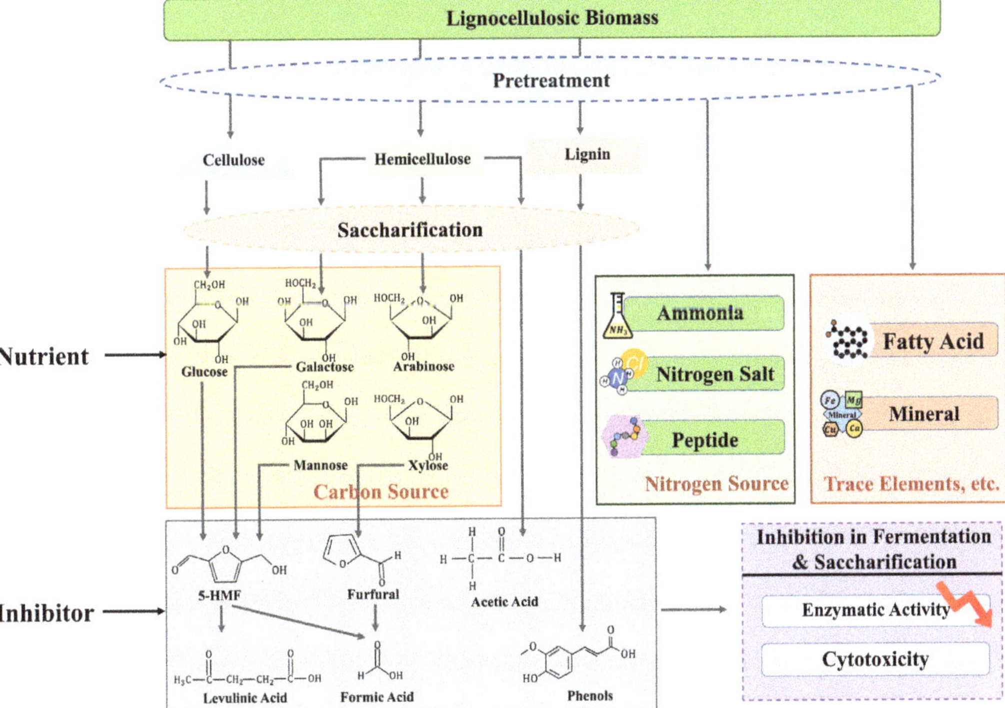

Figure 1. Nutrients and inhibitors present in lignocellulosic hydrolysate.

Lignocellulosic hydrolysate is typically used for further microbial fermentation to produce the final biofuels and bio-based chemicals, but the relatively low price of the end products makes it economically impractical to separate and purify fermentable sugars from lignocellulosic hydrolysate. Therefore, it is crucial that we understand the various chemical compositions (including the nutrients and inhibitors) of the lignocellulosic hydrolysate and their impact on the subsequent fermentation step in depth for the development of economically feasible biorefinery strategies.

In the downstream fermentation process of biorefinery, the nutritional requirements for fermenting microbes include carbon sources, nitrogen sources, trace elements, and other essential nutrients. In most studies, lignocellulosic hydrolysates are used as carbon sources, and whether hexoses (such as glucose) or pentoses (such as xylose) in the hydrolysates can be fully utilized by microbes or not will be taken into account [20] and determine the sugar conversion rate. In fact, due to the complex structural components of lignocellulosic materials themselves and the complexity of the pretreatment and saccharification processes, not only carbon sources but also many other complex components are contained in lignocellulosic hydrolysates. For example, nitrogen sources are necessary for microbial growth and often present in the lignocellulosic hydrolysate. The raw materials themselves, especially those agricultural wastes, naturally contain certain amounts of nitrogen, such as proteins, amino acids, and other nitrogen compounds [15,16,21]. These nitrogen compounds may be retained during certain pretreatment and saccharification processes and, ultimately, remain in lignocellulosic hydrolysates. Furthermore, if ammonia or nitrates are added during the pretreatment or saccharification in some process designs, they may also remain in the lignocellulosic hydrolysates with higher concentrations than the amount of nitrogen possibly required for downstream fermentation [22,23]. Other minor but essential nutrients such as phosphorus, calcium, magnesium, and iron also have significant impacts on growth and production, and all their contents in the lignocellulosic hydrolysates depend on the raw materials, pretreatment methods, and saccharification strategies [24]. Their amounts might directly meet the needs of downstream fermentation production [25,26], be insufficient and require supplementation, or be excessive and have inhibiting effects [27]. Additionally, in some pretreatment and saccharification strategies, especially those whole-cell-based saccharification processes, amino acids, organic acids, vitamins, and other biostimulants might be produced. Although the content of these substances is low, they may have significant impacts on the downstream fermentation process [28].

Nutrients and inhibitory compounds in lignocellulosic hydrolysates are highly dependent on the composition of the raw material, and the methods, strategies, and technologies used in specific pretreatment and saccharification processes [19,29–34]. The pretreatment process aims to open the recalcitrant structure of lignocellulose and separate different components as much as possible for subsequent performance [35]. There are many pretreatment methods, including physical (grinding, microwave, ultrasonic, and pyrolysis), chemical (acid, alkali, ozonolysis, organic solvents, and ionic liquids), physicochemical (hot water, steam explosion, ammonia fiber explosion, wet oxidation, and carbon dioxide blasting), biological, and their combinations [36]. While the dense structure of lignocellulose is broken, different pretreatment processes may involve some additional chemical reactions because of the chemical reagents added, high temperature, and high pressure, and some byproducts may be inhibitors for the subsequent saccharification or fermentation [37,38]. For example, some lignin-derived phenolics may be released or converted into more toxic forms during the pretreatment and saccharification, disrupting the integrity of the microbial cell membranes, interacting with or changing the structure of the enzyme's active sites, thereby inhibiting the enzyme activities [39]. In addition, sugar degradation products (such as furfural) produced during the pretreatment process may directly inhibit enzyme functions and affect the efficiency of the saccharification process [32]. After pretreatment, biomass materials often undergo the washing step for detoxification, but this will significantly increase the cost of pretreatment and the burden of wastewater treatment, so strategies for biomass pretreatment need to be considered comprehensively by combining subsequent saccharification and fermentation processes.

The saccharification process is a key step in biorefinery for producing fermentable sugars. Various saccharification strategies have been developed based on the techno-economic feasibility and difficulty level in implementation. Based on the source of the enzymes used, current saccharification strategies can be divided into off-site and on-site approaches [40]. In the off-site approach, separately produced cellulases are used to convert pretreated lignocellulose into fermentable sugars by enzymatic hydrolysis, such as simultaneous saccharification and fermentation (SSF) which is currently used in the majority of pilot-scale demonstrations and industrial plants. Only enzymes with buffers are usually added in the off-site saccharification approach, and monosaccharides and oligosaccharides are produced, along with other nutrients and inhibitors generated mainly from raw materials and pretreatment processes [41,42]. In the on-site approach, the enzyme production is integrated with the saccharification process in one system, which can significantly reduce the cost of enzyme production and separation, such as consolidated bioprocessing (CBP) and consolidated bio-saccharification (CBS) [40]. Since enzymes are produced directly by microorganisms for saccharification in the on-site approach, metabolic products and partial cell lysates from the enzyme-producing microorganisms are often contained in the resulting lignocellulosic hydrolysate. They can serve as nutrients or inhibitors for downstream fermentation. For example, organic acids may be accumulated by metabolic activities of microorganisms during enzyme production, lead to changes in the pH of the medium, or have toxic effects on the microorganisms, thus affecting the growth of microorganisms and the efficiency of the downstream biorefinery process [39,43–46].

In summary, lignocellulosic hydrolysates contain carbon sources and other nutrients, as well as various inhibitors. The composition and concentration are closely dependent on the lignocellulosic raw material types, pretreatment processes, and saccharification strategies and processes. This article aims to explore the key stages in the biorefinery process of lignocellulose, summarize how various factors in the biomass, pretreatment, saccharification, and fermentation steps affect nutrient supply and inhibitor formation, and offer new insights for designing lignocellulosic biorefinery processes and improving the efficiency and yield in biorefinery.

2. Composition of Lignocellulose Feedstocks

Lignocellulosic biomass contains cellulose, hemicellulose, and lignin as the main components. It also contains small amounts of inorganic elements like potassium, calcium, and magnesium, and various organic compounds such as resins, fats, and waxes that can be extracted with solvents [15,16]. Usually, cellulose and hemicellulose can be hydrolyzed to soluble sugars, which provide carbon and energy sources for microorganisms and are the main nutrients from lignocellulose. Cellulose is a structural homopolymer composed of linear chains constituted by repeating β-D-pyranose glucose units linked by β-(1,4) glycosidic bonds [47]. The cellulose chains are bonded through non-covalent interactions (van der Waals forces and hydrogen bonds), forming rigid and insoluble microfibrils [48]. Depending on the different orientations and different levels of crystallinity, cellulose molecules form amorphous (low-crystallinity) and crystalline (high-crystallinity) regions [49]. A higher crystallinity means that the molecular chains are arranged more tightly and orderly within the crystalline regions, enhancing the material's mechanical hardness and chemical stability [50,51]. Such complicated structures render cellulose resistant to biological and chemical degradation, making it a major barrier in the conversion process of lignocellulosic biomass.

Hemicelluloses are branched and heterogenic polysaccharides composed of pentose (D-xylose and L-arabinose) and hexose (D-glucose, D-mannose, and D-galactose) [52]. These monosaccharide units are linked by β-(1,4)-glycosidic bonds and β-(1,3)-glycosidic bonds [53]. The amorphous, random structural properties and the lower physical strength make hemicelluloses easier to be hydrolyzed than celluloses, but they can act as a physical barrier for cellulases' access to celluloses [38]. Therefore, the removal or separation of hemicelluloses is often necessary for the pretreatment process.

Lignin is an amorphous and highly branched phenolic polymer primarily composed of syringyl (S), guaiacyl (G), and p-hydroxyphenyl (H) subunits. Lignin has hydrogen bonding with cellulose and hemicellulose and is also connected to hemicellulose via various alkyl/aromatic ether linkages, forming lignin–carbohydrate complexes (LCCs). These LCCs prevent enzymes from accessing cellulose during enzymatic hydrolysis [13,48] and cause enzyme deactivation by irreversible enzyme adsorption [36]. Therefore, lignin-derived compounds are major inhibitors of enzymatic reactions and microbial fermentation. Moreover, lignin and its derivatives are difficult to be degraded and assimilated by microorganisms. Therefore, it is often necessary to remove lignin and lignin-derived compounds to eliminate their negative effects through pretreatment to obtain more fermentable sugars [53].

The constitution, structure, and distribution of lignocellulose in the cell wall vary significantly from different biomass sources, depending on the plant species, climate conditions, growth stages, and processing methods [54]. For example, cultivars harvested in summer often have a higher cellulose content than those harvested in autumn [55]. The composition of hemicellulose also varies significantly among different plant species and even within different parts of the same plants (such as the leaves, stalks, and roots) [56]. For example, hemicelluloses in softwoods are constituted by glucomannans, arabinoglucuronoxylans (xylans), arabinogalactans, xyloglucans, and other glucans, while, in hardwoods, hemicelluloses are primarily composed of xylans and glucomannans [57]. Generally, the content of hemicellulose in hardwoods and herbages is usually higher than in softwoods [58]. The composition of lignin also varies among plant types. Hardwood lignin primarily contains S and G subunits, with relative amounts of 45–75% and 25–50%, respectively; softwood lignin is mainly composed of G subunits accounting for about 95%; while the relative contents of H/G/S subunits are about 5–35%, 35–80%, and 20–55%, respectively, for herbaceous plants [37,59]. The total lignin contents are also different: softwoods have the highest lignin content, while herbaceous plants have the lowest [60].

Although the components of lignocellulosic biomass are influenced by various factors, their approximate contents in raw materials are relatively stable: 15–30%, 20–40%, and 35–50% for lignin, hemicellulose, and cellulose, respectively [54]. The compositions of some common lignocellulosic biomass are listed in Table 1.

Table 1. Content of cellulose, hemicellulose, and lignin in common lignocellulosic biomass.

Biomass	Cellulose (%)	Hemicellulose (%)	Lignin (%)	Reference
Sugarcane bagasse	32–55	22–36	14–30	[61]
Sugarcane straw	29	28.8	32.2	[45]
Sorghum straw	26.93	32.57	10.16	[62]
Wheat straw	43.4	26.9	22.2	[63]
Barley straw	35.73–45.73	26.8–32.6	5.3–5.9	[64,65]
Aspen wood	50.7	16.6	13.3	[64]
Oak	43.2	21.9	35.4	[66]
Corn stover	38	23	20	[67]
Switchgrass	50	40	20	[67]
Pine chip	33–44.78	17.56–23.75	20.22–26.29	[68,69]
Spruce	24.7	10.2	35	[70]

These biochemical characteristics of lignocellulosic components help predict the yield of fermentable sugars and understand the native recalcitrance of raw materials. Generally, the higher the proportion of cellulose in raw materials was, the more glucose could be released; the higher the percentage of lignin present was, the more difficult it was for the biomass to be degraded [64]. Moreover, lignin and hemicellulose, along with their degradation products, can suppress cellulose hydrolysis [17,18]. Therefore, a lignocellulosic biomass with lower contents of lignin and hemicellulose is more suitable for biorefinery.

The different types of lignocellulose can influence the variable amount of inhibitors [29]. For example, more acetylation normally happens in hardwoods than in softwoods [71]. Acetic acid is produced by acetyl-group hydrolysis and has significant inhibiting effects on microbial fermentation. Moreover, acetylation on hemicellulose chains can cause surface hydrophobicity changes, inhibit hydrolases, and contribute to biomass recalcitrance [72,73]. Therefore, the lower the acetylation degree in hemicellulose was, the higher the sugar yield obtained. Due to significant variations of three components in different lignocellulose biomasses, similar pretreatment methods may generate different amounts of inhibitors with diverse inhibiting effects [37]. Major inhibitors derived from lignin are phenolic compounds such as 4-hydroxybenzoic acid, vanillin, catechol, ferulic acid, and syringic acid [19]. Inhibitors derived from hemicellulose include furfural and formic acid as the degradation products of xylose and arabinose. Inhibitors derived from cellulose include 5-hydroxy methylfurfural (5-HMF), formic acid, and levulinic acid [74]. Although the concentration of inhibitors may not be high, their inhibiting effects can be significant [45], and the inhibitory effects of the same inhibitors from different biomass types may be discrepant. For example, the concentration of phenolic compounds released from beechwood ranges between 2–21.6 mg/g after steam explosion pretreatment, and the conversion of cellulose to glucose is reduced by 5–26% [75]. In contrast, the concentration of the phenolic from maple is 5.65 mg/g after liquid hot water (LHW) pretreatment, and the hydrolysis rate is reduced by 50% [76].

In summary, the contents and structural characteristics of cellulose, hemicellulose, and lignin vary among various lignocellulosic materials. This physical and chemical nature of biomass materials affects saccharification yields and inhibitor generation in pretreatment, thereby determining biorefinery strategy selection.

3. Pretreatment Process and Its Effects on Nutrients and Inhibitors

Lignocellulose pretreatment is a crucial step for biorefineries, aiming to increase the share of the amorphous region in celluloses, promote hemicellulose degradation, and remove lignins, to enhance the susceptibility of the lignocellulosic biomass to enzymatic degradation [35]. Pretreatment methods are diverse, including acid treatment, alkaline treatment, ionic liquid treatment, deep eutectic solvent (DES) treatment, steam explosion, hydrothermal treatment, physical treatment, and biological treatment, as well as their combinations. When selecting a pretreatment method, it is important to consider the biomass type, the anticipated end products, and the economic benefits [77]. For instance, acid pretreatment is more effective in hemicellulose removal, while alkaline pretreatment is more efficient in delignification [78]. During the delignification and decomposition of hemicellulose, some pretreatment byproducts with fermentation-negative effects are generated due to severe conditions [79]. Qualitative and quantitative assessments of these byproducts are important for determining the suitability of each raw material pretreatment method, including whether hydrolysate detoxification is necessary or not for effective subsequent fermentation.

3.1. Alkaline Pretreatment

The following mechanisms for changing the structure and composition of biomass are mainly involved in alkaline pretreatment: first, exposing more celluloses and hemicelluloses by breaking down ether bonds and carbon–carbon bonds between aromatic rings and dissolving lignins; and, second, improving the accessibility of cellulose fibers to enzymes by cellulose swell, surface area increase, and crystallinity reduction. Besides lignin, alkaline pretreatment can partially dissolve hemicelluloses, although it is not as efficient as acid pretreatment, helping more celluloses be exposed to enzymatic hydrolysis and improving the overall conversion efficiency [80,81]. At last, alkaline pretreatment can cleave the ester linkages in lignocelluloses, including bonds between lignin and hemicelluloses, as well as acetyl and other ester groups in hemicelluloses, thus further reducing the degree of polymerization in lignocellulose [82].

Alkaline pretreatment dissolves only a small amount of cellulose and hemicellulose, so nearly the maximum amount of saccharides can be recovered in subsequent steps [30]. It is reported that over 85% of xylan can be extracted from corn stalks under low-alkali-concentration conditions [83]. Common alkaline reagents include NaOH, Ca(OH)$_2$, KOH, and ammonia. A high proportion of demethylated phenolics can be produced by alkaline pretreatment using NaOH [77]. Alkaline pretreatment is very effective for lignin removal from softwoods and grasses, and more holocellulose can be kept compared to acid treatment [30]. Overall, alkaline pretreatment can retain more nutrients with the formation of relatively fewer new inhibitors.

However, some issues need to be addressed about alkaline pretreatment, such as the difficult chemical recovery [84], and equipment corrosion which may result in a short lifespan and high maintenance costs. Moreover, compared to other pretreatment methods, alkaline pretreatment may require a longer time to break down the lignin structure, influencing production efficiency [85]. Conditions of high temperature and high pressure also lead to the energy consumption being increased. After alkaline pretreatment, the lignin-rich black liquor generally requires solid–liquid separation before subsequent saccharification and utilization [86]. Inhibitors in the sugar solution, besides unwashed components of black liquor, mainly include lignin fragments deposited on celluloses and LCC (lignin–carbohydrate complexes) released from the fragmented cellulose [87,88]. The effective utilization of the separated lignin is the key to the economic viability of the entire biorefinery process.

3.2. Acid Pretreatment

Acid hydrolysis is one of the most common pretreatment methods, mainly relying on inorganic acids (such as sulfuric acid, phosphoric acid, or nitric acid) or organic acids (such as formic acid, maleic acid, or oxalic acid) [61,68,89,90]. Initially, acid molecules cause the breakdown of the glucosidic bonds between cellulose and hemicelluloses, and hydrolyze hemicelluloses partially or completely into monosaccharides or smaller oligosaccharides. Meanwhile, acid pretreatment affects hydrogen bonds between celluloses and hemicelluloses. By altering these hydrogen bonds, acid pretreatment can reduce the crystallinity of celluloses, making them more accessible for enzymatic hydrolysis [82].

However, acidic environments can promote sugar degradation to generate a large number of inhibitors, such as furfural, 5-HMF, and phenolic compounds, affecting subsequent fermentation [91]. In addition to the common inhibitors above, there are other newly discovered substances, such as quinone compounds derived from phenolic compounds, severely inhibiting the growth and fermentability of various typical biorefinery fermentation strains [92]. Typically, the more severe the acid pretreatment is, the more phenolic compounds are generated, especially those with carbonyl groups, as well as acetic acid, furfural, and 5-HMF in the hydrolysate. The concentration of glucose and some oligosaccharides will also decrease due to excessive byproduct conversion when the pretreatment strength is too high [39]. Traditional detoxification methods such as water washing and biological detoxification can remove these inhibitors, but a large amount of fermentable sugars derived from pretreated lignocellulosic biomass are also lost simultaneously. Therefore, biorefinery processes that use acid pretreatment tend to develop methods by improving the inhibitor tolerance of strains for biodetoxification or constructing pathways for quinone biodegradation [93].

In addition, researchers are exploring other alternative acids with less toxicity and easier removal for biomass pretreatment. For example, trifluoroacetic acid (TFA) can obtain soluble sugars such as xylose from hemicelluloses with celluloses undegraded. Due to the easy recyclability of TFA, an additional detoxification stage is not necessary [68]. Levulinic acid (Lev) is another eco-friendly organic acid for pretreatment and can also prevent the lignins from re-polymerization [94]. More than 50% (w/w) solids are discharged in the dry acid pretreatment system [95] and all inhibitors are retained in the pretreated solids without any wastewater generated. The pretreated solids can be introduced to fungus cultures for biodetoxification, followed by simultaneous saccharification and co-fermentation for ethanol or lactic acid fermentation [96–98], significantly reducing wastewater generation during the detoxification process.

3.3. Hydrothermal Pretreatment

Hydrothermal pretreatment primarily utilizes H_3O^+ ions ionized from water under increased temperature and pressure. These autoionization products act as catalysts during pretreatment. Acetyl groups on xylan chains are cleaved and form acetic acid in the solution to trigger hemicellulose depolymerization, leaving most celluloses and lignins remaining in the pretreated solids [99]. Compared to other chemical pretreatments, hydrothermal pretreatment has the advantage of being eco-friendly and using pressurized hot water as the only solvent [100]. However, like acid pretreatment, it releases or generates several soluble inhibitors such as acetic acid, 5-HMF, and phenolics, along with some sugar degradation byproducts such as furfural and oligosaccharides [54]. These compounds significantly impact the enzymatic hydrolysis efficiency, especially phenolics, whose inhibitory effects cannot be completely relieved even with an increased enzyme load [45]. Studies have shown that 5-HMF and furfural can undergo polymerization or condensation to form pseudo-lignin [101]. These structures tend to deposit as droplets on the surface of the pretreated biomass, reduce the effective contact area for cellulase, and inhibit cellulase activity [102]. Additional steps are required for lignin removal [103], and alkali and ammonium sulfite are currently commonly used to assist in hydrothermal pretreatment for delignification [104–108]. On the other hand, the low separation efficiency of hemicellulose is reported to be a main technical challenge for hydrothermal pretreatment, and various methods and technologies have been studied to improve hemicellulose separation efficiency, such as pH pre-control [109] and metal ion catalysis [110].

3.4. High-Pressure Explosion Pretreatment

Steam explosion (SE) is a widely used physicochemical pretreatment method. It treats biomass with high-temperature and high-pressure steam, then rapidly decompresses to break down the lignin–hemicellulose barrier and effectively facilitate subsequent hydrolysis. During the SE pretreatment process, with the temperature increasing, hemicellulose degradation may result from autohydrolysis reactions and inhibitors like furfural and 5-HMF can be generated in side reactions [111–113]. Lignin also partially depolymerizes and melts at high temperatures, similar to that during hydrothermal pretreatment, but these dissolved components may be recondensed or transformed afterward [114].

Compared to SE, nitrogen explosion decompression (NED) offers a different pretreatment approach and is particularly suitable for biomass hydrolysis under mild treatment conditions. It operates at lower temperatures, applies a gentler treatment for biomass, and minimizes the generation of inhibitors. It achieves biomass explosion effects through dissolved nitrogen expanding rapidly, breaks down the lignin–hemicellulose matrix to some extent, and promotes hemicellulose dissolution [65]. NED is characterized by its potential environmental friendliness and lower operation temperatures. As an emerging technology, further evaluation is still required for its technological maturity, cost-effectiveness, and adaptability to various biomasses. Additionally, although fewer inhibitors are produced in NED, detailed generation mechanisms and control strategies are still required in further research to ensure the efficiency and reliability of NED in practical applications.

3.5. Solvent Pretreament

Organic solvent pretreatment often uses methanol, ethanol, acetone, and organic acids, and removes lignins and some hemicelluloses through solubilization. Compared to other chemical pretreatments, its advantage lies in the ability to recover relatively pure lignin [115]. Sometimes, organic acids, inorganic acids, or alkalis are added as catalysts to lower the operating temperature or increase delignification [116]. This method requires balancing the relationship between the cost and inhibitor generation. Low-boiling-point alcohols are easily recovered but require a high-pressure pretreatment process; acetone, although better at recovering saccharides, has a higher overall cost and is not feasible for large-scale production [117]. Organic acid pretreatment can be carried out under atmospheric pressure but may lead to cellulose acetylation and inhibitory components accumulation in the system. Although most organic reagents used in pretreatment can be recovered by distillation, it is challenging to ensure that they do not remain on the pretreated solids and flow into the downstream sugar solution during large-scale applications.

Compared to organic solvents, deep eutectic solvents (DESs) are stable, biodegradable, and recyclable green solvents [118]. They are composed of different molar ratios of a hydrogen bond acceptor (such as choline chloride) and a hydrogen bond donor (like lactic acid, urea, ethylene glycol, etc.) [119,120]. The DES pretreatment can increase both the digestibility and solubility of lignocellulose and reduce its resistance to enzymatic digestion; it can also selectively break ether bonds to separate lignins and celluloses, thus deconstructing plant cell walls effectively [121]. Additionally, DESs can suppress the re-polymerization of depolymerized lignin and reduce the lignin molecular weight [122]. The DES pretreatment is considered a green and cost-effective process, typically with characteristics of non-toxicity and recyclability [123]. Despite the large number of laboratory studies emerging in recent years, the efficiency, stability, recyclability, and biocompatibility of DES are still the major challenges in large-scale pretreatment under industrial conditions [124,125].

Ionic liquids (ILs) are also known as "green solvents" with the typical constitution of organic cations and organic or inorganic anions. In the lignocellulose biochemical processing, ILs improve the convertibility of biomass and increase the efficiency of enzymatic hydrolysis and fermentation by disrupting non-covalent interactions between lignocellulose components, such as hydrogen bonds between polysaccharide chains, and ether/ester bonds between lignin and carbohydrates [126,127]. Although the formation of inhibitors is diminished, the residual small amounts of ILs still have potential toxicity to enzymes and fermenting microbes [37]. Therefore, it is necessary to use excess water or antisolvent washing after ionic liquid pretreatment to remove ionic liquids, lignin, and other derivatives [126].

Generally, the economic viability of solvent-based pretreatment is determined by the solvent recovery. Most inhibitors in the subsequent enzymatic sugar solution may be from the solvent itself after efficient solvent recovery. The residue solvent affects not only operating costs but also the qualities of the produced sugar solution and other downstream products.

3.6. Other Pretreatment Techniques

Physical pretreatment techniques primarily use external mechanical or electrical forces, such as milling, ball milling, extrusion, ultrasonication, and microwave irradiation [128,129]. The mechanical pretreatments disrupt the intrinsic structure of biomass, thereby increasing the surface area, reducing crystallinity, and improving efficiency. However, mechanical forces cannot break down the chemical structures of lignins and have little effect on the degradation of hemicellulose and lignin. Ultrasonication pretreatment is based on the cavitation effect during radiation with ultrasonic energy, which produces both physical forces and chemical effects on the biomass structure. Microwave irradiation provides rapid and uniform heating effects on the lignocellulose, thus generating structural changes in biomass. However, both ultrasonication and microwave irradiation pretreatments have a high energy demand and are difficult to scale up for high-volume applications. Physical pretreatment is often used in conjunction with other pretreatment methods [128,130,131], but new sources of inhibitors are also introduced. Biological pretreatment is mainly carried out by microorganisms utilizing biomass for growth directly or by enzyme mixtures added. No inhibitors form and less energy is consumed during biological pretreatment [82]. However, compared to chemical pretreatment, it takes longer and has limited effects on facilitating subsequent saccharification.

Although numerous pretreatment methods have been developed, none of them can perfectly achieve the separation of the three major components of lignocellulose. Other compounds, such as extractives and ash, are largely lost during pretreatment. The choice of pretreatment process primarily affects the carbon sources for downstream bioconversion and inhibitor formation. Besides the total process cost, there are two more factors highly recommended for pretreatment evaluation: accelerating polysaccharides hydrolysis or not, and reducing side reactions or not for more main carbon sources remaining and fewer inhibitors generated. Generally, more control over inhibitor generation is required during acid pretreatment, while chemical reagent recovery and lignin utilization need to be addressed in the methods based on alkalis and solvents that focus on delignification.

The possible inhibitors from different pretreatments are summarized in Table 2.

Table 2. Summary of nutrient retention and inhibitor formation under various pretreatment methods.

Pretreatment Method	Nutrient Retention	Inhibitor Production	Reference
Alkaline pretreatment	Removal of lignin, partial hemicellulose; less sugar dissolution	Formic acid; acetic acid; hydroxy acid; phenols	[37,77]
Acid pretreatment	Partial or complete removal of hemicellulose; more sugar dissolution	Furfural; 5-HMF; phenols; quinones; acetic acid	[91,92]
Steam explosion	Significant dissolution of hemicellulose, minor dissolution of cellulose; less degradation of sugar	Furfural; 5-HMF; formic acid; acetic acid	[36,111–113]
Nitrogen explosion	Hemicellulose dissolution		[65]
Liquid hot water	More hemicellulose dissolved; higher sugar recovery; less cellulose loss	Furfural; 5-HMF; acetic acid; phenols; pseudo-lignin	[2,36,54]
Organic solvent	Removal of part of the hemicellulose, dissolution of lignin	Furfural; 5-HMF	[132]
Deep eutectic solvents	Removal of hemicellulose and lignin	Furfural; 5-HMF; levulinic acid	[123]
Ionic liquid	High lignin extraction rate, partial degradation of hemicellulose, possibly reduced cellulose crystallinity	Furfural; 5-HMF; weak acid	[37,126,133,134]
Physical pretreatment	Reduced cellulose crystallinity, less sugar degradation	Furfural; phenols	[36,133,135]
Biological pretreatment	High lignin degradation, low cellulose degradation, reduced sugar	Furfural; 5-HMF; organic acids	[36,133]

4. Saccharification Process

Saccharification is a key step in the biotransformation process of lignocellulose, aiming at breaking down complex polysaccharides of cellulose and hemicellulose into fermentable oligosaccharides and monosaccharides through enzymatic hydrolysis [12]. These oligosaccharides and monosaccharides are primary nutrients for subsequent microbial fermentation, providing carbon sources and energy for the production of ethanol, biofuels, or other chemicals [40,136]. The saccharification process is mainly catalyzed by various enzymes, and the efficiency and yield of saccharification are two of the most critical factors for determining the utilization rate of lignocellulosic polysaccharides. The saccharification process involves the synergistic action of multiple enzymes, primarily glycoside hydrolases (GHs) including cellulases and hemicellulases [57,137,138]. The cellulase system primarily includes the following three types of enzymes: endoglucanases (EGs), which break down the crystalline structure of cellulose microfibrils to release individual polysaccharide chains and reduce the polymerization degree [57]; exoglucanases, or cellobiohydrolases (CBHs), that cut cellulose from the free ends of polysaccharides, mainly releasing cellobiose; and β-glucosidases, which can further hydrolyze cellobiose and other oligosaccharides to produce individual glucose molecules. These enzymes work together to break down celluloses into glucose units [12]. Unlike celluloses, the structure of hemicelluloses is very complex and requires a wider variety of enzymes for complete degradation, mainly including xylanases, arabinoxylanases, mannanases, β-xylosidases, and esterases [12,138]. These enzymes work synergistically and degrade polysaccharides specifically according to the chemical composition and different linkages in hemicelluloses, releasing sugar monomers such as xyloses and mannoses [13]. The efficiency of these enzymes is influenced by various factors, including the various biomass sources, the chemical composition after biomass pretreatment, the source of enzymes, the ratios of different enzymes, and the enzyme catalytic activities [13,139,140]. In addition to glycoside hydrolases, recent studies discovered that adding auxiliary enzymes such as lytic polysaccharide monooxygenases (LPMOs) during the cellulose hydrolysis process can significantly improve hydrolysis efficiency and reduce the enzyme loadings required for saccharification [141–144].

Current enzyme systems for biomass saccharification primarily include free enzyme systems and cellulosome systems (Figure 2). Most lignocellulose-degrading bacteria and fungi in nature can secrete a variety of cellulases and hemicellulases. Some fungi, such as *Trichoderma reesei* and *Penicillium oxalicum*, possess a high cellulase secretion system and have been modified to become main production strains of cellulases for industrial-scale use [145–147]. Due to the complex structure of lignocellulose, currently, no microbial enzyme can independently decompose all components for industrial application, so many studies are focusing on designing optimal enzyme mixtures to hydrolyze the pretreated biomass effectively [13]. Most industrial demonstration plants for biomass saccharification currently use processes based on free enzyme formulation [148–151]. Despite the crucial role and multiple advantages of these free enzymes, such as the specific activity, mild reaction conditions, and environmental friendliness, the high enzyme usage in the saccharification process, even with recent significant improvements in enzyme production, still represents a substantial cost in the biorefinery process. This enzyme cost is one of the main factors hindering the economic viability of biorefinery processes [152,153].

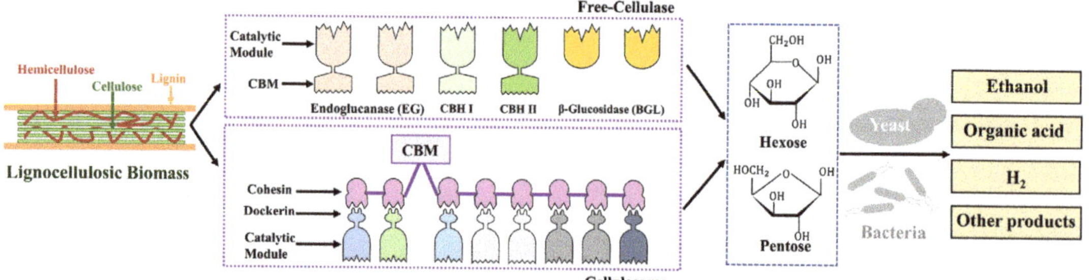

Figure 2. Free-cellulase- and -cellulosome-based saccharification.

In addition to free enzymes produced by fungi and bacteria [154], there exists in nature a large multi-enzyme complex produced by anaerobic microorganisms for degrading lignocellulose, known as cellulosome, composed of non-catalytic proteins (scaffoldins) and a variety of glycoside hydrolases [155,156]. The cellulosome is one of the most efficient lignocellulose degradation systems known in nature, with a far higher degradation efficiency than that of free cellulase systems [154,157–159]. The cellulosome assembles various types of cellulases and hemicellulases into a large complex through non-covalent interactions between scaffoldins and enzymes [160,161], creating synergistic and proximity effects among enzymes. Cellulosome also binds to substrates through carbohydrate-binding modules (CBMs), forming synergistic enzyme-substrate interactions. Furthermore, the cellulosome attaches to the bacterial cell wall through cell-wall-binding modules on scaffoldins, creating synergistic interactions between the enzymes and cells [154,158,162]. These multi-level synergistic actions collectively enhance the efficiency of lignocellulose degradation. The modules within the cellulosome are connected by flexible linkers, allowing the cellulosome to undergo conformational changes according to the substrate, thus degrading the substrate better [162]. Additionally, bacteria express genes of cellulosomal components dynamically based on the type of substrate and adapt to the different substrate compositions [162–166]. Compared to free cellulases, cellulosomes not only exhibit a stronger tolerance to chemical inhibitors such as formate, lactate, and furfural present in the hydrolysate but also show a higher ethanol tolerance and thermostability [157].

The biocatalytic saccharification process can be inhibited by various factors, and the most common one is product inhibition, which is that the accumulation of product sugars inhibits enzyme activity. For instance, cellobiose and cello-oligosaccharide inhibit the activity of various cellulases, and β-glucosidases are susceptible to glucose inhibition [167,168]. Typically, the cellobiose inhibition of cellulases is more severe than the glucose inhibition of β-glucosidases. To relieve feedback inhibition and promote cellulose saccharification, strategies often applied are the direct supplementation of exogenous β-glucosidases with a high glucose tolerance or using recombinant strains secreting these enzymes [169,170]. Cellobiose and xylan also inhibit cellobiohydrolases and cellulases, respectively. Notably, studies have found that, under certain conditions, xylose at low concentrations can stimulate β-glucosidases with doubled hydrolytic activity, while the binding of cellobiose to the active site of the enzyme may be interfered by high xylose concentrations [171]. This phenomenon implies the need for precise control over the effects of different components on enzyme activity for saccharification condition optimization and efficiency improvement. In the cellulosome system, it was found that soluble lignin and arabinoxylan released during lignocellulose hydrolysis can interact with key exoglucanases [137,172]. In biorefinery strategies that integrate the saccharification process with downstream fermentation (such as simultaneous saccharification and fermentation or consolidated bioprocessing), the products of downstream fermentation, such as bioethanol, can significantly inhibit cellulase activity as their concentration increases, sometimes even leading to enzyme deactivation [173].

Since the saccharification process is the core step to generate nutrients for downstream fermentation in lignocellulose biorefining, various current lignocellulose bioconversion strategies have been developed based on biocatalyst production methods and their integration with upstream and downstream processes, including off-site and on-site saccharification [40]. Off-site saccharification strategies are among the earliest proposed biorefining technologies, and separate hydrolysis and fermentation (SHF) and simultaneous saccharification and fermentation (SSF) are the most commonly applied, both of which use free cellulases from fungi as biocatalysts. Considering the joint utilization of pentose and hexose downstream, separate hydrolysis co-fermentation (SHCF) and simultaneous saccharification co-fermentation (SSCF) strategies have been further developed [174]. On-site saccharification strategies are new approaches developed for their low operating cost, especially the enzyme production cost by avoiding enzyme production and separation and integrating the enzyme production and saccharification steps into one single step, mainly including consolidated bioprocessing (CBP) and consolidated bio-saccharification (CBS) [40]. On-site saccharification requires a single enzyme-producing strain to produce all enzymes needed for saccharification, thus demanding a higher capacity of the strain for lignocellulose degradation. In addition, the tolerance of the strain is also required to be higher because of the multi-step integration.

The lignocellulosic hydrolysates produced by on-site and off-site strategies are quite different for downstream fermentation. In off-site saccharification, free enzymes are used as catalysts for the saccharification process, so the nutrients and inhibitors in the hydrolysates mostly come from the lignocellulosic substrates themselves and pretreatment processes, while, in on-site saccharification, the hydrolysates are also a fermentation medium for enzyme-producing strains [40]; therefore, besides the nutrients and inhibitors from the substrates and pretreatment processes, metabolic products from the strains are also contained in hydrolysates, and both nutrients and potential inhibitors may be included for downstream fermentation [3]. The differences among lignocellulose hydrolysis in various saccharification approaches will be further discussed in the next section.

5. Lignocellulose Hydrolysate in Different Biorefinery Strategies

The production cost of biofuels or fermentable sugars as intermediate platform products from biorefinery requires competitiveness with fossil fuels or starch-based sugars on the market. Therefore, reducing the operational cost of biorefining is the primary concern for strategy development. To overcome the techno-economic challenges of biorefinery, several different strategies have been developed, including separated hydrolysis and fermentation (SHF), simultaneous saccharification and fermentation (SSF), consolidated bioprocessing (CBP), and consolidated bio-saccharification (CBS) [139]. These strategies, by separating or combining different steps of biorefinery, result in variations of obtained lignocellulosic hydrolysates (Figure 3). Such differences further affect the performance of downstream fermentation, which is a critical factor that needs to be considered in the designs of downstream fermentation products and processes.

SHF is a strategy in which each step—pretreatment, enzyme production, saccharification, and fermentation—is carried out separately. The advantage of SHF is that the saccharification and fermentation processes can be conducted under their optimal conditions respectively, thus achieving higher yields [175]. However, the main disadvantage is that each step is performed separately, so a high overall process cost is generated owing to specific operational costs for each step, and additional costs are also introduced by the connections between steps. The lignocellulosic hydrolysate produced by SHF, due to the precise control over each step, typically contains high concentrations of fermentable sugars [176,177]. The inhibitors for fermentation primarily originate from raw materials and the pretreatment process, allowing the generation of nutrients and inhibitors to be better controlled. High sugar yields can be achieved in SHF, with most of the pentoses from hemicellulose retained in the fermentable sugars after saccharification. Therefore, the main consideration for downstream fermentation in SHF is the joint utilization of pentoses and hexoses for the polysaccharide nutrients in lignocelluloses to be fully converted and utilized [42]. The strategy that enables the joint utilization of pentoses and hexoses in SHF is also known as separate hydrolysis co-fermentation (SHCF) [178].

SSF integrates saccharification and fermentation simultaneously in a single bioreactor, reducing operating costs while avoiding the accumulation of high concentrations of sugars in the hydrolysate and related inhibitory effects. However, the optimal conditions for saccharification and fermentation are usually different, especially those of temperature and pH. The optimum temperature for enzymatic hydrolysis is usually higher than that for fermentation, thus typically leading to reduced overall efficiency. Meanwhile, metabolites generated during the fermentation process may inhibit enzyme activities in the saccharification process [31]. To address these issues, many works of research are focusing on enzyme optimization and fermentative microorganism screening to ensure both of them can work efficiently under the same conditions [179–183]. Co-cultures of multiple microorganism strains are also often applied in SSF [184–186]. This strategy is advantageous for more a complex metabolite production and is beneficial for strain growth and fermentation processes due to the synergistic interactions between different strains [28]. However, some studies have also shown that the co-culture strategy in SSF may lead to reduced yields due to various reasons (such as nutrient competition and the accumulation of metabolic products) [27].

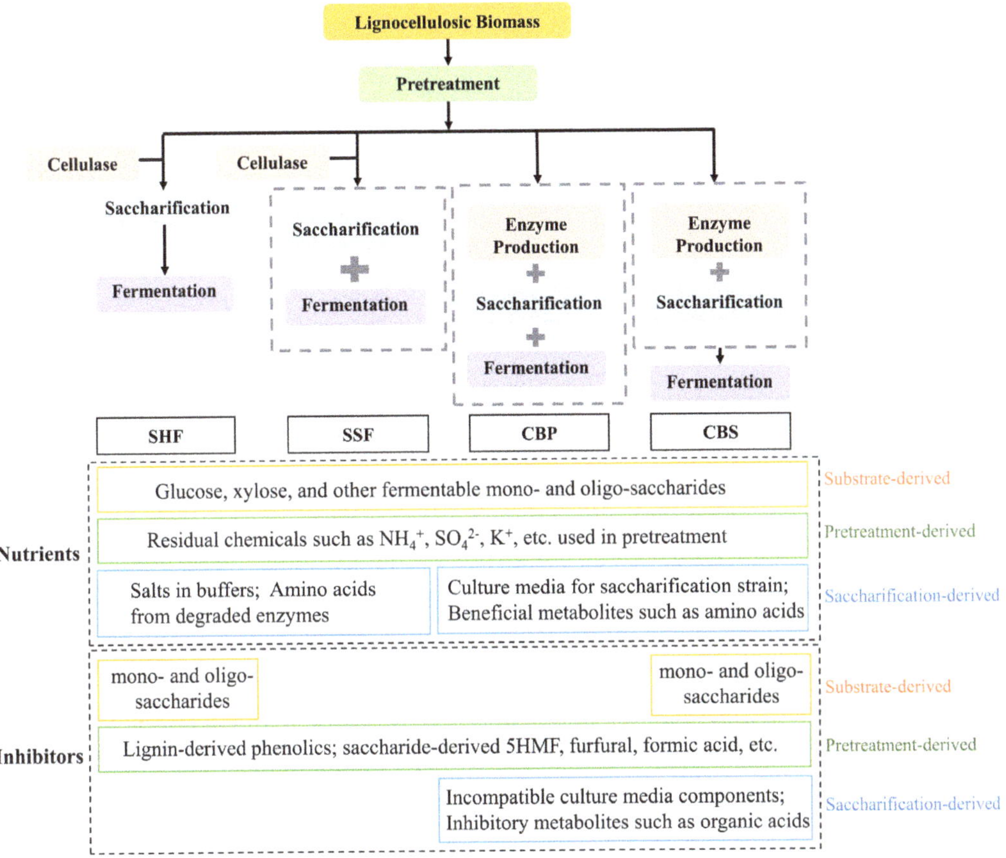

Figure 3. Nutrients and inhibitors in different biorefining strategies. SHF: separate hydrolysis and fermentation; SSF: simultaneous saccharification and fermentation; CBP: consolidated bioprocessing; and CBS: consolidated bio-saccharification.

In SSF, due to the fermentation step integrated, small molecule sugars produced from the hydrolysis of celluloses and hemicelluloses are directly utilized by fermentation strains; thus, lignocellulosic hydrolysates cannot be generated with high sugar concentrations. However, like SHF, SSF also faces issues of inhibitor production during the pretreatment process and joint utilization of pentoses and hexoses. Regarding inhibitors, some studies focus on improving pretreatment methods or using inhibitor-resistant or inhibitor-utilizing strains [187]. Other processes such as simultaneous saccharification and co-fermentation (SSCF) have been developed based on the SSF to achieve effective xylose utilization and less product inhibition [10,188,189]. The approach known as fed-batch delayed SSF (fed-batch dSSF) has also been developed: dSSF (also known as PSSF, pre-hydrolysis SSF [190]) is performed in the first bioreactor by using cellulases for pre-saccharification and followed by simultaneous saccharification and fermentation, while, in the second bioreactor, a cultivation medium was used only for enzymatic hydrolysis. When glucose is depleted in the first bioreactor, the medium is fed from the second bioreactor. This process is primarily designed to avoid early carbon deficiency in SSF and to enhance the hydrolysis rate of cellulase into glucose by integrating a pre-saccharification step at the optimum temperature for cellulose decomposition [41].

Consolidated bioprocessing (CBP) integrates the enzyme production, saccharification, and fermentation steps into a single system, thus significantly reducing the costs of enzymes and the overall process. The core of CBP is to develop microorganisms with the capability of degrading lignocellulose and conducting downstream fermentation simultaneously [191]. Similar to SSF, hydrolysates with high sugar concentrations do not exist in CBP, but issues about inhibitors produced from substrates and the pretreatment process, as well as inhibitory effects of fermentation products on microorganisms and enzymes, have to be faced in CBP. The whole CBP process is carried out by living microbial cells, and the derived inhibitors from pretreatment and raw materials may interfere with the host cell membrane integrity, protein synthesis, cell growth, and target product production [32]. In the CBP system, since all steps occur in the same bioreactor, the efficient consumption of the sugar mixture or the biomass hydrolysate is particularly critical. For instance, the hydrolysis of hemicelluloses produces xyloses, arabinoses, galactoses, and rhamnoses, but many host microorganisms in CBP cannot consume these sugars, leading to a carbon catabolite repression [192]. In such cases, CBP hosts are required to be modified by genetic engineering or metabolic engineering [32,193,194]. Another approach is the co-culture strategy by introducing other strains to metabolize these sugars and fully utilizing various carbohydrates in the lignocellulosic hydrolysates, thus improving the hydrolysis efficiency of upstream strains [191,195,196]. However, this approach has its challenges, as the growth conditions in the co-culture system should meet the requirements for all different microorganisms (such as pH, oxygen, and temperature) and the growth of one species does not have toxic or inhibitory effects on others [197]. In lignocellulose biorefinery, most cellulases exhibit optimal enzymatic activity at higher temperatures; thus, downstream strains are required to adapt to such a high-temperature environment for growth and fermentation. Cellulosomes of *Clostridium thermocellum* exhibit a higher efficiency in degrading woody and herbaceous cellulose materials than commercial fungal cellulases [198], although their fermenting property is unsatisfactory. However, the co-culture of *Clostridium thermocellum* DSM1313 and *Thermoanaerobacterium thermosaccharolyticum* MJ1 can produce more hydrogen [196,199]. This is due to relieving the inhibition of DSM1313, improving substrate degradation, and enhancing electron transfer activity. Additionally, besides substrate utilization, the co-culture system can also improve the low tolerance of organic acids (such as acetic acid, formic acid, and lactic acid) and ethanol produced during fermentation with *Clostridium thermocellum* [33,34]. Overall, more complex modifications of strains are required in the CBP strategy to make full use of hydrolysis products and tolerate various inhibitors produced during pretreatment and fermentation.

Consolidated bio-saccharification (CBS) is a strategy that separates the fermentation step in CBP while keeping the enzyme production and saccharification process completed by a single strain [40]. The key advantage of the CBS strategy is that it integrates enzyme production with saccharification to minimize the production cost of cellulases, overcoming the major bottleneck in lignocellulosic biorefinery. Furthermore, it also separates the cellulose hydrolysis process from the downstream microbial fermentation process, avoiding the compromise of different reaction conditions required in both processes. Since the fermentation process is separated, the CBS strategy, like the SHF strategy, can produce hydrolysates containing high concentrations of fermentable sugars, while cellulases may suffer from the feedback inhibition of monosaccharides or oligosaccharides in the hydrolysates. Therefore, as in the SHF strategy, the product feedback inhibition by cellobiose and cello-oligosaccharide can be released by the addition of β-glucosidases (BGLs) with both a high enzymatic activity and high glucose tolerance in the CBS strategy [170]. However, whole-cell catalysts are used for the saccharification process in the CBS strategy, which is different from the SHF strategy, so the resulted lignocellulosic hydrolysates are also the fermentation broth of the CBS strain, containing more complex components than those of the SHF strategy. These components mainly come from the fermentation medium and bacterial metabolites, so most of them may act as nutrients for downstream fermentation and may be beneficial for the downstream fermentation process. For example, Liu et al. [26] successfully achieved the fermentation of pullulan using condensed CBS sugar liquor without any nutrient supplementation. Similarly, lactic acids were produced by directly inoculating lactic acid bacteria (LAB) strains into the CBS hydrolysates to initiate the fermentation process [3]. On the other hand, since the strains used in CBS are anaerobic *Clostridium thermocellum* [200], certain amounts of metabolic intermediates such as lactic acid and acetic acid are produced, which may inhibit some anaerobic fermentation processes for bioenergy production, such as acetone–butanol–ethanol (ABE) fermentation [157,201]. Therefore, lignocellulosic hydrolysates in the CBS strategy require further analysis of their components, integration with downstream fermentation processes, nutrients needed for downstream fermentation, and potential inhibitors, thereby optimizing the whole CBS process and downstream fermentation processes for the best-matched optimal process.

In this section, we have analyzed the advantages and limitations of various biorefining strategies in terms of nutrient retention and inhibitor control. Overall, SHF allows enzyme production, saccharification, and fermentation under optimal conditions, which can significantly preserve the nutrients from lignocellulose (sugar yield) and independently optimize each step for minimizing inhibitor accumulation or residues. However, the operational cost is high, making techno-economic viability the primary challenge. The SSF strategy effectively integrates the saccharification and fermentation processes, reducing process costs, but it also faces challenges such as high enzyme costs and difficulties in matching saccharification with fermentation. The CBP strategy, theoretically, can achieve lower costs but has extremely high requirements for microbial strains, and there is still a long way to go to develop technologically and economically viable CBP strains. The CBS strategy combines the low-cost advantage of CBP with the high yield of SHF, but the produced saccharification liquid has complex components, and it is necessary to develop compatible downstream strains.

The overall consideration of the biorefinery strategy and process to maximize nutrient retention and control inhibitors at a low cost is critical for the techno-economic assessment (TEA) and the life cycle assessment (LCA) of different biorefinery scenarios [202–205]. Due to the long research history and the establishment of many demonstration pilot plants of off-site biorefinery approaches, there are many TEA and LCA analyses of SHF and SSF strategies, as well as the specific steps of them such as feedstocks and pretreatments [204,206–213]. The TEA and LCA analyses of on-site approaches are relatively less, but limited studies have shown that the CBP strategy has significant advantages in feasibility and sustainability compared with SHF/SSF strategies [214–216]. As we analyzed in this paper, both nutrients and inhibitors will run through the entire biorefinery process, so any improvements to a single step will require a further TEA and LCA analysis of the entire biorefinery, which needs to be greatly strengthened in future research.

6. Conclusions and Perspective

This article provides an in-depth analysis of how different steps in lignocellulose biotransformation affect the nutrients and inhibitors in lignocellulosic hydrolysates, highlighting the differences in raw material selection, pretreatment methods, and biorefinery strategies in terms of nutrient production and inhibitor control. We found that, although each strategy has its unique advantages, it also faces various challenges such as cost, efficiency, inhibitory product generation, and difficulties in strain development. A vast amount of research focuses on nutrient retention (increasing fermentable sugar yields or conversion rate), the control or removal of inhibitors, and strain development for specific strategies, but there are only limited studies that considered the overall compatibility of steps in the entire lignocellulosic bioconversion process. Our analysis shows that both nutrients and inhibitors can be produced throughout all biorefinery stages, and, at the same time, each subsequent step is closely related to the nutrient retention and inhibitor generation in the previous step. To achieve the techno-economic viability of lignocellulosic biorefining, it is necessary to consider all steps comprehensively, design the most suitable biorefining strategy based on the characteristics of the raw materials and final target products, and optimize the whole process to maximize nutrient retention and conversion and minimize inhibitor generation.

Therefore, we propose that future research on lignocellulosic biorefinery should focus more on matching the overall biorefinery strategy and process, especially the compatibility of the pretreatment, enzyme production, and fermentation strain development, to maximize nutrient retention and control inhibitors at a lower cost. For example, pretreatment technologies should be developed based on the requirements of downstream saccharification and fermentation processes. It is necessary to analyze the role of residual chemicals or byproducts during pretreatment more meticulously and thoroughly because they can be inhibitors for subsequent saccharification and fermentation, act as activators for enzyme activity, or provide nutrients for fermentation. For instance, nitrogen sources required for the growth of downstream fermentation microorganisms often need to be added additionally. The ammonium salts, such as ammonia or ammonium sulfite, from nitrogen-containing pretreatment processes, may move into the microbial conversion stage with residual lignin, but their stimulative or inhibitory effects on fermentation still require further studies. Most previous research on nutrient retention, inhibitor control, or removal focused on off-site strategies (SHF and SSF), while very limited studies target more recent on-site strategies (CBP and CBS). *Clostridium thermocellum* is currently used in on-site strategies, whose core of the saccharification process is their secreted cellulosomes. However, research about the effects of various sugars, residues, and byproducts from pretreatment on the activity and stability of *Clostridium thermocellum* with their cellulosomes is still very few and should be focused on in the future. For the CBS strategy first developed in our lab, we are committed to the investigation of the overall compatibility of steps in the entire lignocellulosic bioconversion process. For example, the composition of the produced saccharification liquid is complex, and it should be developed in collaboration with downstream processes. The medium used in the CBS process for enzyme production and saccharification contains various nutrients required by *Clostridium thermocellum* growing, and most of them can remain in the final saccharification liquid in various forms, along with some products generated from *Clostridium thermocellum* metabolism. More detailed studies need to be carried out about the suitability between these components and downstream fermentation strains, which are ongoing in our lab for the CBS strategy. Moreover, the salts in the CBS hydrolysates come from not only the medium but also ash in the substrate dissolved during pretreatment and saccharification. These salts have potentially significant effects on downstream fermentation equipment and processes. The choice of medium in an on-site saccharification strategy not only determines the production status of the enzyme-producing microorganisms but also affects crucial carbon source supply. Moreover, the inhibitory effects of metabolites and the changes in nutrient components during the enzyme production process of the on-site saccharification strategy also require an overall assessment of their impacts on the performance of production strains. More TEA and LCA studies for different biorefinery approaches, particularly the more recent on-site strategies (CBP and CBS), with improved nutrient retention and inhibitor control are needed to validate their feasibility and sustainability.

Author Contributions: Conceptualization, Y.F. and J.X.; writing—original draft preparation, Y.W., Y.Z. and Y.F.; writing—review and editing, Q.C., Y.F. and J.X.; funding acquisition, Y.Z., Q.C., Y.F. and J.X. All authors have read and agreed to the published version of the manuscript.

Funding: This work was funded by the National Key Research and Development Program of China (2023YFC3402300 to Y.F. and 2023YFB4203502 to Y.Z.), the Training Program for Young Teaching Backbone Talents, USTB (2302020JXGGRC-005 to J.X.), the Major Education and Teaching Reform

Research Project, USTB (grant number JG2021ZD01 and JG2023ZD05 to J.X.), the National Natural Science Foundation of China (32070125 to Y.F. and 32170051 to Q.C.), QIBEBT (QIBEBT/SEI/QNESL S202302 to Y.F.), the QIBEBT International Cooperation Project (QIBEBT ICP202304 to Y.F.), and the State Key Laboratory of Microbial Technology Open Projects Fund (M2022-01 to Y.F.).

Institutional Review Board Statement: Not applicable.

Informed Consent Statement: Not applicable.

Data Availability Statement: No new data were created or analyzed in this study. Data sharing is not applicable to this article.

Conflicts of Interest: The authors declare no conflicts of interest.

References

1. Abraham, A.; Mathew, A.K.; Park, H.; Choi, O.; Sindhu, R.; Parameswaran, B.; Pandey, A.; Park, J.H.; Sang, B.-I. Pretreatment Strategies for Enhanced Biogas Production from Lignocellulosic Biomass. *Bioresour. Technol.* **2020**, *301*, 122725. [CrossRef]
2. Basak, B.; Kumar, R.; Bharadwaj, A.V.S.L.S.; Kim, T.H.; Kim, J.R.; Jang, M.; Oh, S.-E.; Roh, H.-S.; Jeon, B.-H. Advances in Physicochemical Pretreatment Strategies for Lignocellulose Biomass and Their Effectiveness in Bioconversion for Biofuel Production. *Bioresour. Technol.* **2023**, *369*, 128413. [CrossRef]
3. Liu, Y.-J.; Zhang, Y.; Chi, F.; Chen, C.; Wan, W.; Feng, Y.; Song, X.; Cui, Q. Integrated Lactic Acid Production from Lignocellulosic Agricultural Wastes under Thermal Conditions. *J. Environ. Manag.* **2023**, *342*, 118281. [CrossRef]
4. Jia, Q.; Zhang, H.; Zhao, A.; Qu, L.; Xiong, W.; Alam, M.A.; Miao, J.; Wang, W.; Li, F.; Xu, J.; et al. Produce D-Allulose from Non-Food Biomass by Integrating Corn Stalk Hydrolysis with Whole-Cell Catalysis. *Front. Bioeng. Biotechnol.* **2023**, *11*, 1156953. [CrossRef]
5. He, T.; Zhang, D.; Chen, W.; Liu, Z.; Zhao, R.; Li, J.; Wu, J.; Wang, Z.; Wu, J. Synergistic Oxidation-Reforming of Biomass for High Quality Syngas Production Based on a Bifunctional Catalyst. *Green Carbon* **2024**, *2*, 118–123. [CrossRef]
6. Seo, M.W.; Lee, S.H.; Nam, H.; Lee, D.; Tokmurzin, D.; Wang, S.; Park, Y.-K. Recent Advances of Thermochemical Conversion Processes for Biorefinery. *Bioresour. Technol.* **2022**, *343*, 126109. [CrossRef]
7. Dhakal, N.; Acharya, B. Syngas Fermentation for the Production of Bio-Based Polymers: A Review. *Polymers* **2021**, *13*, 3917. [CrossRef]
8. Harahap, B.M.; Ahring, B.K. Acetate Production from Syngas Produced from Lignocellulosic Biomass Materials along with Gaseous Fermentation of the Syngas: A Review. *Microorganisms* **2023**, *11*, 995. [CrossRef]
9. Yang, Z.; Leero, D.D.; Yin, C.; Yang, L.; Zhu, L.; Zhu, Z.; Jiang, L. Clostridium as Microbial Cell Factory to Enable the Sustainable Utilization of Three Generations of Feedstocks. *Bioresour. Technol.* **2022**, *361*, 127656. [CrossRef]
10. Patel, A.; Shah, A.R. Integrated Lignocellulosic Biorefinery: Gateway for Production of Second Generation Ethanol and Value Added Products. *J. Bioresour. Bioprod.* **2021**, *6*, 108–128. [CrossRef]
11. Xu, L.-H.; Ma, C.-Y.; Wang, P.-F.; Xu, Y.; Shen, X.-J.; Wen, J.-L.; Yuan, T.-Q. Conversion of Control and Genetically-Modified Poplar into Multi-Scale Products Using Integrated Pretreatments. *Bioresour. Technol.* **2023**, *385*, 129415. [CrossRef]
12. Lu, H.; Xue, M.; Nie, X.; Luo, H.; Tan, Z.; Yang, X.; Shi, H.; Li, X.; Wang, T. Glycoside Hydrolases in the Biodegradation of Lignocellulosic Biomass. *3 Biotech* **2023**, *13*, 402. [CrossRef]
13. Adsul, M.; Sandhu, S.K.; Singhania, R.R.; Gupta, R.; Puri, S.K.; Mathur, A. Designing a Cellulolytic Enzyme Cocktail for the Efficient and Economical Conversion of Lignocellulosic Biomass to Biofuels. *Enzym. Microb. Technol.* **2020**, *133*, 109442. [CrossRef]
14. Huang, C.; Jiang, X.; Shen, X.; Hu, J.; Tang, W.; Wu, X.; Ragauskas, A.; Jameel, H.; Meng, X.; Yong, Q. Lignin-Enzyme Interaction: A Roadblock for Efficient Enzymatic Hydrolysis of Lignocellulosics. *Renew. Sustain. Energy Rev.* **2022**, *154*, 111822. [CrossRef]
15. Milagres, A.M.F.; Carvalho, W.; Ferraz, A. Topochemistry, Porosity and Chemical Composition Affecting Enzymatic Hydrolysis of Lignocellulosic Materials. In *Routes to Cellulosic Ethanol*; Buckeridge, M.S., Goldman, G.H., Eds.; Springer: New York, NY, USA, 2010; pp. 53–72, ISBN 978-0-387-92740-4.
16. da Silva Perez, D.; Dupont, C.; Guillemain, A.; Jacob, S.; Labalette, F.; Briand, S.; Marsac, S.; Guerrini, O.; Broust, F.; Commandre, J.-M. Characterisation of the Most Representative Agricultural and Forestry Biomasses in France for Gasification. *Waste Biomass Valor.* **2015**, *6*, 515–526. [CrossRef]
17. Lai, C.; Yang, C.; Jia, Y.; Xu, X.; Wang, K.; Yong, Q. Lignin Fractionation to Realize the Comprehensive Elucidation of Structure-Inhibition Relationship of Lignins in Enzymatic Hydrolysis. *Bioresour. Technol.* **2022**, *355*, 127255. [CrossRef]
18. Malgas, S.; Kwanya Minghe, V.M.; Pletschke, B.I. The Effect of Hemicellulose on the Binding and Activity of Cellobiohydrolase I, Cel7A, from *Trichoderma reesei* to Cellulose. *Cellulose* **2020**, *27*, 781–797. [CrossRef]
19. Cunha, J.T.; Romaní, A.; Costa, C.E.; Sá-Correia, I.; Domingues, L. Molecular and Physiological Basis of *Saccharomyces cerevisiae* Tolerance to Adverse Lignocellulose-Based Process Conditions. *Appl. Microbiol. Biotechnol.* **2019**, *103*, 159–175. [CrossRef]
20. Wang, Y.; Cao, W.; Luo, J.; Wan, Y. Exploring the Potential of Lactic Acid Production from Lignocellulosic Hydrolysates with Various Ratios of Hexose versus Pentose by *Bacillus coagulans* IPE22. *Bioresour. Technol.* **2018**, *261*, 342–349. [CrossRef]
21. Han, Y.; Chen, H. Synergism between Corn Stover Protein and Cellulase. *Enzym. Microb. Technol.* **2007**, *41*, 638–645. [CrossRef]

22. Jin, C.; Li, J.; Huang, Z.; Han, X.; Bao, J. Engineering *Corynebacterium glutamicum* for Synthesis of Poly(3-Hydroxybutyrate) from Lignocellulose Biomass. *Biotechnol. Bioeng.* **2022**, *119*, 1598–1613. [CrossRef]
23. Latif, A.A.; Harun, S.; Shaiful, S.M.; Markom, M.; Jahim, J. Ammonia-Based Pretreatment for Ligno-Cellulosic Biomass Conversion—An Overview. *J. Eng. Sci. Technol.* **2018**, *13*, 1595–1620.
24. Xu, F. Chapter 2—Structure, Ultrastructure, and Chemical Composition. In *Cereal Straw as a Resource for Sustainable Biomaterials and Biofuels*; Sun, R.-C., Ed.; Elsevier: Amsterdam, The Netherlands, 2010; pp. 9–47, ISBN 978-0-444-53234-3.
25. He, Q.; Hemme, C.L.; Jiang, H.; He, Z.; Zhou, J. Mechanisms of Enhanced Cellulosic Bioethanol Fermentation by Co-Cultivation of *Clostridium* and *Thermoanaerobacter* spp. *Bioresour. Technol.* **2011**, *102*, 9586–9592. [CrossRef]
26. Liu, G.; Zhao, X.; Chen, C.; Chi, Z.; Zhang, Y.; Cui, Q.; Chi, Z.; Liu, Y.-J. Robust Production of Pigment-Free Pullulan from Lignocellulosic Hydrolysate by a New Fungus Co-Utilizing Glucose and Xylose. *Carbohydr. Polym.* **2020**, *241*, 116400. [CrossRef]
27. Xia, M.; Zhang, X.; Xiao, Y.; Sheng, Q.; Tu, L.; Chen, F.; Yan, Y.; Zheng, Y.; Wang, M. Interaction of Acetic Acid Bacteria and Lactic Acid Bacteria in Multispecies Solid-State Fermentation of Traditional Chinese Cereal Vinegar. *Front. Microbiol.* **2022**, *13*, 964855. [CrossRef]
28. Intasit, R.; Cheirsilp, B.; Louhasakul, Y.; Thongchul, N. Enhanced Biovalorization of Palm Biomass Wastes as Biodiesel Feedstocks through Integrated Solid-State and Submerged Fermentations by Fungal Co-Cultures. *Bioresour. Technol.* **2023**, *380*, 129105. [CrossRef]
29. Menegon, Y.A.; Gross, J.; Jacobus, A.P. How Adaptive Laboratory Evolution Can Boost Yeast Tolerance to Lignocellulosic Hydrolyses. *Curr. Genet.* **2022**, *68*, 319–342. [CrossRef]
30. Sharma, S.; Nandal, P.; Arora, A. Ethanol Production from NaOH Pretreated Rice Straw: A Cost Effective Option to Manage Rice Crop Residue. *Waste Biomass Valor.* **2019**, *10*, 3427–3434. [CrossRef]
31. Zhang, W.; Zhang, X.; Lei, F.; Jiang, J. Co-Production Bioethanol and Xylooligosaccharides from Sugarcane Bagasse via Autohydrolysis Pretreatment. *Renew. Energy* **2020**, *162*, 2297–2305. [CrossRef]
32. Sharma, J.; Kumar, V.; Prasad, R.; Gaur, N.A. Engineering of *Saccharomyces cerevisiae* as a Consolidated Bioprocessing Host to Produce Cellulosic Ethanol: Recent Advancements and Current Challenges. *Biotechnol. Adv.* **2022**, *56*, 107925. [CrossRef]
33. Awate, B.; Steidl, R.J.; Hamlischer, T.; Reguera, G. Stimulation of Electro-Fermentation in Single-Chamber Microbial Electrolysis Cells Driven by Genetically Engineered Anode Biofilms. *J. Power Sources* **2017**, *356*, 510–518. [CrossRef]
34. Singh, N.; Mathur, A.S.; Tuli, D.K.; Gupta, R.P.; Barrow, C.J.; Puri, M. Cellulosic Ethanol Production via Consolidated Bioprocessing by a Novel Thermophilic Anaerobic Bacterium Isolated from a Himalayan Hot Spring. *Biotechnol. Biofuels* **2017**, *10*, 73. [CrossRef] [PubMed]
35. Cantero, D.; Jara, R.; Navarrete, A.; Pelaz, L.; Queiroz, J.; Rodríguez-Rojo, S.; Cocero, M.J. Pretreatment Processes of Biomass for Biorefineries: Current Status and Prospects. *Annu. Rev. Chem. Biomol. Eng.* **2019**, *10*, 289–310. [CrossRef] [PubMed]
36. Alvira, P.; Tomás-Pejó, E.; Ballesteros, M.; Negro, M.J. Pretreatment Technologies for an Efficient Bioethanol Production Process Based on Enzymatic Hydrolysis: A Review. *Bioresour. Technol.* **2010**, *101*, 4851–4861. [CrossRef] [PubMed]
37. Jönsson, L.J.; Martín, C. Pretreatment of Lignocellulose: Formation of Inhibitory by-Products and Strategies for Minimizing Their Effects. *Bioresour. Technol.* **2016**, *199*, 103–112. [CrossRef] [PubMed]
38. Bhatia, S.K.; Jagtap, S.S.; Bedekar, A.A.; Bhatia, R.K.; Patel, A.K.; Pant, D.; Rajesh Banu, J.; Rao, C.V.; Kim, Y.-G.; Yang, Y.-H. Recent Developments in Pretreatment Technologies on Lignocellulosic Biomass: Effect of Key Parameters, Technological Improvements, and Challenges. *Bioresour. Technol.* **2020**, *300*, 122724. [CrossRef] [PubMed]
39. Jiang, X.; Zhai, R.; Leng, Y.; Deng, Q.; Jin, M. Understanding the Toxicity of Lignin-Derived Phenolics towards Enzymatic Saccharification of Lignocellulose for Rationally Developing Effective in-Situ Mitigation Strategies to Maximize Sugar Production from Lignocellulosic Biorefinery. *Bioresour. Technol.* **2022**, *349*, 126813. [CrossRef] [PubMed]
40. Liu, Y.-J.; Li, B.; Feng, Y.; Cui, Q. Consolidated Bio-Saccharification: Leading Lignocellulose Bioconversion into the Real World. *Biotechnol. Adv.* **2020**, *40*, 107535. [CrossRef] [PubMed]
41. Paulová, L.; Patáková, P.; Rychtera, M. High Solid Fed-Batch SSF with Delayed Inoculation for Improved Production of Bioethanol from Wheat Straw. *Fuel* **2014**, *122*, 294–300. [CrossRef]
42. Liu, C.-G.; Xiao, Y.; Xia, X.-X.; Zhao, X.-Q.; Peng, L.; Srinophakun, P.; Bai, F.-W. Cellulosic Ethanol Production: Progress, Challenges and Strategies for Solutions. *Biotechnol. Adv.* **2019**, *37*, 491–504. [CrossRef]
43. Suo, Y.; Li, W.; Wan, L.; Luo, L.; Liu, S.; Qin, S.; Wang, J. Transcriptome Analysis Reveals Reasons for the Low Tolerance of *Clostridium tyrobutyricum* to Furan Derivatives. *Appl. Microbiol. Biotechnol.* **2023**, *107*, 327–339. [CrossRef] [PubMed]
44. Shabbir, S.; Wang, W.; Nawaz, M.; Boruah, P.; Kulyar, M.F.-A.; Chen, M.; Wu, B.; Liu, P.; Dai, Y.; Sun, L.; et al. Molecular Mechanism of Engineered *Zymomonas mobilis* to Furfural and Acetic Acid Stress. *Microb. Cell Fact.* **2023**, *22*, 88. [CrossRef] [PubMed]
45. Michelin, M.; Ximenes, E.; Polizeli, M.d.L.T.M.; Ladisch, M.R. Inhibition of Enzyme Hydrolysis of Cellulose by Phenols from Hydrothermally Pretreated Sugarcane Straw. *Enzym. Microb. Technol.* **2023**, *166*, 110227. [CrossRef] [PubMed]
46. Zeng, L.; Si, Z.; Zhao, X.; Feng, P.; Huang, J.; Long, X.; Yi, Y. Metabolome Analysis of the Response and Tolerance Mechanisms of *Saccharomyces cerevisiae* to Formic Acid Stress. *Int. J. Biochem. Cell Biol.* **2022**, *148*, 106236. [CrossRef] [PubMed]
47. Ranzi, E.; Debiagi, P.E.A.; Frassoldati, A. Mathematical Modeling of Fast Biomass Pyrolysis and Bio-Oil Formation. Note I: Kinetic Mechanism of Biomass Pyrolysis. *ACS Sustain. Chem. Eng.* **2017**, *5*, 2867–2881. [CrossRef]

48. Zhang, B.; Li, J.; Guo, L.; Chen, Z.; Li, C. Photothermally Promoted Cleavage of β-1,4-Glycosidic Bonds of Cellulosic Biomass on Ir/HY Catalyst under Mild Conditions. *Appl. Catal. B Environ.* **2018**, *237*, 660–664. [CrossRef]
49. Ling, Z.; Chen, S.; Zhang, X.; Xu, F. Exploring Crystalline-Structural Variations of Cellulose during Alkaline Pretreatment for Enhanced Enzymatic Hydrolysis. *Bioresour. Technol.* **2017**, *224*, 611–617. [CrossRef] [PubMed]
50. Gnanasekaran, L.; Priya, A.K.; Thanigaivel, S.; Hoang, T.K.A.; Soto-Moscoso, M. The Conversion of Biomass to Fuels via Cutting-Edge Technologies: Explorations from Natural Utilization Systems. *Fuel* **2023**, *331*, 125668. [CrossRef]
51. Zheng, Y.; Zhao, J.; Xu, F.; Li, Y. Pretreatment of Lignocellulosic Biomass for Enhanced Biogas Production. *Prog. Energy Combust. Sci.* **2014**, *42*, 35–53. [CrossRef]
52. Varriale, L.; Ulber, R. Fungal-Based Biorefinery: From Renewable Resources to Organic Acids. *ChemBioEng Rev.* **2023**, *10*, 272–292. [CrossRef]
53. Kim, D. Physico-Chemical Conversion of Lignocellulose: Inhibitor Effects and Detoxification Strategies: A Mini Review. *Molecules* **2018**, *23*, 309. [CrossRef] [PubMed]
54. Sun, D.; Lv, Z.-W.; Rao, J.; Tian, R.; Sun, S.-N.; Peng, F. Effects of Hydrothermal Pretreatment on the Dissolution and Structural Evolution of Hemicelluloses and Lignin: A Review. *Carbohydr. Polym.* **2022**, *281*, 119050. [CrossRef] [PubMed]
55. Bals, B.; Rogers, C.; Jin, M.; Balan, V.; Dale, B. Evaluation of Ammonia Fibre Expansion (AFEX) Pretreatment for Enzymatic Hydrolysis of Switchgrass Harvested in Different Seasons and Locations. *Biotechnol. Biofuels* **2010**, *3*, 1. [CrossRef]
56. Feng, J.; Techapun, C.; Phimolsiripol, Y.; Phongthai, S.; Khemacheewakul, J.; Taesuwan, S.; Mahakuntha, C.; Porninta, K.; Htike, S.L.; Kumar, A.; et al. Utilization of Agricultural Wastes for Co-Production of Xylitol, Ethanol, and Phenylacetylcarbinol: A Review. *Bioresour. Technol.* **2024**, *392*, 129926. [CrossRef]
57. Zhang, Z.; Donaldson, A.A.; Ma, X. Advancements and Future Directions in Enzyme Technology for Biomass Conversion. *Biotechnol. Adv.* **2012**, *30*, 913–919. [CrossRef]
58. Shen, X.; Sun, R. Recent Advances in Lignocellulose Prior-Fractionation for Biomaterials, Biochemicals, and Bioenergy. *Carbohydr. Polym.* **2021**, *261*, 117884. [CrossRef] [PubMed]
59. Dharmaraja, J.; Shobana, S.; Arvindnarayan, S. Lignocellulosic Biomass Conversion via Greener Pretreatment Methods towards Biorefinery Applications. *Bioresour. Technol.* **2023**, *369*, 128328. [CrossRef]
60. Kumar, V.; Yadav, S.K.; Kumar, J.; Ahluwalia, V. A Critical Review on Current Strategies and Trends Employed for Removal of Inhibitors and Toxic Materials Generated during Biomass Pretreatment. *Bioresour. Technol.* **2020**, *299*, 122633. [CrossRef] [PubMed]
61. Risanto, L.; Adi, D.T.N.; Fajriutami, T.; Teramura, H.; Fatriasari, W.; Hermiati, E.; Kahar, P.; Kondo, A.; Ogino, C. Pretreatment with Dilute Maleic Acid Enhances the Enzymatic Digestibility of Sugarcane Bagasse and Oil Palm Empty Fruit Bunch Fiber. *Bioresour. Technol.* **2023**, *369*, 128382. [CrossRef]
62. Hernández-Beltrán, J.U.; Hernández-Escoto, H. Enzymatic Hydrolysis of Biomass at High-Solids Loadings through Fed-Batch Operation. *Biomass Bioenergy* **2018**, *119*, 191–197. [CrossRef]
63. Shah, T.A.; Ullah, R. Pretreatment of Wheat Straw with Ligninolytic Fungi for Increased Biogas Productivity. *Int. J. Environ. Sci. Technol.* **2019**, *16*, 7497–7508. [CrossRef]
64. Sjulander, N.; Kikas, T. Two-Step Pretreatment of Lignocellulosic Biomass for High-Sugar Recovery from the Structural Plant Polymers Cellulose and Hemicellulose. *Energies* **2022**, *15*, 8898. [CrossRef]
65. Raud, M.; Krennhuber, K.; Jäger, A.; Kikas, T. Nitrogen Explosive Decompression Pre-Treatment: An Alternative to Steam Explosion. *Energy* **2019**, *177*, 175–182. [CrossRef]
66. Yu, J.; Paterson, N.; Blamey, J.; Millan, M. Cellulose, Xylan and Lignin Interactions during Pyrolysis of Lignocellulosic Biomass. *Fuel* **2017**, *191*, 140–149. [CrossRef]
67. Sriariyanun, M.; Gundupalli, M.P.; Phakeenuya, V.; Phusamtisampan, T.; Cheng, Y.-S.; Venkatachalam, P. Biorefinery Approaches for Production of Cellulosic Ethanol Fuel Using Recombinant Engineered Microorganisms. *J. Appl. Sci. Eng.* **2024**, *27*, 1985–2005. [CrossRef]
68. Piedrahita-Rodríguez, S.; Baumberger, S.; Cézard, L.; Poveda-Giraldo, J.A.; Alzate-Ramírez, A.F.; Cardona Alzate, C.A. Comparative Analysis of Trifluoracetic Acid Pretreatment for Lignocellulosic Materials. *Materials* **2023**, *16*, 5502. [CrossRef]
69. García-Velásquez, C.A.; Cardona, C.A. Comparison of the Biochemical and Thermochemical Routes for Bioenergy Production: A Techno-Economic (TEA), Energetic and Environmental Assessment. *Energy* **2019**, *172*, 232–242. [CrossRef]
70. Gao, J.; Chen, L.; Yuan, K.; Huang, H.; Yan, Z. Ionic Liquid Pretreatment to Enhance the Anaerobic Digestion of Lignocellulosic Biomass. *Bioresour. Technol.* **2013**, *150*, 352–358. [CrossRef]
71. Sjulander, N.; Kikas, T. Origin, Impact and Control of Lignocellulosic Inhibitors in Bioethanol Production—A Review. *Energies* **2020**, *13*, 4751. [CrossRef]
72. Wang, Z.; Pawar, P.M.-A.; Derba-Maceluch, M.; Hedenström, M.; Chong, S.-L.; Tenkanen, M.; Jönsson, L.J.; Mellerowicz, E.J. Hybrid Aspen Expressing a Carbohydrate Esterase Family 5 Acetyl Xylan Esterase Under Control of a Wood-Specific Promoter Shows Improved Saccharification. *Front. Plant Sci.* **2020**, *11*, 380. [CrossRef]
73. Zhang, Y.; Mu, X.; Wang, H.; Li, B.; Peng, H. Combined Deacetylation and PFI Refining Pretreatment of Corn Cob for the Improvement of a Two-Stage Enzymatic Hydrolysis. *J. Agric. Food Chem.* **2014**, *62*, 4661–4667. [CrossRef] [PubMed]
74. Świątek, K.; Gaag, S.; Klier, A.; Kruse, A.; Sauer, J.; Steinbach, D. Acid Hydrolysis of Lignocellulosic Biomass: Sugars and Furfurals Formation. *Catalysts* **2020**, *10*, 437. [CrossRef]

75. Brethauer, S.; Antczak, A.; Balan, R.; Zielenkiewicz, T.; Studer, M.H. Steam Explosion Pretreatment of Beechwood. Part 2: Quantification of Cellulase Inhibitors and Their Effect on Avicel Hydrolysis. *Energies* **2020**, *13*, 3638. [CrossRef]

76. Kim, Y.; Ximenes, E.; Mosier, N.S.; Ladisch, M.R. Soluble Inhibitors/Deactivators of Cellulase Enzymes from Lignocellulosic Biomass. *Enzym. Microb. Technol.* **2011**, *48*, 408–415. [CrossRef] [PubMed]

77. Chen, X.; Zhai, R.; Li, Y.; Yuan, X.; Liu, Z.-H.; Jin, M. Understanding the Structural Characteristics of Water-Soluble Phenolic Compounds from Four Pretreatments of Corn Stover and Their Inhibitory Effects on Enzymatic Hydrolysis and Fermentation. *Biotechnol. Biofuels* **2020**, *13*, 44. [CrossRef] [PubMed]

78. Gundupalli, M.P.; Cheenkachorn, K.; Chuetor, S.; Kirdponpattara, S.; Gundupalli, S.P.; Show, P.-L.; Sriariyanun, M. Assessment of Pure, Mixed and Diluted Deep Eutectic Solvents on Napier Grass (*Cenchrus purpureus*): Compositional and Characterization Studies of Cellulose, Hemicellulose and Lignin. *Carbohydr. Polym.* **2023**, *306*, 120599. [CrossRef] [PubMed]

79. Mikulski, D.; Kłosowski, G. High-Pressure Microwave-Assisted Pretreatment of Softwood, Hardwood and Non-Wood Biomass Using Different Solvents in the Production of Cellulosic Ethanol. *Biotechnol. Biofuels Bioprod.* **2023**, *16*, 19. [CrossRef]

80. Awoyale, A.A.; Lokhat, D. Experimental Determination of the Effects of Pretreatment on Selected Nigerian Lignocellulosic Biomass in Bioethanol Production. *Sci. Rep.* **2021**, *11*, 557. [CrossRef]

81. Vu, H.P.; Nguyen, L.N.; Vu, M.T.; Johir, M.A.H.; McLaughlan, R.; Nghiem, L.D. A Comprehensive Review on the Framework to Valorise Lignocellulosic Biomass as Biorefinery Feedstocks. *Sci. Total Environ.* **2020**, *743*, 140630. [CrossRef]

82. Baruah, J.; Nath, B.K.; Sharma, R.; Kumar, S.; Deka, R.C.; Baruah, D.C.; Kalita, E. Recent Trends in the Pretreatment of Lignocellulosic Biomass for Value-Added Products. *Front. Energy Res.* **2018**, *6*, 141. [CrossRef]

83. Liu, Q.; Fan, H. Preparation and Characterization of Xylan by an Efficient Approach with Mechanical Pretreatments. *Ind. Crops Prod.* **2021**, *165*, 113420. [CrossRef]

84. Chen, Y.; Stevens, M.A.; Zhu, Y.; Holmes, J.; Xu, H. Understanding of Alkaline Pretreatment Parameters for Corn Stover Enzymatic Saccharification. *Biotechnol. Biofuels* **2013**, *6*, 8. [CrossRef] [PubMed]

85. Ho, M.C.; Ong, V.Z.; Wu, T.Y. Potential Use of Alkaline Hydrogen Peroxide in Lignocellulosic Biomass Pretreatment and Valorization—A Review. *Renew. Sustain. Energy Rev.* **2019**, *112*, 75–86. [CrossRef]

86. Wang, W.; Chen, X.; Tan, X.; Wang, Q.; Liu, Y.; He, M.; Yu, Q.; Qi, W.; Luo, Y.; Zhuang, X.; et al. Feasibility of Reusing the Black Liquor for Enzymatic Hydrolysis and Ethanol Fermentation. *Bioresour. Technol.* **2017**, *228*, 235–240. [CrossRef] [PubMed]

87. Łukajtis, R.; Rybarczyk, P.; Kucharska, K.; Konopacka-Łyskawa, D.; Słupek, E.; Wychodnik, K.; Kamiński, M. Optimization of Saccharification Conditions of Lignocellulosic Biomass under Alkaline Pre-Treatment and Enzymatic Hydrolysis. *Energies* **2018**, *11*, 886. [CrossRef]

88. Feng, N.; She, S.; Tang, F.; Zhao, X.; Chen, J.; Wang, P.; Wu, Q.; Rojas, O.J. Formation and Identification of Lignin-Carbohydrate Complexes in Pre-Hydrolysis Liquors. *Biomacromolecules* **2023**, *24*, 2541–2548. [CrossRef] [PubMed]

89. Wang, Y.; Zhan, P.; Shao, L.; Zhang, L.; Qing, Y. Effects of Inhibitors Generated by Dilute Phosphoric Acid Plus Steam-Exploded Poplar on *Saccharomyces cerevisiae* Growth. *Microorganisms* **2022**, *10*, 1456. [CrossRef]

90. Liu, B.; Liu, L.; Deng, B.; Huang, C.; Zhu, J.; Liang, L.; He, X.; Wei, Y.; Qin, C.; Liang, C.; et al. Application and Prospect of Organic Acid Pretreatment in Lignocellulosic Biomass Separation: A Review. *Int. J. Biol. Macromol.* **2022**, *222*, 1400–1413. [CrossRef] [PubMed]

91. Domínguez-Gómez, C.X.; Nochebuena-Morando, L.E.; Aguilar-Uscanga, M.G.; López-Zamora, L. Statistical Optimization of Dilute Acid and H_2O_2 Alkaline Pretreatment Using Surface Response Methodology and Tween 80 for the Enhancement of the Enzymatic Hydrolysis of Corncob. *Biomass Conv. Bioref.* **2023**, *13*, 6185–6196. [CrossRef]

92. Yan, Z.; Gao, X.; Gao, Q.; Bao, J. Mechanism of Tolerance to the Lignin-Derived Inhibitor p-Benzoquinone and Metabolic Modification of Biorefinery Fermentation Strains. *Appl. Environ. Microbiol.* **2019**, *85*, e01443-19. [CrossRef]

93. Qiu, Z.; Fang, C.; He, N.; Bao, J. An Oxidoreductase Gene ZMO1116 Enhances the P-Benzoquinone Biodegradation and Chiral Lactic Acid Fermentability of *Pediococcus acidilactici*. *J. Biotechnol.* **2020**, *323*, 231–237. [CrossRef] [PubMed]

94. Li, J.; Liu, B.; Liu, L.; Luo, Y.; Zeng, F.; Qin, C.; Liang, C.; Huang, C.; Yao, S. Pretreatment of Poplar with Eco-Friendly Levulinic Acid to Achieve Efficient Utilization of Biomass. *Bioresour. Technol.* **2023**, *376*, 128855. [CrossRef] [PubMed]

95. Han, T.; Zhang, B.; Yang, H.; Liu, X.; Bao, J. Changes in pH Values Allow for a Visible Detection of the End Point in Submerged Liquid Biodetoxification during Biorefinery Processing. *ACS Sustain. Chem. Eng.* **2023**, *11*, 16608–16617. [CrossRef]

96. Liu, G.; Zhang, Q.; Li, H.; Qureshi, A.S.; Zhang, J.; Bao, X.; Bao, J. Dry Biorefining Maximizes the Potentials of Simultaneous Saccharification and Co-Fermentation for Cellulosic Ethanol Production. *Biotechnol. Bioeng.* **2018**, *115*, 60–69. [CrossRef]

97. He, Y.; Zhang, J.; Bao, J. Dry Dilute Acid Pretreatment by Co-Currently Feeding of Corn Stover Feedstock and Dilute Acid Solution without Impregnation. *Bioresour. Technol.* **2014**, *158*, 360–364. [CrossRef] [PubMed]

98. Han, X.; Bao, J. General Method to Correct the Fluctuation of Acid Based Pretreatment Efficiency of Lignocellulose for Highly Efficient Bioconversion. *ACS Sustain. Chem. Eng.* **2018**, *6*, 4212–4219. [CrossRef]

99. Gomes, D.G.; Michelin, M.; Romaní, A.; Domingues, L.; Teixeira, J.A. Co-Production of Biofuels and Value-Added Compounds from Industrial Eucalyptus Globulus Bark Residues Using Hydrothermal Treatment. *Fuel* **2021**, *285*, 119265. [CrossRef]

100. Batista, G.; Souza, R.B.A.; Pratto, B.; dos Santos-Rocha, M.S.R.; Cruz, A.J.G. Effect of Severity Factor on the Hydrothermal Pretreatment of Sugarcane Straw. *Bioresour. Technol.* **2019**, *275*, 321–327. [CrossRef]

101. Shinde, S.D.; Meng, X.; Kumar, R.; Ragauskas, A.J. Recent Advances in Understanding the Pseudo-Lignin Formation in a Lignocellulosic Biorefinery. *Green Chem.* **2018**, *20*, 2192–2205. [CrossRef]

102. He, J.; Huang, C.; Lai, C.; Jin, Y.; Ragauskas, A.; Yong, Q. Investigation of the Effect of Lignin/Pseudo-Lignin on Enzymatic Hydrolysis by Quartz Crystal Microbalance. *Ind. Crops Prod.* **2020**, *157*, 112927. [CrossRef]
103. Liu, Z.C.; Wang, Z.W.; Gao, S.; Tong, Y.X.; Le, X.; Hu, N.W.; Yan, Q.S.; Zhou, X.G.; He, Y.R.; Wang, L. Isolation and Fractionation of the Tobacco Stalk Lignin for Customized Value-Added Utilization. *Front. Bioeng. Biotechnol.* **2021**, *9*, 811287. [CrossRef] [PubMed]
104. Camargo, F.P.; Sakamoto, I.K.; Duarte, I.C.S.; Varesche, M.B.A. Influence of Alkaline Peroxide Assisted and Hydrothermal Pretreatment on Biodegradability and Bio-Hydrogen Formation from Citrus Peel Waste. *Int. J. Hydrog. Energy* **2019**, *44*, 22888–22903. [CrossRef]
105. Rocha, G.J.M.; Silva, V.F.N.; Martín, C.; Gonçalves, A.R.; Nascimento, V.M.; Souto-Maior, A.M. Effect of Xylan and Lignin Removal by Hydrothermal Pretreatment on Enzymatic Conversion of Sugarcane Bagasse Cellulose for Second Generation Ethanol Production. *Sugar Technol.* **2013**, *15*, 390–398. [CrossRef]
106. Singh, R.D.; Bhuyan, K.; Banerjee, J.; Muir, J.; Arora, A. Hydrothermal and Microwave Assisted Alkali Pretreatment for Fractionation of Arecanut Husk. *Ind. Crops Prod.* **2017**, *102*, 65–74. [CrossRef]
107. Sun, S.-F.; Yang, H.-Y.; Yang, J.; Shi, Z.-J. Structural Characterization of Poplar Lignin Based on the Microwave-Assisted Hydrothermal Pretreatment. *Int. J. Biol. Macromol.* **2021**, *190*, 360–367. [CrossRef] [PubMed]
108. Yu, G.; Liu, S.; Feng, X.; Zhang, Y.; Liu, C.; Liu, Y.-J.; Li, B.; Cui, Q.; Peng, H. Impact of Ammonium Sulfite-Based Sequential Pretreatment Combinations on Two Distinct Saccharifications of Wheat Straw. *RSC Adv.* **2020**, *10*, 17129–17142. [CrossRef] [PubMed]
109. Yao, S.; Nie, S.; Yuan, Y.; Wang, S.; Qin, C. Efficient Extraction of Bagasse Hemicelluloses and Characterization of Solid Remainder. *Bioresour. Technol.* **2015**, *185*, 21–27. [CrossRef] [PubMed]
110. Hou, Y.; Wang, S.; Deng, B.; Ma, Y.; Long, X.; Qin, C.; Liang, C.; Huang, C.; Yao, S. Selective Separation of Hemicellulose from Poplar by Hydrothermal Pretreatment with Ferric Chloride and pH Buffer. *Int. J. Biol. Macromol.* **2023**, *251*, 126374. [CrossRef]
111. Zhao, X.; Moates, G.K.; Wilson, D.R.; Ghogare, R.J.; Coleman, M.J.; Waldron, K.W. Steam Explosion Pretreatment and Enzymatic Saccharification of Duckweed (*Lemna Minor*) Biomass. *Biomass Bioenergy* **2015**, *72*, 206–215. [CrossRef]
112. Cui, W.; Wang, Y.; Sun, Z.; Cui, C.; Li, H.; Luo, K.; Cheng, A. Effects of Steam Explosion on Phenolic Compounds and Dietary Fiber of Grape Pomace. *LWT* **2023**, *173*, 114350. [CrossRef]
113. Ghoreishi, S.; Løhre, C.; Hermundsgård, D.H.; Molnes, J.L.; Tanase-Opedal, M.; Brusletto, R.; Barth, T. Identification and Quantification of Valuable Platform Chemicals in Aqueous Product Streams from a Preliminary Study of a Large Pilot-Scale Steam Explosion of Woody Biomass Using Quantitative Nuclear Magnetic Resonance Spectroscopy. *Biomass Conv. Bioref.* **2024**, *14*, 3331–3349. [CrossRef]
114. Hoang, A.T.; Nguyen, X.P.; Duong, X.Q.; Ağbulut, Ü.; Len, C.; Nguyen, P.Q.P.; Kchaou, M.; Chen, W.-H. Steam Explosion as Sustainable Biomass Pretreatment Technique for Biofuel Production: Characteristics and Challenges. *Bioresour. Technol.* **2023**, *385*, 129398. [CrossRef] [PubMed]
115. Tang, C.; Shan, J.; Chen, Y.; Zhong, L.; Shen, T.; Zhu, C.; Ying, H. Organic Amine Catalytic Organosolv Pretreatment of Corn Stover for Enzymatic Saccharification and High-Quality Lignin. *Bioresour. Technol.* **2017**, *232*, 222–228. [CrossRef] [PubMed]
116. Chin, D.W.K.; Lim, S.; Pang, Y.L.; Lim, C.H.; Shuit, S.H.; Lee, K.M.; Chong, C.T. Effects of Organic Solvents on the Organosolv Pretreatment of Degraded Empty Fruit Bunch for Fractionation and Lignin Removal. *Sustainability* **2021**, *13*, 6757. [CrossRef]
117. Zhao, X.; Cheng, K.; Liu, D. Organosolv Pretreatment of Lignocellulosic Biomass for Enzymatic Hydrolysis. *Appl. Microbiol. Biotechnol.* **2009**, *82*, 815–827. [CrossRef] [PubMed]
118. Satlewal, A.; Agrawal, R.; Bhagia, S.; Sangoro, J.; Ragauskas, A.J. Natural Deep Eutectic Solvents for Lignocellulosic Biomass Pretreatment: Recent Developments, Challenges and Novel Opportunities. *Biotechnol. Adv.* **2018**, *36*, 2032–2050. [CrossRef] [PubMed]
119. Saratale, R.G.; Ponnusamy, V.K.; Piechota, G.; Igliński, B.; Shobana, S.; Park, J.-H.; Saratale, G.D.; Shin, H.S.; Banu, J.R.; Kumar, V.; et al. Green Chemical and Hybrid Enzymatic Pretreatments for Lignocellulosic Biorefineries: Mechanism and Challenges. *Bioresour. Technol.* **2023**, *387*, 129560. [CrossRef] [PubMed]
120. Li, P.; Lu, Y.; Li, X.; Ren, J.; Jiang, Z.; Jiang, B.; Wu, W. Comparison of the Degradation Performance of Seven Different Choline Chloride-Based DES Systems on Alkaline Lignin. *Polymers* **2022**, *14*, 5100. [CrossRef] [PubMed]
121. Zhou, X.; Li, F.; Li, C.; Li, Y.; Jiang, D.; Zhang, T.; Lu, C.; Zhang, Q.; Jing, Y. Effect of Deep Eutectic Solvent Pretreatment on Biohydrogen Production from Corncob: Pretreatment Temperature and Duration. *Bioengineered* **2023**, *14*, 2252218. [CrossRef]
122. Xiao, T.; Hou, M.; Guo, X. Recent Progress in Deep Eutectic Solvent (DES) Fractionation of Lignocellulosic Components: A Review. *Renew. Sustain. Energy Rev.* **2024**, *192*, 114243. [CrossRef]
123. Kumar, B.; Bhardwaj, N.; Agrawal, K.; Chaturvedi, V.; Verma, P. Current Perspective on Pretreatment Technologies Using Lignocellulosic Biomass: An Emerging Biorefinery Concept. *Fuel Process. Technol.* **2020**, *199*, 106244. [CrossRef]
124. Tan, Y.T.; Chua, A.S.M.; Ngoh, G.C. Deep Eutectic Solvent for Lignocellulosic Biomass Fractionation and the Subsequent Conversion to Bio-Based Products—A Review. *Bioresour. Technol.* **2020**, *297*, 122522. [CrossRef] [PubMed]
125. Lobato-Rodríguez, Á.; Gullón, B.; Romaní, A.; Ferreira-Santos, P.; Garrote, G.; Del-Río, P.G. Recent Advances in Biorefineries Based on Lignin Extraction Using Deep Eutectic Solvents: A Review. *Bioresour. Technol.* **2023**, *388*, 129744. [CrossRef] [PubMed]
126. Zhao, J.; Wilkins, M.R.; Wang, D. A Review on Strategies to Reduce Ionic Liquid Pretreatment Costs for Biofuel Production. *Bioresour. Technol.* **2022**, *364*, 128045. [CrossRef] [PubMed]
127. Colussi, F.; Rodríguez, H.; Michelin, M.; Teixeira, J.A. Challenges in Using Ionic Liquids for Cellulosic Ethanol Production. *Molecules* **2023**, *28*, 1620. [CrossRef] [PubMed]

128. Gallego-García, M.; Moreno, A.D.; Manzanares, P.; Negro, M.J.; Duque, A. Recent Advances on Physical Technologies for the Pretreatment of Food Waste and Lignocellulosic Residues. *Bioresour. Technol.* **2023**, *369*, 128397. [CrossRef] [PubMed]
129. Zhang, Y.; Ding, Z.; Shahadat Hossain, M.; Maurya, R.; Yang, Y.; Singh, V.; Kumar, D.; Salama, E.-S.; Sun, X.; Sindhu, R.; et al. Recent Advances in Lignocellulosic and Algal Biomass Pretreatment and Its Biorefinery Approaches for Biochemicals and Bioenergy Conversion. *Bioresour. Technol.* **2023**, *367*, 128281. [CrossRef] [PubMed]
130. Ling, Z.; Tang, W.; Su, Y.; Shao, L.; Wang, P.; Ren, Y.; Huang, C.; Lai, C.; Yong, Q. Promoting Enzymatic Hydrolysis of Aggregated Bamboo Crystalline Cellulose by Fast Microwave-Assisted Dicarboxylic Acid Deep Eutectic Solvents Pretreatments. *Bioresour. Technol.* **2021**, *333*, 125122. [CrossRef] [PubMed]
131. Dai, L.; He, C.; Wang, Y.; Liu, Y.; Yu, Z.; Zhou, Y.; Fan, L.; Duan, D.; Ruan, R. Comparative Study on Microwave and Conventional Hydrothermal Pretreatment of Bamboo Sawdust: Hydrochar Properties and Its Pyrolysis Behaviors. *Energy Convers. Manag.* **2017**, *146*, 1–7. [CrossRef]
132. Farmanbordar, S.; Amiri, H.; Karimi, K. Simultaneous Organosolv Pretreatment and Detoxification of Municipal Solid Waste for Efficient Biobutanol Production. *Bioresour. Technol.* **2018**, *270*, 236–244. [CrossRef]
133. Guo, H.; Zhao, Y.; Chang, J.-S.; Lee, D.-J. Inhibitor Formation and Detoxification during Lignocellulose Biorefinery: A Review. *Bioresour. Technol.* **2022**, *361*, 127666. [CrossRef] [PubMed]
134. Silveira, M.H.L.; Morais, A.R.C.; da Costa Lopes, A.M.; Olekszyszen, D.N.; Bogel-Łukasik, R.; Andreaus, J.; Pereira Ramos, L. Current Pretreatment Technologies for the Development of Cellulosic Ethanol and Biorefineries. *ChemSusChem* **2015**, *8*, 3366–3390. [CrossRef] [PubMed]
135. Capolupo, L.; Faraco, V. Green Methods of Lignocellulose Pretreatment for Biorefinery Development. *Appl. Microbiol. Biotechnol.* **2016**, *100*, 9451–9467. [CrossRef] [PubMed]
136. Vasić, K.; Knez, Ž.; Leitgeb, M. Bioethanol Production by Enzymatic Hydrolysis from Different Lignocellulosic Sources. *Molecules* **2021**, *26*, 753. [CrossRef] [PubMed]
137. Xiao, M.; Liu, Y.-J.; Bayer, E.A.; Kosugi, A.; Cui, Q.; Feng, Y. Cellulosomal Hemicellulases: Indispensable Players for Ensuring Effective Lignocellulose Bioconversion. *Green Carbon* **2024**, *2*, 57–69. [CrossRef]
138. You, C.; Liu, Y.-J.; Cui, Q.; Feng, Y. Glycoside Hydrolase Family 48 Cellulase: A Key Player in Cellulolytic Bacteria for Lignocellulose Biorefinery. *Fermentation* **2023**, *9*, 204. [CrossRef]
139. Singh, N.; Singhania, R.R.; Nigam, P.S.; Dong, C.-D.; Patel, A.K.; Puri, M. Global Status of Lignocellulosic Biorefinery: Challenges and Perspectives. *Bioresour. Technol.* **2022**, *344*, 126415. [CrossRef] [PubMed]
140. Ubando, A.T.; Felix, C.B.; Chen, W.-H. Biorefineries in Circular Bioeconomy: A Comprehensive Review. *Bioresour. Technol.* **2020**, *299*, 122585. [CrossRef] [PubMed]
141. Rani Singhania, R.; Dixit, P.; Kumar Patel, A.; Shekher Giri, B.; Kuo, C.-H.; Chen, C.-W.; Di Dong, C. Role and Significance of Lytic Polysaccharide Monooxygenases (LPMOs) in Lignocellulose Deconstruction. *Bioresour. Technol.* **2021**, *335*, 125261. [CrossRef]
142. Müller, G.; Várnai, A.; Johansen, K.S.; Eijsink, V.G.H.; Horn, S.J. Harnessing the Potential of LPMO-Containing Cellulase Cocktails Poses New Demands on Processing Conditions. *Biotechnol. Biofuels* **2015**, *8*, 187. [CrossRef]
143. Moon, M.; Lee, J.-P.; Park, G.W.; Lee, J.-S.; Park, H.J.; Min, K. Lytic Polysaccharide Monooxygenase (LPMO)-Derived Saccharification of Lignocellulosic Biomass. *Bioresour. Technol.* **2022**, *359*, 127501. [CrossRef] [PubMed]
144. Lopes, A.M.; Ferreira Filho, E.X.; Moreira, L.R.S. An Update on Enzymatic Cocktails for Lignocellulose Breakdown. *J. Appl. Microbiol.* **2018**, *125*, 632–645. [CrossRef] [PubMed]
145. Zhao, S.; Zhang, T.; Hasunuma, T.; Kondo, A.; Zhao, X.-Q.; Feng, J.-X. Every Road Leads to Rome: Diverse Biosynthetic Regulation of Plant Cell Wall-Degrading Enzymes in Filamentous Fungi *Penicillium oxalicum* and *Trichoderma reesei*. *Crit. Rev. Biotechnol.* **2023**; *online ahead of print*. [CrossRef]
146. Ma, C.; Liu, J.; Tang, J.; Sun, Y.; Jiang, X.; Zhang, T.; Feng, Y.; Liu, Q.; Wang, L. Current Genetic Strategies to Investigate Gene Functions in *Trichoderma reesei*. *Microb. Cell Fact.* **2023**, *22*, 97. [CrossRef] [PubMed]
147. Pant, S.; Ritika; Nag, P.; Ghati, A.; Chakraborty, D.; Maximiano, M.R.; Franco, O.L.; Mandal, A.K.; Kuila, A. Employment of the CRISPR/Cas9 System to Improve Cellulase Production in *Trichoderma reesei*. *Biotechnol. Adv.* **2022**, *60*, 108022. [CrossRef] [PubMed]
148. Qi, W.; Feng, Q.; Wang, W.; Zhang, Y.; Hu, Y.; Shakeel, U.; Xiao, L.; Wang, L.; Chen, H.; Liang, C. Combination of Surfactants and Enzyme Cocktails for Enhancing Woody Biomass Saccharification and Bioethanol Production from Lab-Scale to Pilot-Scale. *Bioresour. Technol.* **2023**, *384*, 129343. [CrossRef] [PubMed]
149. Ramos, M.D.N.; Sandri, J.P.; Claes, A.; Carvalho, B.T.; Thevelein, J.M.; Zangirolami, T.C.; Milessi, T.S. Effective Application of Immobilized Second Generation Industrial *Saccharomyces cerevisiae* Strain on Consolidated Bioprocessing. *New Biotechnol.* **2023**, *78*, 153–161. [CrossRef] [PubMed]
150. Qiao, J.; Cui, H.; Wang, M.; Fu, X.; Wang, X.; Li, X.; Huang, H. Integrated Biorefinery Approaches for the Industrialization of Cellulosic Ethanol Fuel. *Bioresour. Technol.* **2022**, *360*, 127516. [CrossRef] [PubMed]
151. Haldar, D.; Dey, P.; Thomas, J.; Singhania, R.R.; Patel, A.K. One Pot Bioprocessing in Lignocellulosic Biorefinery: A Review. *Bioresour. Technol.* **2022**, *365*, 128180. [CrossRef] [PubMed]
152. Ilić, N.; Milić, M.; Beluhan, S.; Dimitrijević-Branković, S. Cellulases: From Lignocellulosic Biomass to Improved Production. *Energies* **2023**, *16*, 3598. [CrossRef]

153. Nargotra, P.; Sharma, V.; Lee, Y.-C.; Tsai, Y.-H.; Liu, Y.-C.; Shieh, C.-J.; Tsai, M.-L.; Dong, C.-D.; Kuo, C.-H. Microbial Lignocellulolytic Enzymes for the Effective Valorization of Lignocellulosic Biomass: A Review. *Catalysts* **2023**, *13*, 83. [CrossRef]
154. Singh, N.; Mathur, A.S.; Gupta, R.P.; Barrow, C.J.; Tuli, D.K.; Puri, M. Enzyme Systems of Thermophilic Anaerobic Bacteria for Lignocellulosic Biomass Conversion. *Int. J. Biol. Macromol.* **2021**, *168*, 572–590. [CrossRef] [PubMed]
155. Béguin, P.; Lemaire, M. The Cellulosome: An Exocellular, Multiprotein Complex Specialized in Cellulose Degradation. *Crit. Rev. Biochem. Mol. Biol.* **1996**, *31*, 201–236. [CrossRef] [PubMed]
156. Feng, Y.; Liu, Y.; Cui, Q. Research Progress in Cellulosomes and Their Applications in Synthetic Biology. *Synth. Biol. J.* **2022**, *3*, 138. [CrossRef]
157. Xu, C.; Qin, Y.; Li, Y.; Ji, Y.; Huang, J.; Song, H.; Xu, J. Factors Influencing Cellulosome Activity in Consolidated Bioprocessing of Cellulosic Ethanol. *Bioresour. Technol.* **2010**, *101*, 9560–9569. [CrossRef] [PubMed]
158. Alves, V.D.; Fontes, C.M.G.A.; Bule, P. Cellulosomes: Highly Efficient Cellulolytic Complexes. *Subcell. Biochem.* **2021**, *96*, 323–354. [CrossRef] [PubMed]
159. Fierer, J.O.; Tovar-Herrera, O.E.; Weinstein, J.Y.; Kahn, A.; Moraïs, S.; Mizrahi, I.; Bayer, E.A. Affinity-Induced Covalent Protein-Protein Ligation via the SpyCatcher-SpyTag Interaction. *Green Carbon* **2023**, *1*, 33–42. [CrossRef]
160. Yao, X.; Chen, C.; Wang, Y.; Dong, S.; Liu, Y.-J.; Li, Y.; Cui, Z.; Gong, W.; Perrett, S.; Yao, L.; et al. Discovery and Mechanism of a pH-Dependent Dual-Binding-Site Switch in the Interaction of a Pair of Protein Modules. *Sci. Adv.* **2020**, *6*, eabd7182. [CrossRef]
161. Chen, C.; Yang, H.; Dong, S.; You, C.; Moraïs, S.; Edward, B.; Liu, Y.-J.; Xuan, J.; Cui, Q.; Mizrahi, I.; et al. A Cellulosomal Double-dockerin Module from *Clostridium thermocellum* Shows Distinct Structural and Cohesin-binding Features. *Protein Sci.* **2024**, *33*, e4937. [CrossRef]
162. Artzi, L.; Bayer, E.A.; Moraïs, S. Cellulosomes: Bacterial Nanomachines for Dismantling Plant Polysaccharides. *Nat. Rev. Microbiol.* **2017**, *15*, 83–95. [CrossRef]
163. Nataf, Y.; Bahari, L.; Kahel-Raifer, H.; Borovok, I.; Lamed, R.; Bayer, E.A.; Sonenshein, A.L.; Shoham, Y. *Clostridium thermocellum* Cellulosomal Genes Are Regulated by Extracytoplasmic Polysaccharides via Alternative Sigma Factors. *Proc. Natl. Acad. Sci. USA* **2010**, *107*, 18646–18651. [CrossRef]
164. Wei, Z.; Chen, C.; Liu, Y.-J.; Dong, S.; Li, J.; Qi, K.; Liu, S.; Ding, X.; Ortiz de Ora, L.; Muñoz-Gutiérrez, I.; et al. Alternative σI/Anti-σI Factors Represent a Unique Form of Bacterial σ/Anti-σ Complex. *Nucleic Acids Res.* **2019**, *47*, 5988–5997. [CrossRef] [PubMed]
165. Chen, C.; Dong, S.; Yu, Z.; Qiao, Y.; Li, J.; Ding, X.; Li, R.; Lin, J.; Bayer, E.A.; Liu, Y.-J.; et al. Essential Autoproteolysis of Bacterial Anti-σ Factor RsgI for Transmembrane Signal Transduction. *Sci. Adv.* **2023**, *9*, eadg4846. [CrossRef] [PubMed]
166. Li, J.; Zhang, H.; Li, D.; Liu, Y.-J.; Bayer, E.A.; Cui, Q.; Feng, Y.; Zhu, P. Structure of the Transcription Open Complex of Distinct σI Factors. *Nat. Commun.* **2023**, *14*, 6455. [CrossRef]
167. Nhim, S.; Waeonukul, R.; Uke, A.; Baramee, S.; Ratanakhanokchai, K.; Tachaapaikoon, C.; Pason, P.; Liu, Y.-J.; Kosugi, A. Biological Cellulose Saccharification Using a Coculture of *Clostridium thermocellum* and *Thermobrachium celere* Strain A9. *Appl. Microbiol. Biotechnol.* **2022**, *106*, 2133. [CrossRef] [PubMed]
168. Konar, S.; Sinha, S.K.; Datta, S.; Ghorai, P.K. Probing the Effect of Glucose on the Activity and Stability of β-Glucosidase: An All-Atom Molecular Dynamics Simulation Investigation. *ACS Omega* **2019**, *4*, 11189–11196. [CrossRef]
169. Zhang, J.; Liu, S.; Li, R.; Hong, W.; Xiao, Y.; Feng, Y.; Cui, Q.; Liu, Y. J. Efficient Whole-Cell-Catalyzing Cellulose Saccharification Using Engineered *Clostridium thermocellum*. *Biotechnol. Biofuels* **2017**, *10*, 124. [CrossRef] [PubMed]
170. Qi, K.; Chen, C.; Yan, F.; Feng, Y.; Bayer, E.A.; Kosugi, A.; Cui, Q.; Liu, Y.-J. Coordinated β-Glucosidase Activity with the Cellulosome Is Effective for Enhanced Lignocellulose Saccharification. *Bioresour. Technol.* **2021**, *337*, 125441. [CrossRef] [PubMed]
171. Erkanli, M.E.; El-Halabi, K.; Kim, J.R. Exploring the Diversity of β-Glucosidase: Classification, Catalytic Mechanism, Molecular Characteristics, Kinetic Models, and Applications. *Enzym. Microb. Technol.* **2024**, *173*, 110363. [CrossRef] [PubMed]
172. Chen, C.; Qi, K.; Chi, F.; Song, X.; Feng, Y.; Cui, Q.; Liu, Y.-J. Dissolved Xylan Inhibits Cellulosome-Based Saccharification by Binding to the Key Cellulosomal Component of *Clostridium thermocellum*. *Int. J. Biol. Macromol.* **2022**, *207*, 784–790. [CrossRef]
173. Qiao, J.; Sheng, Y.; Wang, M.; Li, A.; Li, X.; Huang, H. Evolving Robust and Interpretable Enzymes for the Bioethanol Industry. *Angew. Chem. Int. Ed.* **2023**, *62*, e202300320. [CrossRef]
174. Zhao, X.-Q.; Liu, C.-G.; Bai, F.-W. Making the Biochemical Conversion of Lignocellulose More Robust. *Trends Biotechnol.* **2024**, *42*, 418–430. [CrossRef] [PubMed]
175. Lu, J.; Li, J.; Gao, H.; Zhou, D.; Xu, H.; Cong, Y.; Zhang, W.; Xin, F.; Jiang, M. Recent Progress on Bio-Succinic Acid Production from Lignocellulosic Biomass. *World J. Microbiol. Biotechnol.* **2021**, *37*, 16. [CrossRef] [PubMed]
176. Ask, M.; Olofsson, K.; Di Felice, T.; Ruohonen, L.; Penttilä, M.; Lidén, G.; Olsson, L. Challenges in Enzymatic Hydrolysis and Fermentation of Pretreated Arundo Donax Revealed by a Comparison between SHF and SSF. *Process Biochem.* **2012**, *47*, 1452–1459. [CrossRef]
177. Jesus, M.S.; Romaní, A.; Genisheva, Z.; Teixeira, J.A.; Domingues, L. Integral Valorization of Vine Pruning Residue by Sequential Autohydrolysis Stages. *J. Clean. Prod.* **2017**, *168*, 74–86. [CrossRef]
178. Mohapatra, S.; Ranjan Mishra, R.; Nayak, B.; Chandra Behera, B.; Das Mohapatra, P.K. Development of Co-Culture Yeast Fermentation for Efficient Production of Biobutanol from Rice Straw: A Useful Insight in Valorization of Agro Industrial Residues. *Bioresour. Technol.* **2020**, *318*, 124070. [CrossRef]

179. Chavan, S.; Shete, A.; Mirza, Y.; Dharne, M.S. Investigation of Cold-Active and Mesophilic Cellulases: Opportunities Awaited. *Biomass Conv. Bioref.* **2023**, *13*, 8829–8852. [CrossRef]

180. Zhang, Q.; Bao, J. Industrial Cellulase Performance in the Simultaneous Saccharification and Co-Fermentation (SSCF) of Corn Stover for High-Titer Ethanol Production. *Bioresour. Bioprocess.* **2017**, *4*, 17. [CrossRef]

181. Sóti, V.; Lenaerts, S.; Cornet, I. Of Enzyme Use in Cost-Effective High Solid Simultaneous Saccharification and Fermentation Processes. *J. Biotechnol.* **2018**, *270*, 70–76. [CrossRef]

182. Zhang, K.; Jiang, Z.; Li, X.; Wang, D.; Hong, J. Enhancing Simultaneous Saccharification and Co-Fermentation of Corncob by Kluyveromyces Marxianus through Overexpression of Putative Transcription Regulator. *Bioresour. Technol.* **2024**, *399*, 130627. [CrossRef]

183. Olofsson, K.; Bertilsson, M.; Lidén, G. A Short Review on SSF—An Interesting Process Option for Ethanol Production from Lignocellulosic Feedstocks. *Biotechnol. Biofuels* **2008**, *1*, 7. [CrossRef]

184. Suriyachai, N.; Weerasaia, K.; Laosiripojana, N.; Champreda, V.; Unrean, P. Optimized Simultaneous Saccharification and Co-Fermentation of Rice Straw for Ethanol Production by *Saccharomyces cerevisiae* and *Scheffersomyces stipitis* Co-Culture Using Design of Experiments. *Bioresour. Technol.* **2013**, *142*, 171–178. [CrossRef]

185. Golias, H.; Dumsday, G.J.; Stanley, G.A.; Pamment, N.B. Evaluation of a Recombinant *Klebsiella oxytoca* Strain for Ethanol Production from Cellulose by Simultaneous Saccharification and Fermentation: Comparison with Native Cellobiose-Utilising Yeast Strains and Performance in Co-Culture with Thermotolerant Yeast and *Zymomonas mobilis*. *J. Biotechnol.* **2002**, *96*, 155–168. [CrossRef] [PubMed]

186. Miah, R.; Siddiqa, A.; Chakraborty, U.; Tuli, J.F.; Barman, N.K.; Uddin, A.; Aziz, T.; Sharif, N.; Dey, S.K.; Yamada, M.; et al. Development of High Temperature Simultaneous Saccharification and Fermentation by Thermosensitive *Saccharomyces cerevisiae* and *Bacillus amyloliquefaciens*. *Sci. Rep.* **2022**, *12*, 3630. [CrossRef]

187. Wei, N.; Oh, E.J.; Million, G.; Cate, J.H.D.; Jin, Y.-S. Simultaneous Utilization of Cellobiose, Xylose, and Acetic Acid from Lignocellulosic Biomass for Biofuel Production by an Engineered Yeast Platform. *ACS Synth. Biol.* **2015**, *4*, 707–713. [CrossRef] [PubMed]

188. Toor, M.; Kumar, S.S.; Malyan, S.K.; Bishnoi, N.R.; Mathimani, T.; Rajendran, K.; Pugazhendhi, A. An Overview on Bioethanol Production from Lignocellulosic Feedstocks. *Chemosphere* **2020**, *242*, 125080. [CrossRef]

189. Haokok, C.; Lunprom, S.; Reungsang, A.; Salakkam, A. Efficient Production of Lactic Acid from Cellulose and Xylan in Sugarcane Bagasse by Newly Isolated *Lactiplantibacillus plantarum* and *Levilactobacillus brevis* through Simultaneous Saccharification and Co-Fermentation Process. *Heliyon* **2023**, *9*, e17935. [CrossRef]

190. Tareen, A.K.; Punsuvon, V.; Sultan, I.N.; Khan, M.W.; Parakulsuksatid, P. Cellulase Addition and Pre-Hydrolysis Effect of High Solid Fed-Batch Simultaneous Saccharification and Ethanol Fermentation from a Combined Pretreated Oil Palm Trunk. *ACS Omega* **2021**, *6*, 26119–26129. [CrossRef]

191. Liu, Y.; Tang, Y.; Gao, H.; Zhang, W.; Jiang, Y.; Xin, F.; Jiang, M. Challenges and Future Perspectives of Promising Biotechnologies for Lignocellulosic Biorefinery. *Molecules* **2021**, *26*, 5411. [CrossRef] [PubMed]

192. Dev, C.; Jilani, S.B.; Yazdani, S.S. Adaptation on Xylose Improves Glucose–Xylose Co-Utilization and Ethanol Production in a Carbon Catabolite Repression (CCR) Compromised Ethanologenic Strain. *Microb. Cell Fact.* **2022**, *21*, 154. [CrossRef]

193. Xin, F.; Dong, W.; Zhang, W.; Ma, J.; Jiang, M. Biobutanol Production from Crystalline Cellulose through Consolidated Bioprocessing. *Trends Biotechnol.* **2019**, *37*, 167–180. [CrossRef]

194. Zhou, S.; Zhang, M.; Zhu, L.; Zhao, X.; Chen, J.; Chen, W.; Chang, C. Hydrolysis of Lignocellulose to Succinic Acid: A Review of Treatment Methods and Succinic Acid Applications. *Biotechnol. Biofuels* **2023**, *16*, 1. [CrossRef] [PubMed]

195. Froese, A.G.; Nguyen, T.-N.; Ayele, B.T.; Sparling, R. Digestibility of Wheat and Cattail Biomass Using a Co-Culture of Thermophilic Anaerobes for Consolidated Bioprocessing. *Bioenerg. Res.* **2020**, *13*, 325–333. [CrossRef]

196. Mai, J.; Hu, B.-B.; Zhu, M.-J. Metabolic Division of Labor between *Acetivibrio thermocellus* DSM 1313 and *Thermoanaerobacterium thermosaccharolyticum* MJ1 Enhanced Hydrogen Production from Lignocellulose. *Bioresour Technol.* **2023**, *390*, 129871. [CrossRef] [PubMed]

197. Roell, G.W.; Zha, J.; Carr, R.R.; Koffas, M.A.; Fong, S.S.; Tang, Y.J. Engineering Microbial Consortia by Division of Labor. *Microb. Cell Factories* **2019**, *18*, 35. [CrossRef]

198. Holwerda, E.K.; Worthen, R.S.; Kothari, N.; Lasky, R.C.; Davison, B.H.; Fu, C.; Wang, Z.-Y.; Dixon, R.A.; Biswal, A.K.; Mohnen, D.; et al. Multiple Levers for Overcoming the Recalcitrance of Lignocellulosic Biomass. *Biotechnol. Biofuels* **2019**, *12*, 15. [CrossRef] [PubMed]

199. Beri, D.; York, W.S.; Lynd, L.R.; Peña, M.J.; Herring, C.D. Development of a Thermophilic Coculture for Corn Fiber Conversion to Ethanol. *Nat. Commun.* **2020**, *11*, 1937. [CrossRef] [PubMed]

200. Xiao, Y.; Liu, Y.; Feng, Y.; Cui, Q. Progress in Synthetic Biology Research of *Clostridium thermocellum* for Biomass Energy Applications. *Synth. Biol. J.* **2023**, *4*, 1055. [CrossRef]

201. Buehler, E.A.; Mesbah, A. Kinetic Study of Acetone-Butanol-Ethanol Fermentation in Continuous Culture. *PLoS ONE* **2016**, *11*, e0158243. [CrossRef] [PubMed]

202. Ceaser, R.; Montané, D.; Constantí, M.; Medina, F. Current Progress on Lignocellulosic Bioethanol Including a Technological and Economical Perspective. *Environ. Dev. Sustain.* **2024**, *online ahead of print*. [CrossRef]

203. Banerjee, S.; Pandit, C.; Gundupalli, M.P.; Pandit, S.; Rai, N.; Lahiri, D.; Chaubey, K.K.; Joshi, S.J. Life Cycle Assessment of Revalorization of Lignocellulose for the Development of Biorefineries. *Environ. Dev. Sustain.* **2023**, *online ahead of print*. [CrossRef]
204. Hakeem, I.G.; Sharma, A.; Sharma, T.; Sharma, A.; Joshi, J.B.; Shah, K.; Ball, A.S.; Surapaneni, A. Techno-Economic Analysis of Biochemical Conversion of Biomass to Biofuels and Platform Chemicals. *Biofuels Bioprod. Biorefining* **2023**, *17*, 718–750. [CrossRef]
205. Bilal, M.; Iqbal, H.M.N. Recent Advancements in the Life Cycle Analysis of Lignocellulosic Biomass. *Curr. Sustain. Renew. Energy Rep.* **2020**, *7*, 100–107. [CrossRef]
206. Unrean, P. Techno-Economic Assessment of Bioethanol Production from Major Lignocellulosic Residues Under Different Process Configurations. In *Sustainable Biotechnology—Enzymatic Resources of Renewable Energy*; Singh, O.V., Chandel, A.K., Eds.; Springer International Publishing: Cham, Germany, 2018; pp. 177–204, ISBN 978-3-319-95480-6.
207. Fernández-Rodríguez, J.; Erdocia, X.; Alriols, M.G.; Labidi, J. Techno-Economic Analysis of Different Integrated Biorefinery Scenarios Using Lignocellulosic Waste Streams as Source for Phenolic Alcohols Production. *J. Clean. Prod.* **2021**, *285*, 124829. [CrossRef]
208. Sassner, P.; Galbe, M.; Zacchi, G. Techno-Economic Evaluation of Bioethanol Production from Three Different Lignocellulosic Materials. *Biomass Bioenergy* **2008**, *32*, 422–430. [CrossRef]
209. Mu, D.; Seager, T.; Rao, P.S.; Zhao, F. Comparative Life Cycle Assessment of Lignocellulosic Ethanol Production: Biochemical Versus Thermochemical Conversion. *Environ. Manag.* **2010**, *46*, 565–578. [CrossRef]
210. Prasad, A.; Sotenko, M.; Blenkinsopp, T.; Coles, S.R. Life Cycle Assessment of Lignocellulosic Biomass Pretreatment Methods in Biofuel Production. *Int. J. Life Cycle Assess.* **2016**, *21*, 44–50. [CrossRef]
211. Wingren, A.; Galbe, M.; Zacchi, G. Techno-Economic Evaluation of Producing Ethanol from Softwood: Comparison of SSF and SHF and Identification of Bottlenecks. *Biotechnol. Prog.* **2003**, *19*, 1109–1117. [CrossRef] [PubMed]
212. Kadhum, H.J.; Mahapatra, D.M.; Murthy, G.S. A Comparative Account of Glucose Yields and Bioethanol Production from Separate and Simultaneous Saccharification and Fermentation Processes at High Solids Loading with Variable PEG Concentration. *Bioresour. Technol.* **2019**, *283*, 67–75. [CrossRef] [PubMed]
213. Khajeeram, S.; Unrean, P. Techno-Economic Assessment of High-Solid Simultaneous Saccharification and Fermentation and Economic Impacts of Yeast Consortium and on-Site Enzyme Production Technologies. *Energy* **2017**, *122*, 194–203. [CrossRef]
214. Dempfle, D.; Kröcher, O.; Studer, M.H.-P. Techno-Economic Assessment of Bioethanol Production from Lignocellulose by Consortium-Based Consolidated Bioprocessing at Industrial Scale. *New Biotechnol.* **2021**, *65*, 53–60. [CrossRef] [PubMed]
215. Raftery, J.P.; Karim, M.N. Economic Viability of Consolidated Bioprocessing Utilizing Multiple Biomass Substrates for Commercial-Scale Cellulosic Bioethanol Production. *Biomass Bioenergy* **2017**, *103*, 35–46. [CrossRef]
216. Andhalkar, V.V.; Foong, S.Y.; Kee, S.H.; Lam, S.S.; Chan, Y.H.; Djellabi, R.; Bhubalan, K.; Medina, F.; Constantí, M. Integrated Biorefinery Design with Techno-Economic and Life Cycle Assessment Tools in Polyhydroxyalkanoates Processing. *Macromol. Mater. Eng.* **2023**, *308*, 2300100. [CrossRef]

Review

Challenges and Opportunities in the Catalytic Synthesis of Diphenolic Acid and Evaluation of Its Application Potential

Sara Fulignati [1,2], Nicola Di Fidio [1,2], Claudia Antonetti [1,2], Anna Maria Raspolli Galletti [1,2,*] and Domenico Licursi [1,2]

[1] Department of Chemistry and Industrial Chemistry, University of Pisa, Via Giuseppe Moruzzi 13, 56124 Pisa, Italy; sara.fulignati@unipi.it (S.F.); nicola.difidio@unipi.it (N.D.F.); claudia.antonetti@unipi.it (C.A.); domenico.licursi@unipi.it (D.L.)

[2] Consorzio Interuniversitario Reattività Chimica e Catalisi (CIRCC), Via Celso Ulpiani 27, 70126 Bari, Italy

* Correspondence: anna.maria.raspolli.galletti@unipi.it; Tel.: +39-050-2219290

Citation: Fulignati, S.; Di Fidio, N.; Antonetti, C.; Raspolli Galletti, A.M.; Licursi, D. Challenges and Opportunities in the Catalytic Synthesis of Diphenolic Acid and Evaluation of Its Application Potential. *Molecules* **2024**, *29*, 126. https://doi.org/10.3390/molecules29010126

Academic Editors: José A.P. Coelho and Roumiana P. Stateva

Received: 23 November 2023
Revised: 21 December 2023
Accepted: 22 December 2023
Published: 24 December 2023

Abstract: Diphenolic acid, or 4,4-bis(4-hydroxyphenyl)pentanoic acid, represents one of the potentially most interesting bio-products obtainable from the levulinic acid supply-chain. It represents a valuable candidate for the replacement of bisphenol A, which is strongly questioned for its toxicological issues. Diphenolic acid synthesis involves the condensation reaction between phenol and levulinic acid and requires the presence of a Brønsted acid as a catalyst. In this review, the state of the art related to the catalytic issues of its synthesis have been critically discussed, with particular attention to the heterogeneous systems, the reference benchmark being represented by the homogeneous acids. The main opportunities in the field of heterogeneous catalysis are deeply discussed, as well as the bottlenecks to be overcome to facilitate diphenolic acid production on an industrial scale. The regioselectivity of the reaction is a critical point because only the p,p'-isomer is of industrial interest; thus, several strategies aiming at the improvement of the selectivity towards this isomer are considered. The future potential of adopting alkyl levulinates, instead of levulinic acid, as starting materials for the synthesis of new classes of biopolymers, such as new epoxy and phenolic resins and polycarbonates, is also briefly considered.

Keywords: diphenolic acid; levulinic acid; phenol derivatives; homogeneous catalysis; heterogeneous catalysis; epoxy resins; polycarbonates

1. Introduction

Levulinic acid (LA) is a well-known platform chemical that is obtainable nowadays from cheap and waste cellulosic biomasses, thus allowing its industrial production on a larger scale [1,2]. The growth of LA production is allowing the progressive reduction of its manufacturing costs, thus favoring the research and development of new added-value bio-products derived from its supply-chain. At the state of the art, many technological improvements have been made by GF Biochemicals [3] to lower the LA production costs, aiming at a target of USD 1/kg, which is highly competitive in the market [4]. One of the most important added-value LA derivatives is 4,4-bis(4-hydroxyphenyl)pentanoic acid, also named diphenolic acid (DPA), for which the main physico-chemical properties are reported in Table 1.

Due to the structural similarities, DPA has been identified as a potential replacement for bisphenol A (BPA), the latter representing one of the most adopted monomers for the synthesis of epoxy resins and polycarbonates [5,6], which are widely exploited for the formulation of paints, cosmetics, surfactants, plasticizers, textile chemicals, etc. [7,8]. However, BPA is well-known for its dangerous toxicological properties, which mime the effects of estrogens and thus disrupt the endocrine system, even at very low concentrations (<1 ng L^{-1}) [9,10]. Moreover, BPA use has been associated with the onset of health issues, such as cancer, obesity, and diabetes [11,12], and its use, especially in foodstuffs, is strictly regulated by European Food Safety Authority (EFSA) legislation [13]. In this context, DPA represents an excellent substitute for BPA, and the progressive lowering of the LA production costs makes further R&D on this topic particularly valuable. Moreover, DPA can be envisaged as a completely bio-based molecule if the phenol production from renewable lignin sources, mainly through pyrolysis or reductive depolymerization, will be implemented on an industrial scale [14,15]. Remarkably, such pathways could be even better tailored to give substituted phenols (catechol, resorcinol, cresol, guaiacol, etc.), thus further broadening the application opportunities of the final specialty biomaterials. Whilst BPA is produced through a condensation reaction between acetone and two moles of phenol, in the DPA synthesis, acetone is replaced by LA, and both reactions require the presence of a proper Brønsted catalyst. A comparison between the reactions involved in the BPA and DPA syntheses is shown in Figure 1.

Table 1. Physico-chemical properties of DPA.

Physical Property	Details
Formula	$C_{17}H_{18}O_4$
Chemical structure	
Molecular weight (g mol^{-1})	286.33
Melting point (°C)	167–170
Boiling point (°C)	507
Flash point (°C)	208
Density (g mL^{-1})	1.30–1.32
Refractive index	1.675
Solubility	Slightly soluble in water; soluble in acetic acid, acetone, ethanol, isopropanol, and methyl ethyl ketone; and insoluble in benzene

According to Rahaman et al. [16], the reaction mechanism of DPA synthesis (Figure 2) involves the preliminary formation of the mono-phenolic acid (MPA) intermediate via activation of the LA ketone group, followed by the subsequent nucleophilic attack of the phenol and finally through the acid-catalyzed loss of a water molecule. The obtained carbocation intermediate (MPA) undergoes the nucleophilic attack of another phenol molecule to obtain the target DPA.

Just as for BPA, the synthesis of DPA is traditionally carried out in the presence of strong mineral acids, such as concentrated HCl or H_2SO_4, which are generally considered benchmarks for the development of new investigations. Remarkably, an isolated yield of about 60 mol% (evaluated with respect to the starting LA) was claimed in Bader's patent [17], typically employing an LA/phenol molar ratio of 1/2 and working at 25 °C for 20 h with concentrated H_2SO_4 (about 80 wt%). Such molar yields were also claimed employing the same LA/phenol molar ratio and a lower concentration of HCl (about 25 wt%) but working at a higher reaction temperature (about 90 °C) for 6 h. The use of HCl is generally preferred over H_2SO_4 due to the simpler work-up operations, which are required for its recovery. More recently, the use of concentrated mineral acids (37 wt% HCl or 96 wt% H_2SO_4), excess of phenol (LA/phenol molar ratio within the range 1/2–1/4), mild reaction temperatures (up to 60 °C), and long reaction times (up to 48 h) have generally been proposed, thus achieving a maximum DPA yield of about 70 mol% [18–22]. Remarkably, the use of a thiol (namely, methyl mercaptan) as a reaction co-catalyst led to significant improvement of the DPA yield up to a maximum of about 93 mol%, working at 55 °C for 16 h, in the presence of 37 wt% HCl as the main catalyst, and adopting the LA/phenol molar ratio of 1/2 [23]. The chemistry involved in such reactions was not explained in detail by the authors, and claims of such patents were mainly based on experimental evidence. Moreover, the reaction is mostly performed under solvent-free conditions, e.g., without any addition of external water and/or other solvents, thus exploiting the use of an excess of phenol. To specifically assess issues related to the selectivity of the DPA synthesis, it is necessary to consider more recent works, including, in some cases, the chemistry of BPA as the only reference, which admits the formation of different degradation products, mainly including triphenols and chromans [24]. In any case, the identification of the corresponding by-products in the DPA synthesis has not been specifically discussed in the literature, and scarce information is available on a few reaction by-products. In this context, according to the reaction mechanism proposed in Figure 2, the nucleophilic attack of the second molecule of phenol may lead to the formation of the *o,p′*-DPA isomer (Figure 3), and this isomer is considered an unwanted reaction by-product because it has no industrial interest and cannot be recycled within the process. In such condensation reactions involving unsubstituted phenol, the formation of the *o,o′*-isomer is almost negligible for steric hindrance, whilst those of the *p,p′*- and *o,p′*- ones are both thoroughly allowed [25]. The presence of the *o,p′*-DPA isomer significantly complicates the purification procedures of the *p,p′*-DPA, which represents the only industrially useful product. On this basis, the *p,p′*/*o,p′* molar ratio represents a key parameter for the optimization of the DPA synthesis process due to the difficult separation of the two isomers through crystallization. For the BPA synthesis, an isomerization reaction was planned at the industrial scale to advantageously convert the *o,p′*-BPA isomer to the *p,p′*-BPA one [26], but this isomerization is extremely difficult in the case of DPA isomers; thus, this crucial aspect requires the adoption of more appropriate strategies already during the synthesis step.

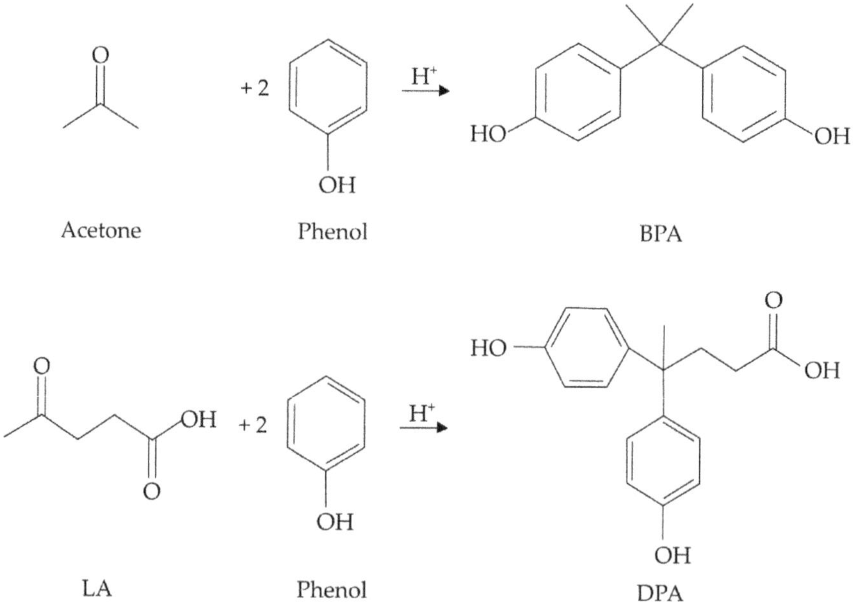

Figure 1. Acid-catalyzed syntheses of BPA and DPA.

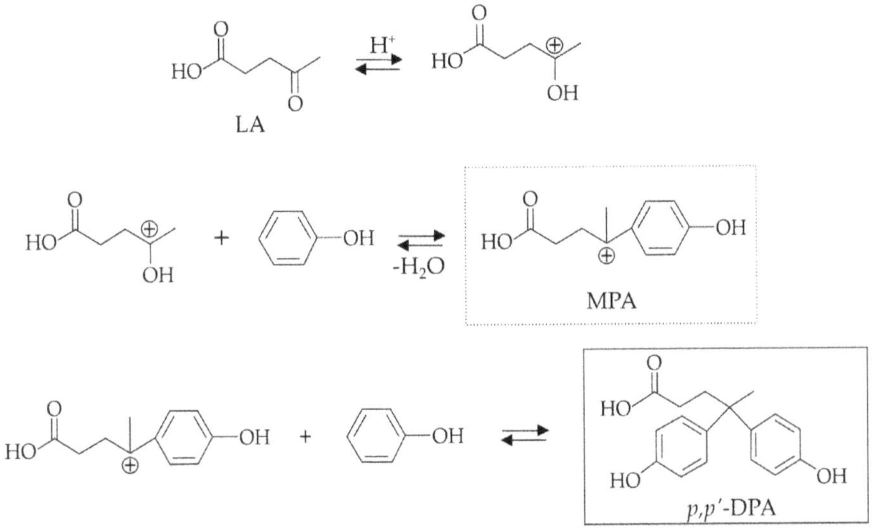

Figure 2. General reaction mechanism proposed for the synthesis of DPA (adapted from [16]).

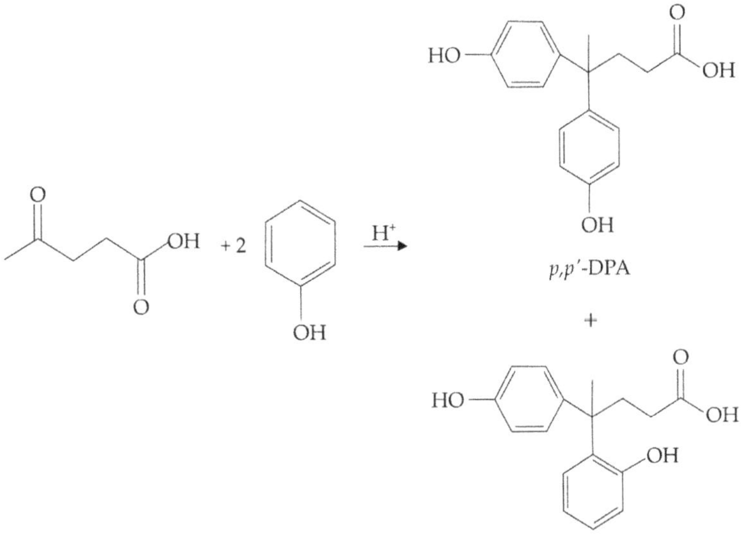

Figure 3. Main DPA isomers originating from the acid-catalyzed reaction occurring between LA and phenol.

The p,p'-DPA/o,p'-DPA molar ratio also has important effects on the properties of the final polymers, such as the color stability, crystallinity, and intermolecular attractive forces between the polymer chains. Thus, the p,p'-DPA/o,p'-DPA molar ratio should be as high as possible to improve these properties [27]. Remarkably, the available data based on Bader's patents, such as those of reference in this discussion [17], seem to confirm the exclusive formation of the p,p'-DPA isomer, but these data must be much more carefully considered due to the instrumental limitations at that time. In fact, the most recent literature admits the formation of the o,p'-DPA isomer starting from phenol [16], and, in this regard, different approaches have been proposed to improve the selectivity towards the p,p'-isomer. Remarkably, 2,6-xylenols, such as 2,6-dimethylphenol, were proposed as a phenolic source for the acid-catalyzed condensation with LA to the corresponding bis-xylenol product [28], thus completely addressing the reaction selectivity towards the corresponding p,p'-isomer (that, is 4,4-bis(3,5-dimethyl-4-hydroxyphenyl)pentanoic acid). Even more advantageously, natural phenols deriving from lignin sources, such as catechol and resorcinol, could be smartly exploited as valuable replacements for unsubstituted phenol [29]. In this case, in addition to the para-orientating effect, the presence of an additional hydroxyl group in the appropriate position of the phenolic structure is advantageous because it adds another reactive group to the DPA derivative, which could be better exploited to produce even more added-value products, such as glycidyl ethers [29].

Regarding the catalytic issues, although mineral acids are the traditional ones, their use has some well-known and important drawbacks, including corrosivity, difficult separation, and expensive treatment of the process's wastewater. These general issues are stimulating the development of new catalysts to perform DPA synthesis with the intention of improving the activity and the selectivity towards the p,p'-DPA isomer and, above all, overcoming the corrosion and separation problems. In this context, sulfonated acid catalysts [19,27,30,31], heteropolyacids [16,18,20,30,32], zeolites and oxides [18,27], as well as ionic liquids [21,22,33] have been proposed for DPA synthesis in the search for further improvement of the selectivity to the p,p'-DPA isomer. In this context, the use of a thiolic derivative as a co-catalyst was proposed by Van De Vyver et al. [30,31] to further improve the selectivity to the p,p'-DPA isomer. According to the reaction mechanism proposed by the authors for explaining the behavior of sulfonic-acid-functionalized catalysts, the thiol additive acts as a promoter by improving the reactivity of the carbonyl group of LA, thus making it more prone to react with phenol. Therefore, the thiol would react with the carbonyl of the LA to give an intermediate that is more electrophilic towards the reaction with phenol (Figure 2). The authors found that thiols with small substituents, such as ethanethiol, were particularly effective towards the p,p'-DPA formation, preferentially working at low LA conversion and therefore under kinetic control [31]. On the other hand, at higher LA conversions, the formation of the o,p'-DPA isomer became relevant according to a regime of thermodynamic control.

Starting from the above milestones, the state of the art related to the production of DPA from LA is critically discussed in this review by focusing on the development of new acid catalysts, with particular attention to the heterogeneous systems and considering the homogeneous ones as the benchmark. Subsequently, the main available applications of p,p'-DPA are summarized and discussed while highlighting the bottlenecks to be solved for further improving the research on this topic.

2. DPA Synthesis

2.1. DPA Synthesis with Homogeneous Catalysts

The condensation reaction between LA and phenol to give DPA has been traditionally carried out in the presence of strong Brønsted mineral acids, mainly HCl and H_2SO_4. The most interesting results reported in the literature with these mineral acids and other homogeneous systems, mainly referring to sulfonic acids, are reported in Table 2.

Table 2. DPA synthesis catalyzed using inorganic mineral acids and homogeneous sulfonic acids.

Entry	LA/Phenol (mol/mol)	Catalyst (g_{LA}/g_{cat})	Additive ($mol_{LA}/mol_{additive}$)	T (°C)	t (h)	C_{LA} (mol%)	$Y_{p,p'\text{-DPA}}$ (mol%)	p,p'-DPA/o,p'-DPA Molar Ratio	Ref.
1	1/3	HCl (8.0)	/	100	6	46	19	2.2	[18]
2	1/7	HCl (8.0)	/	100	24	n.a. [a]	56	1.8	[20]
3	1/4	HCl (6.3)	Ethanethiol (100)	60	24	65	41	2.0	[21]
4	1/4	H_2SO_4 (11.8)	/	60	24	62	42	11.2	[19]
5	1/3.7	H_2SO_4 (0.9)	/	75	6	n.a. [a]	74 [b]	n.a. [a]	[33]
6	1/4	H_2SO_4 (2.3)	/	60	48	n.a. [a]	61	24.0	[22]
7	1/4	H_2SO_4 (2.3)	Mercaptoacetic acid (20.0)	60	48	n.a. [a]	68	9.0	[22]
8	1/4	p-TSA (1.3)	Ethanethiol (100)	60	24	25	18	3.9	[21]
9	1/3.7	p-TSA (0.9)	/	75	6	n.a. [a]	70 [b]	n.a. [a]	[33]
10	1/4	NH_2SO_3H (2.4)	Ethanethiol (100)	60	24	58	51	7.4	[21]
11	1/3.7	CH_3SO_3H (0.9)	/	75	6	90	37(66 [b])	1.2	[33]
12	1/9.2	CH_3SO_3H (0.9)	/	75	6	n.a. [a]	86 [b]	n.a. [a]	[33]
13	1/4	CF_3SO_3H (1.5)	/	60	48	n.a. [a]	43	25.0	[22]
14	1/4	$HS(CH_2)_3SO_3H$ (1.5)	/	60	48	n.a. [a]	77	30.0	[22]

[a] n.a. = not available; [b] yield of p,p'-DPA + o,p'-DPA.

Very different reaction conditions in terms of the LA/phenol molar ratio, LA/catalyst weight ratio, reaction temperature, and time have been proposed when working with mineral acids and, in some cases, adding thiols to improve the selectivity towards the desired p,p'-DPA isomer [21]. However, H_2SO_4 generally led to higher yields and p,p'-DPA/o,p'-DPA molar ratios compared to HCl, regardless of the adopted reaction conditions (entries 1–7, Table 2). Thus, H_2SO_4 results are more promising for this application. On this basis, it is possible to suppose that the sulfonic group plays a key role in the reaction. For this reason, many other sulfonic acids have been proposed, including p-toluenesulfonic acid (entries 8 and 9, Table 2), sulfamic acid (entry 10, Table 2), methanesulfonic acid (entries 11 and 12, Table 2), triflic acid (entry 13, Table 2), and 3-mercaptopropyl sulfonic acid (entry 14, Table 2). In this context, Shen et al. [21] compared the catalytic performances of p-toluenesulfonic acid and sulfamic acid with those of HCl (entries 8, 10, and 3, Table 2) working under the same reaction conditions, including the amount of catalyst and ethanethiol (as an additive). The authors found that HCl led to the highest LA conversion (entry 3, Table 2, 65 mol%) but also to the lowest p,p'-DPA/o,p'-DPA molar ratio (2.0), whilst the sulfamic acid resulted in the most promising catalyst (entry 10, Table 2), showing comparable conversion but with a much higher selectivity towards the p,p'-DPA isomer (p,p'-DPA/o,p'-DPA molar ratio of 7.4) and the yield of 51 mol%. Mthembu et al. [33] optimized the Response Surface Methodology (Box–Behnken) for the condensation reaction between commercial LA and phenol by employing methanesulfonic acid as the homogeneous catalyst. The effects of the reaction time, temperature, and acid loading were investigated while keeping constant the LA/phenol molar ratio (3.7). The highest overall DPA yield of 66 mol%, corresponding to a yield of the p,p'-DPA isomer of 37 mol%, was achieved by working for 6 h at 75 °C with a LA/catalyst ratio of 0.9 wt/wt (entry 11, Table 2). The same authors also tested H_2SO_4 (entry 5, Table 2) and p-toluenesulfonic acid (entry 9, Table 2), in both cases achieving higher overall DPA yields than that reported for the methanesulfonic acid while working under the same reaction conditions. However, the authors did not report the p,p'-DPA/o,p'-DPA molar ratio reached with H_2SO_4 and p-toluenesulfonic acid; thus, further specific information on the p,p'-DPA yield is not available. The increment of the amount of phenol up to an LA/phenol ratio of 1/9.2 mol/mol was effective for improving the overall DPA yield from 66 to 86 mol% (entry 12, Table 2) but, in this case as well, no further information was given on the p,p'-DPA isomer yield. Moreover, the authors synthesized LA from depithed sugarcane bagasse as a real waste biomass to obtain an LA-rich liquor, working for 7 h at 100 °C in the presence of [EMim][HSO$_4$] as both solvent and catalyst, thus achieving an LA yield of 55 mol%. The crude LA-rich mixture further reacted with phenol under the same reaction conditions (75 °C, 6 h, and 1/3.7 as the LA/phenol molar ratio), thus achieving a total DPA yield of 65 mol% and proving that even crude LA derived from waste biomass can be directly used for DPA production. Liu et al. compared the catalytic activity of H$_2$SO$_4$ (entry 6, Table 2), triflic acid (entry 13, Table 2), and 3-mercaptopropyl sulfonic acid (entry 14, Table 2) by keeping constant the reaction conditions (60 °C for 48 h and 1/4 as the LA/phenol molar ratio), reaching the highest p,p'-DPA yield and p,p'-DPA/o,p'-DPA molar ratio by working with 3-mercaptopropyl sulfonic acid (entry 14, Table 2) [22]. Considering the promoting effect of the thiol groups, the mercaptoacetic acid was employed as an additive for the H$_2$SO$_4$-catalyzed reaction (entry 7, Table 2), which only modestly improved the p,p'-DPA yield from 61 to 68 mol%, but a reduction of the p,p'-DPA/o,p'-DPA molar ratio was also ascertained. According to the available literature [31], the formation of the o,p'-DPA isomer becomes relevant at high LA conversion under a regime of thermodynamic control, whereas the use of thiol additives for improving the p,p'-DPA production can be advantageous at low LA conversions, therefore working under kinetic control.

Isolation of DPA from a crude reaction mixture is a challenging topic, and only few examples are available in the literature, which mainly apply to DPA synthesized in the presence of a mineral acid (H$_2$SO$_4$ and HCl) as the catalyst. According to Bader [17], the isolation of DPA from such a reaction environment can be realized by diluting the crude reaction mixture with water and, next, extraction with ethyl acetate. The extract (containing phenol and DPA) can be further extracted with an aqueous solution of sodium bicarbonate to selectively deprotonate the DPA, which becomes soluble in the water phase. The latter is acidified and extracted with ether, and the corresponding extract is stripped in vacuo to yield the isolated DPA. Depending on the required purity degree, crude DPA can be further crystallized, generally with organic solvents, such as aromatic hydrocarbons (toluene), but also with water or ethanol.

2.2. DPA Synthesis with Heterogeneous Sulfonated Systems

Starting from the previous discussion, the use of heterogeneous catalysts that include sulfonic groups has represented the smartest choice to be immediately developed. In this context, many authors have proposed the use of heterogeneous systems, including the sulfonic acid group, to perform the DPA synthesis. The main available results based on such commercial and ad hoc synthesized catalysts are reported in Table 3.

Among the commercial, heterogeneous, sulfonic-acid-based systems, Amberlyst-15 is the preferred choice according to entries 15–17 of Table 3, with the highest p,p'-DPA yield (55 mol%) and p,p'-DPA/o,p'-DPA molar ratio (15.8), which were ascertained while working at 60 °C for 24 h and employing an LA/phenol molar ratio of 1/4 and LA/catalyst weight ratio of 4.4. Remarkably, comparing the results obtained with the same LA/phenol ratio (1/4, according to entries 15 and 16), it is evident that a lower reaction temperature should be preferred while compensating with a greater amount of the catalyst to obtain a similar conversion of LA but significantly better selectivity control towards the p,p'-DPA formation. Van De Vyver et al. [30] compared the catalytic activity of the Amberlyst-15 with that of Nafion NR50 (entry 18, Table 3), highlighting that the latter led to a higher p,p'-DPA/o,p'-DPA molar ratio while achieving similar LA conversion under the same reaction conditions. Moreover, the authors synthesized a new class of sulfonated hyperbranched poly(arylene oxindole)s (SHPAOs), which were tested for DPA synthesis. The macromolecular structure of these hyperbranched polymers positively acts on the increase in the functional group density, which is a desirable aspect for developing catalytic applications. When such catalysts were employed for the DPA synthesis, better performances than those of Amberlyst-15 and Nafion NR50 were achieved, mainly in terms of higher conversion and selectivity to the p,p'-DPA isomer (entry 19, Table 3). In addition to the desired p,p'-DPA product, the authors reported the formation of the o,p'-DPA isomer, as well as oligomeric phenol derivatives and LA dimers [34]. Such condensation reactions are typical of the acid catalysis approach, such as those occurring during BPA synthesis [35]. These side-reactions involving LA, which represents the limiting reagent of DPA synthesis, must be avoided, as they negatively impact p,p'-DPA selectivity. To selectively favor p,p'-DPA production, Van De Vyver et al. [30] proposed the use of thiol as the additive, thus showing an improvement in LA conversion and selectivity towards the p,p'-DPA isomer (entries 20–25, Table 3). By adopting thiols with different chain lengths, the authors found that the phenol condensation was the rate-determining step, and the condensation rate decreased in the following order: ethanthiol > 1-propanethiol ≈ benzylthiol > 1-butanthiol > 2-propanethiol > 2-methyl-2-propanethiol. This suggests that the reaction rate is significantly affected by the thiol steric hindrance. These results agree with the work of Margelefsky et al. [36], who proposed that the formation of the electrophilic sulfonium intermediate accelerates the condensation rate towards the target product, thus preferentially shifting the regioselectivity towards the p,p'-DPA isomer thanks to the introduction of minor steric hindrance of the thiol side chain. In this context, the increase in the thiol steric hindrance significantly worsens the catalytic activity and p,p'-DPA selectivity, making necessary the employment of an accessible thiol group [30]. Moreover, it is well-known that the p,p'-DPA can isomerize to the o,p'-DPA isomer in the presence of acid catalysts; thus, Van De Vyver et al. investigated in greater depth the effect of thiol additives on this isomerization reaction [31]. They found that thiols scarcely affected the isomerization step and the second condensation with phenol to give DPA, whilst such additives play a key role in determining the regioselectivity of the first condensation step of LA with phenol to give MPA. An appropriate balance between the acidity of the catalyst and the steric hindrance and the amount of the thiol additive represents the right solution to achieve high p,p'-DPA yields. In particular, the choice of a low SO_3H/SH ratio is effective in favoring the reaction rate and increasing the p,p'-DPA/o,p'-DPA molar ratio (compare entry 23 with entry 26, Table 3). Based on such promising results, the authors also investigated the post-synthetic modification of SHPAOs by incorporating aminothiols, such as 2-mercaptoethylamine (SHPAOs-MEA) and 4-(2-thioethtyl)-pyridine (SHPAOs-TEP) (entries 27 and 28, Table 3). Both of these catalysts led to a significant improvement in the LA conversion and regioselectivity control to the p,p'-DPA when compared with the absence of the thiol additive (entry 19, Table 3). More recently, Zhu et al. [19] prepared magnetic catalysts to be employed in the DPA synthesis from LA. In this context, an N-doped carbon nanotube modified with -SO_3H groups encapsulating Fe, Ni, or Co nanoparticles (Fe@NC-SO_3H, Ni@NC-SO_3H, Co@NC-SO_3H) were synthesized. All of the investigated heterogeneous catalysts gave the best performances in terms of higher LA conversions, p,p'-DPA/o,p'-DPA molar ratios, and p,p'-DPA yields (entries 29–31, Table 3) compared to those achieved with H_2SO_4 under the same reaction conditions (entry 4, Table 2). Remarkably, Co@NC-SO_3H was the most promising catalyst, leading to the highest p,p'-DPA yield of 63 mol% (entry 31, Table 3). The authors attributed these valuable catalytic performances to the regular microstructure and larger specific surface area of Co@NC-SO_3H, which resulted in a more uniform distribution and easier accessibility of the SO_3H groups on the catalyst's surface. At the end of the reaction, the catalyst was magnetically separated and employed in recycling tests, showing good performances after four cycles. The authors attributed the high stability of this catalyst to its encapsulated structure, which protects the internal metal nanoparticles from unwanted degradation processes, such as leaching. The stability of the recovered nanotubes was experimentally confirmed by verifying the absence of Co in the reaction medium, the uniform distribution of -SO_3H groups on the catalyst's surface, and the presence of encapsulated Co nanoparticles. In addition, Zhu et al. [19] studied the role of thiol in the improvement of p,p'-DPA production by adopting the same Co@NC-SO_3H system. Again, a small thiol (namely, mercaptoacetic acid) was effective in improving both LA conversion and the p,p'-DPA/o,p'-DPA molar ratio (entry 32, Table 3). The adoption of longer reaction times (from 24 to 48 h) improved the LA conversion and the p,p'-DPA yield (91 mol%), keeping almost constant the p,p'-DPA/o,p'-DPA molar ratio (entry 33, Table 3). The authors also investigated the catalytic performances of mercaptoacetic acid and Co@NC alone to exclude any relevant catalytic activity due to their individual contributions. In both cases, the LA conversion was lower than 5 mol% and no DPA was detected, thus proving that mercaptoacetic acid had a too-low acidity and that the -SO$_3$H groups were the effective active sites of the catalyst. An overall evaluation of the data related to sulfonated catalysts highlights good catalytic performances for the Amberlyst-15 sulfonic resin and, even better, excellent performances for the synthesized Co@NC catalytic system, which is easily recoverable and thermally stable.

Table 3. Heterogeneous catalysts with sulfonic acids applied to the synthesis of DPA.

Entry	LA/Phenol (mol/mol)	Catalyst (g$_{LA}$/g$_{cat}$)	Additive (mol$_{LA}$/mol$_{additive}$)	T (°C)	t (h)	C$_{LA}$ (mol%)	Y$_{p,p'\text{-DPA}}$ (mol%)	p,p'-DPA/o,p'-DPA Molar Ratio	Ref.
15	1/4	Amberlyst-15 (6.1)	/	120	24	64	6	0.4	[27]
16	1/4	Amberlyst-15 (4.4)	/	60	24	70	55	15.8	[19]
17	1/3	Amberlyst-15 (9.0)	/	100	16	34	14	4.0	[30]
18	1/3	Nafion NR50 (1.7)	/	100	16	36	24	5.8	[30]
19	1/3	SHPAOs [a] (7.6)	/	100	16	40	35	7.6	[30]
20	1/3	SHPAOs [a] (7.6)	Benzylthiol (15.5)	100	16	65	42	15.6	[30]
21	1/3	SHPAOs [a] (7.6)	Ethanethiol (15.5)	100	16	70	53	19.5	[30]
22	1/3	SHPAOs [a] (7.6)	1-Propanethiol (15.5)	100	16	65	49	17.6	[30]
23	1/3	SHPAOs [a] (7.6)	1-Butanethiol (15.5)	100	16	60	38	14.0	[30]
24	1/3	SHPAOs [a] (7.6)	2-Propanethiol (15.5)	100	16	54	38	12.0	[30]
25	1/3	SHPAOs [a] (7.6)	2-Methyl-2-propanethiol (15.5)	100	16	39	26	10.5	[30]
26	1/3	SHPAOs [a] (7.6)	1-Butanethiol (3.9)	100	32	93	n.a. [d]	20.0	[31]
27	1/3	SHPAO$_s$-MEA [b] (8.2)	/	100	16	59	38	9.8	[30]
28	1/3	SHPAOs-TEP [c] (10.1)	/	100	16	57	35	15.5	[30]
29	1/4	Fe@NC-SO$_3$H [d] (1.2)	/	60	24	69	49	17.5	[19]
30	1/4	Ni@NC-SO$_3$H [d] (1.9)	/	60	24	72	57	17.2	[19]
31	1/4	Co@NC-SO$_3$H [d] (2.1)	/	60	24	74	63	16.9	[19]
32	1/4	Co@NC-SO$_3$H [d] (2.1)	Mercaptoacetic acid (5.0)	60	24	82	76	24.4	[19]
33	1/4	Co@NC-SO$_3$H [d] (2.1)	Mercaptoacetic acid (5.0)	60	48	98	91	23.7	[19]

[a] Sulfonated hyperbranched poly(arylene oxindole)s; [b] Sulfonated hyperbranched poly(arylene oxindole)s modified with 2-mercaptoethylamine; [c] Sulfonated hyperbranched poly(arylene oxindole)s modified with 4-(2-thioethyl)-pyridine; [d] Sulfonated N-doped carbon nanotube.

2.3. DPA Synthesis with Heteropolyacids

The heteropolyacids are another class of acid catalysts that have been employed in the synthesis of DPA. They have a higher acid strength than traditional solid acids (oxides and zeolites) and are less corrosive than mineral ones [16]. On this basis, several types of heteropolyacids have been synthesized and studied for DPA synthesis (Table 4) while working in the absence of thiol additives.

Table 4. Heteropolyacids applied to the synthesis of DPA.

Entry	LA/Phenol (mol/mol)	Catalyst (g_{LA}/g_{cat})	T (°C)	t (h)	C_{LA} (mol%)	$Y_{p,p'\text{-DPA}}$ (mol%)	p,p'-DPA/o,p'-DPA Molar Ratio	Ref.
34	1/4	$H_3PW_{12}O_{40}$ (4.0)	100	6	55	46 [a]	n.a. [b]	[16]
35	1/4	$H_4SiW_{12}O_{40}$ (4.0)	100	6	69	25 [a]	n.a. [b]	[16]
36	1/4	$H_4SiMo_{12}O_{40}$ (6.4)	100	6	79	17 [a]	n.a. [b]	[16]
37	1/4	$H_3PW_{12}O_{40}$ (4.0)	140	6	87	82(85 [a])	28.0	[16]
38	1/7	$H_3PW_{12}O_{40}$ (7.9)	100	24	n.a. [b]	1	2.4	[20]
39	1/3	$H_3PW_{12}O_{40}$ (4.2)	100	16	55	31	8.7	[30]
40	1/3	$H_3PW_{12}O_{40}$ (8.0)	100	6	33	18	1.5	[18]
41	1/3	$H_6P_2W_{18}O_{62}$ (8.0)	100	6	39	29	3.5	[18]
42	1/4	$H_3PW_{12}O_{40}/SiO_2$-E-4.0 [c] (8.0)	100	8	24	1	3.6	[32]
43	1/4	$H_3PW_{12}O_{40}/SiO_2$-E-7.5 [c] (8.0)	100	8	31	2	2.8	[32]
44	1/4	$H_3PW_{12}O_{40}/SiO_2$-E-14.5 [c] (8.0)	100	8	75	21	3.0	[32]
45	1/4	$H_3PW_{12}O_{40}/SiO_2$-E-17.5 [c] (8.0)	100	8	80	27	2.8	[32]
46	1/4	$H_3PW_{12}O_{40}/SiO_2$-C-7.5 [c] (8.0)	100	8	19	1	2.1	[32]
47	1/7	$H_3PW_{12}O_{40}/SiO_2$-3D$_{hex}$-C-7.5 [d] (7.9)	100	24	n.a. [b]	4	1.6	[20]
48	1/7	$H_3PW_{12}O_{40}/SiO_2$-3D$_{hex}$-C-11.1 [d] (7.9)	100	24	n.a. [b]	14	1.8	[20]
49	1/7	$H_3PW_{12}O_{40}/SiO_2$-3D$_{hex}$-C-14.5 [d] (7.9)	100	24	n.a. [b]	13	1.3	[20]
50	1/7	$H_3PW_{12}O_{40}/SiO_2$-3D$_{hex}$-C-15.8 [d] (7.9)	100	24	n.a. [b]	12	1.1	[20]
51	1/7	$H_3PW_{12}O_{40}/SiO_2$-3D$_{hex}$-C-29.9 [d] (7.9)	100	24	n.a. [b]	12	1.7	[20]
52	1/7	$H_3PW_{12}O_{40}/SiO_2$-3D$_{hex}$-C-35.2 [d] (7.9)	100	24	n.a. [b]	9	1.7	[20]
53	1/7	$H_3PW_{12}O_{40}/SiO_2$-3D$_{hex}$-C-43.6 [d] (7.9)	100	24	n.a. [b]	9	1.7	[20]
54	1/7	$H_3PW_{12}O_{40}/SiO_2$-2D$_{hex}$-C-11.7 [d] (7.9)	100	24	n.a. [b]	3	3.3	[20]
55	1/7	$H_3PW_{12}O_{40}/SiO_2$-2D$_{hex}$-E-11.1 [d] (7.9)	100	24	n.a. [b]	6	3.3	[20]
56	1/3	$Cs_{1.5}H_{4.5}P_2W_{18}O_{62}$ (8.0)	100	6	36	31	7.3	[18]
57	1/3	$Cs_{2.5}H_{0.5}PW_{12}O_{40}$ (8.0)	100	6	28	22	4.0	[18]
58	1/9	$Cs_{1.5}H_{4.5}P_2W_{18}O_{62}$ (4.0) [e]	150	10	n.a. [b]	62	7.3	[18]
59	1/4	$Cs_{2.5}H_{0.5}PW_{12}O_{40}$ (4.0) [e]	150	10	n.a. [b]	37	4.9	[18]

[a] Yield of p,p'-DPA + o,p'-DPA; [b] n.a.= Not available; [c] Calcined(C)/extracted(E)$H_3PW_{12}O_{40}$-silica materials with x wt% of $H_3PW_{12}O_{40}$; [d] Hexagonal(hex) calcined(C)/extracted(E) 3D/2D $H_3PW_{12}O_{40}$-silica materials with x wt% of $H_3PW_{12}O_{40}$; [e] The test was carried out with a stirring rate of 1200 rpm.

Rahaman et al. [16] prepared different Keggin-type heteropolyacids ($[XM_{12}O_{40}]^{n-}$), including phosphotungstic acid ($H_3PW_{12}O_4$), silicotungstic acid ($H_4SiW_{12}O_{40}$), and silicomolybdic acid ($H_4SiMo_{12}O_4$), and compared their catalytic performances under the same reaction conditions (entry 34–36, Table 4). The authors proved that all of the tested catalysts were active in LA conversion, which was always higher than 55 mol%, but only $H_3PW_{12}O_4$ gave a good DPA yield (46 mol%) while at the same time showing higher selectivity to the p,p'-DPA. Because of the higher productivity obtained with $H_3PW_{12}O_4$, it was employed for further optimization of the reaction. Regarding the catalyst amount, the authors found that its increased sped up LA conversion, but it did not affect the selectivity towards DPA. On the other hand, the temperature increase was effective for improving the LA conversion and the p,p'-DPA/o,p'-DPA molar ratio, thus proving that the regioselectivity control of this reaction greatly depends on the temperature. The highest p,p'-DPA/o,p'-DPA molar ratio (28.3) was reached at 140 °C, together with LA conversion of 87 mol% and a p,p'-DPA yield of 82 mol% (entry 37, Table 4). At the state of the art, $H_3PW_{12}O_4$ is the most studied heteropolyacid for the synthesis of DPA, and it was reported as the catalyst of reference in the works of Li et al. [20], Van De Vyver et al. [30], and Yu et al. [18]. However, lower LA conversions and p,p'-DPA yields were reported in all of these studies (entries 38–40, Table 4) when compared with the results of Rahaman et al. (entry 37, Table 4) [16]. Interestingly, Yu et al. [18] also employed Wells–Dawson phosphotungstic acid ($H_6P_2W_{18}O_{62}$), claiming higher LA conversion, p,p'-DPA/o,p'-DPA molar ratio, and p,p'-DPA yield (entry 41, Table 4) with respect to the Keggin phosphotungstic acid ($H_3PW_{12}O_4$). Some bottlenecks still limit the employment of heteropolyacids, including small specific surface area (1–5 m^2/g) and good solubilities in polar solvents, which limit their recovery/reuse. To overcome these issues, some smart strategies have been proposed, such as their immobilization on solid supports and the exchange of protons with large alkali cations, to form insoluble salts. The first approach was proposed by Guo et al. [32], who prepared some mesostructured silica-supported $H_3PW_{12}O_4$ catalysts with different heteropolyacids loadings. This goal was achieved through the contemporary hydrolysis and condensation of tetraethoxysilane (TEOS) with $H_3PW_{12}O_4$ in the presence of a template surfactant (Pluronic P123), followed by the hydrothermal treatment and template removal through calcination or extraction with boiling ethanol. This one-pot synthesis is smarter than the traditional post-synthesis grafting method, because the latter results in poor control of the heteropolyacid loadings and not negligible leaching phenomena. Moreover, the employment of a non-ionic surfactant (Pluronic P123) is effective for weakening the interactions between the template and the inorganic walls; thus, it can be easily removed (through solvent extraction or calcination at low temperatures) without damaging the catalyst structure. The authors investigated the effect of the $H_3PW_{12}O_4$ loading, finding that its increment was effective for improving the LA conversion and p,p'-DPA yield, but a slight decrease in the p,p'-DPA/o,p'-DPA molar ratio was also observed (entries 42–45, Table 4). Moreover, the catalytic activities of the catalysts having the same $H_3PW_{12}O_4$ loading but resulting from different template removal were compared, demonstrating that the extracted catalyst ($H_3PW_{12}O_{40}$/SiO_2-E-7.5, entry 43, Table 4) led to higher LA conversion and p,p'-DPA/o,p'-DPA molar ratio than the calcined one ($H_3PW_{12}O_{40}$/SiO_2-C-7.5, entry 46, Table 4). The authors ascribed this different catalytic behavior to the different textural properties, and, in particular, to the narrower pore sizes of $H_3PW_{12}O_{40}$/SiO_2-C-7.5 (6.0 nm) with respect to $H_3PW_{12}O_{40}$/SiO_2-E-7.5 (7.2 nm). Given the large dimensions of the DPA molecule, a catalyst with a large pore size is more suitable for allowing the accessibility of the reactants on the acid sites, as well as the next desorption of the produced DPA from its surface. On the other hand, the narrow pores may cause an increment of the internal mass transfer limitation and, consequently, slower substrate conversion. Moreover, the calcination of the catalyst at 420 °C caused the loss of some acid sites, leading to a lower activity of $H_3PW_{12}O_{40}$/SiO_2-C-7.5 with respect to $H_3PW_{12}O_{40}$/SiO_2-E-7.5. Finally, the authors investigated the recyclability of the prepared catalysts. At the end of the reaction, the catalyst was separated through filtration and then washed with water and ethanol and calcined at 120 °C under vacuum for 1 h. This spent catalyst was employed within two following reactions, partially losing the catalytic performances after the first recycle run and completely losing it after the second one. This evident worsening aspect was not due to the leaching of $H_3PW_{12}O_{40}$ (its absence in the reaction medium was verified through ICP-AES analysis), but rather the strong adsorption of DPA on the catalyst surface. This hypothesis was confirmed by restoring the initial activity through calcination of the spent catalyst at higher temperatures. The same authors deeply investigated the synthesis of these $H_3PW_{12}O_{40}$-silica materials and, properly adjusting the composition ratios of the precursors, temperature, and acidity of the solution, they prepared well-defined ordered mesoporous catalysts with a 2D or 3D hexagonal structure and tested them in the synthesis of DPA [20]. First, they studied the influence of the LA/phenol molar ratio; in principle, an excess of phenol was employed, but molar ratios that are too low could limit DPA formation due to the dilution effect of phenol towards LA. Once having identified the best LA/phenol molar ratio (fixed at 1/7), the authors studied the influence of the $H_3PW_{12}O_{40}$ loading supported on the 3D hexagonal structure (entries 47–53, Table 4). The highest p,p'-DPA yield of 14 mol% was achieved with the 11.1 wt% loading of $H_3PW_{12}O_{40}$, highlighting that lower loading gave insufficient active sites and that the reaction was particularly slow. On the other hand, higher loadings led to the aggregation of heteropolyacid units at the catalyst surface, thus reducing the available surface of the active sites. Moreover, considering the catalysts with similar $H_3PW_{12}O_{40}$ loading obtained through the same template removal method, the 3D hexagonal silica material led to higher catalytic activity than the 2D hexagonal one (compare entries 48 and 54, Table 4). In fact, the 3D morphology allows for more efficient transport of the reactant molecules in many more directions through an easier diffusion than the 2D morphology. Lastly, in the same work, the authors confirmed that the template removal through the extraction route should be preferred, as it allows for a higher structural ordering and catalytic activity (compare entries 54 and 55, Table 4). The recyclability of the best-performing catalyst ($H_3PW_{12}O_{40}/SiO_2$-3D$_{hex}$-C-11.1) was investigated and, as in the previous case, a bulk loss of activity after three runs was ascertained, which was mainly attributed to the strong adsorption of the DPA molecules on the catalyst surface. However, after calcination at 420 °C for 1 h, the pristine catalytic activity was almost completely restored. In developing a different approach, protons of heteropolyacids are often exchanged with large alkali cations, thus obtaining the corresponding insoluble salts in the polar reaction medium and improving the catalyst recovery. In this context, Yu et al. [18] partly substituted the protons of Keggin ($Cs_xH_{3-x}PW_{12}O_{40}$ with x = 1.0–2.5) and Wells–Dawson ($Cs_xH_{6-x}P_2W_{18}O_{62}$ with x = 1.5–4.5) heteropolyacids with cesium, and they found that the latter led to higher p,p'-DPA yields and p,p'-DPA/o,p'-DPA molar ratios than the Keggin-type catalysts when tested under the same reaction conditions (entries 56 and 57, Table 4). This different behavior depended on the different reaction mechanisms of Wells–Dawson and Keggin-type catalysts. In fact, the adsorption ability of Wells–Dawson heteropolyacids strongly depended on the content of cesium, which gradually decreased by increasing the cesium content, whilst the adsorption ability of Keggin heteropolyacids was related to their specific surface area. These properties suggested that with the Wells–Dawson catalysts, the reaction proceeded in the pseudo-liquid phase; thus, the polar molecules as LA can be rapidly adsorbed into the catalyst and react there. This pseudo-liquid phase mechanism often allows for reaching high catalytic activity and promising selectivity. On the other hand, with the Keggin heteropolyacids, the reaction proceeded according to a surface-type mechanism, where the adsorption–desorption step was generally slow, thus justifying the more interesting results achieved with the Wells–Dawson catalysts. The authors selected the most active catalysts for the Wells–Dawson and Keggin-type heteropolyacids (namely, $Cs_{1.5}H_{4.5}P_2W_{18}O_{62}$ and $Cs_{2.5}H_{0.5}PW_{12}O_{40}$, respectively) and studied the influence of several reaction parameters. They found that the increment of phenol improved the DPA yield without any appreciable change in selectivity to the p,p'-DPA isomer. On this basis, the optimal LA/phenol molar ratio was fixed to 1/9 and 1/4 for the $Cs_{1.5}H_{4.5}P_2W_{18}O_{62}$ and $Cs_{2.5}H_{0.5}PW_{12}O_{40}$, respectively. Analogously, the increment in the catalyst amount promoted the DPA yield, together with a negligible effect on the reaction selectivity. On the contrary, the temperature affected not only the DPA yield but also the p,p'-DPA selectivity. Moving from 80 to 150 °C, the p,p'-DPA selectivity increased from 70 to 85 mol% for the Keggin-type heteropolyacid and from 76 to 89 mol% for the Wells–Dawson one. Moreover, the authors proved the presence of external diffusion limitations with both catalysts by changing the stirring speed from 800 to 1200 rpm, with the latter further improving the DPA yield. Once having optimized the reaction conditions, the highest p,p'-DPA yields obtained with Wells–Dawson and Keggin-type heteropolyacids were 62 and 37 mol%, respectively (entries 58 and 59, Table 4). The recyclability of these heteropolyacids was investigated, and, for this purpose, the recovered catalysts were washed with ethanol and calcined at 100 °C for 1 h. The catalytic performances did not significantly change during three recycle tests, and the ICP-AES analysis of the reaction mixtures confirmed the absence of leaching phenomena, thus demonstrating the catalyst's stability. In conclusion, considering the available data on heteropolyacids, the best catalytic performances are achieved with cesium-based systems combining high yields/selectivity to the p,p'-DPA, with an excellent advantage for its recovery/reuse.

2.4. DPA Synthesis with Other Heterogeneous Systems: Zeolites and Modified Metal Silicas

Acid zeolites and metal oxides are often exploited in the field of acid catalysis due to their mild acidity, which is effective for reducing corrosion issues, as well as being more easily recoverable from the reaction medium and, in many cases, easily recyclable. Their catalytic performances strongly depend on the nature, number, and distribution of acid sites, differentiating them from Brønsted and Lewis ones. Such heterogeneous catalysts have also been employed for DPA synthesis (Table 5) working in the absence of thiol additives.

Table 5. Zeolites and modified metal oxides applied to the synthesis of DPA.

Entry	LA/Phenol (mol/mol)	Catalyst (g_{LA}/g_{cat})	T (°C)	t (h)	C_{LA} (mol%)	$Y_{p,p'\text{-DPA}}$ (mol%)	p,p'-DPA/o,p'-DPA Molar Ratio	Ref.
60	1/4	Pr-SO$_3$H-SBA-15 [c] (6.1)	120	24	38	2	0.4	[27]
61	1/4	Ar-SO$_3$H-SBA-15 [d] (6.1)	120	24	41	3	0.3	[27]
62	1/4	Nafion-SBA-15 (6.1)	120	24	38	5	0.5	[27]
63	1/4	n-ZSM-5 (6.1)	120	24	26	1	4.9	[27]
64	1/4	H-USY (6.1)	120	24	37	8	2.0	[27]
65	1/4	H-Beta 12.5 (6.1)	120	24	44	33	99	[27]
66	1/4	H-Beta 19 (6.1)	120	24	48	40 [a]	n.a. [b]	[27]
67	1/4	H-Beta 75 (6.1)	120	24	57	45 [a]	n.a. [b]	[27]
68	1/4	H-Beta 180 (6.1)	120	24	49	27 [a]	n.a. [b]	[27]
69	1/6	H-Beta 19 (4.6)	140	72	77	69	99	[27]

[a] Yield of p,p'-DPA + o,p'-DPA; [b] n.a. = not available; [c] Propylsulfonic-acid functionalized mesostructured silica; [d] Arenesulfonic-acid functionalized mesostructured silica.

Morales et al. [27] compared the catalytic activity of different modified metal oxides (propylsulfonic-acid mesostructured silica, arenesulfonic-acid functionalized mesostructured silica, and Nafion supported mesostructured silica, entries 60–62, Table 5), zeolites (n-ZSM-5, H-USY and H-Beta, entries 63–65, Table 5), and sulfonic-acid-based resins (Amberlyst-15, entry 15, Table 3) while keeping constant the reaction conditions. Amberlyst-15 was found to be the most active catalyst, leading to the highest LA conversion (64 mol%) but to a very low p,p'-DPA yield (6 mol%), thus highlighting a very poor selectivity. Similarly, the modified SBA-15 silicas showed modest LA conversions and very low DPA yields. These results can be attributed to the presence of strong Brønsted acid sites that promoted undesired pathways, thus causing the LA conversion to by-products. However, the authors did not provide further insight into such by-products. On the other hand, among the tested zeolites, the H-Beta allowed for the achievement of the highest LA conversion of 44 mol%, a p,p'-DPA yield of 33 mol%, and a p,p'-DPA/o,p'-DPA molar ratio of 99 when performing the reaction at 120 °C for 24 h (entry 65, Table 5), thus exploiting the appropriate combination of pore size, structure, type, and strength of acid sites. Moreover, a marked difference between the selectivity of the sulfonic catalysts and zeolites was ascertained: the former preferentially led to the formation of the o,p'-DPA, whereas the latter led to the desired p,p'-DPA. Because the best catalytic performances were achieved with the H-Beta zeolite, the authors focused on the investigation of the influence of the Si/Al ratio for this type of system by increasing it from 12.5 to 180 (entries 65–68, Table 5). The H-Beta 75 was the most active catalyst, thus leading to the highest LA conversion and DPA yield, but the selectivity, equal to 78 mol%, was lower than that achieved with the H-Beta 19 (84 mol%). The authors assigned the best catalytic performances of the H-Beta 19 due to the right Brønsted/Lewis acid sites ratio, which was even more thoroughly evaluated by considering the strong acid sites (strong Brønsted/strong Lewis acid sites ratio equal to about 2.0). After the optimization of the reaction using the Response Surface Methodology, the authors claimed that all of the investigated independent variables (namely, temperature, H-Beta 19 loading, and LA/phenol molar ratio) positively affected the LA conversion and overall DPA yield. The highest overall DPA yield of 70 mol%, with the corresponding LA conversion of 77 mol%, was achieved after 72 h at 140 °C by employing the LA/phenol molar ratio of 1/6. It is noteworthy that the excellent p,p'-DPA/o,p'-DPA molar ratio of 99 was kept under these reaction conditions, corresponding to a p,p'-DPA yield of 69 mol% (entry 69, Table 5). The recyclability of the H-Beta 19 catalyst was investigated in three subsequent runs after washing with acetone and drying at room temperature overnight. The recycled catalyst showed a decrease in DPA yield up to 28 mol% after the third run, with a negligible change in the LA conversion, thus highlighting a loss of selectivity. The authors proved the presence of organic compounds on the catalyst surface, which are responsible for the deactivation issues. However, the pristine activity was almost totally restored through calcination in air at 550 °C for 5 h.

On the basis of the above data, the use of zeolites for this condensation reaction is promising, considering that the zeolite pore size and structure must be adequate for the molecules of interest, as well as the type and strength of the acidity. In this regard, a moderate strength of acid sites should be preferred for p,p'-DPA synthesis in order to avoid undesirable side-reactions, which typically occur in the presence of strong Brønsted acid catalysts. For this purpose, the Si/Al ratio represents the key parameter affecting the acid properties of these catalysts, determining not only the amount and concentration of acid sites but also their nature (Brønsted and Lewis) and strength. Therefore, Beta zeolite with a moderate aluminum content (H-Beta 19, Si/Al = 23) represents the best catalyst to perform the solvent-free condensation between LA and phenol owing to the shape selectivity conferred by its structure and the adequate balance of acidity (Al content and speciation).

2.5. DPA Synthesis with Other Catalysts: Ionic Liquids

Ionic liquids are well-known, robust, and designable organic salts typically composed of large cations and small anions that can find applications as solvents and catalysts for carrying out the syntheses of several bio-based compounds [37]. In the field of DPA synthesis, acid ionic liquids have been proposed for the condensation reaction between LA and phenol. The most significant results available for such catalysts are summarized in Table 6.

Shen et al. [21] tested different ionic liquids, again proposing the use of ethanethiol as the additive for further improving the catalytic performances. Remarkably, it was found that [BSMim]CF$_3$SO$_3$ was the most active and selective towards the p,p'-DPA isomer, reaching the maximum LA conversion of 81 mol% and the p,p'-DPA yield of 79 mol% (entry 70, Table 6), a promising result attributed to the in situ formation of HF, which is mainly responsible for the acid catalysis. [BSMim]OAc and [BSMim]HSO$_4$ were also active, but the former led to only moderate results because of its lower acidity, thus demonstrating the key role of the anion (entries 71 and 72, Table 6). However, the cation also strongly influenced the catalytic activity of the ionic liquids, as shown by [BSMim]HSO$_4$ and [BPy]HSO$_4$ systems (entries 72 and 73, Table 6), with the former leading to higher LA conversion and p,p'-DPA yield due to the presence of the sulphonic acid in the [BSMim]HSO$_4$ cation. The importance of -SO$_3$H groups in the cation/anion of the ionic liquid is also highlighted by the lower LA conversion, p,p'-DPA yield, and p,p'-DPA/o,p'-DPA molar ratio achieved with [AMim]Br and [BMim]Cl (entries 74 and 75, Table 6). Moreover, the authors investigated the effect of the additive (ethanethiol) for the improvement of the catalysis with

[BSMim]CF$_3$SO$_3$, which was identified as the most active catalyst (entry 70, Table 6). For the test carried out in the absence of the additive (entry 76, Table 6), a significant worsening of the p,p'-DPA/o,p'-DPA molar ratio was ascertained, thus confirming that the regioselectivity of the reaction depends on the synergy between the type of ionic liquid and additive. However, although [BSMim]CF$_3$SO$_3$ was identified as the best catalyst, its synthesis is difficult and expensive; thus, the optimization of the DPA synthesis was developed employing [BSMim]HSO$_4$ as the catalyst, which allowed for achieving similar promising results (entry 72, Table 6). Temperature, LA/phenol molar ratio, LA/catalyst weight ratio, and time were optimized (60 °C, 1/4.5, 0.3, and 30 h, respectively), thus improving the p,p'-DPA yield up to 93 mol% without any marked drop in selectivity (entry 77, Table 6). At the end of the reaction, the ionic liquid was regenerated through ethyl acetate extraction, and it was recycled up to four runs. Only a slight decrease in the p,p'-DPA yield up to the value of 85 mol% was observed, which was mainly attributed to some loss of the ionic liquid that occurred during the separation step. Mthembu et al. [33] employed [EMIM][OTs] and [BMim]HSO$_4$, considering the catalysis with H$_2$SO$_4$ (entry 5, Table 2), p-TSA (entry 9, Table 2), and CH$_3$SO$_3$H (entry 11, Table 2), for comparison purposes. Maximum DPA yields of 59 and 68 mol% were obtained with such ionic liquids (entries 78 and 79, Table 6), which are comparable to those already reported for the homogeneous catalysts. In addition, Liu et al. [22] tested several ionic liquids with Brønsted acidity (for which the corresponding chemical structures are reported in Appendix A), highlighting an unclear relationship between their acid strength and the corresponding catalytic activity, which decreased in the order of 4b > 1a > 4a = 2 > 5 > 6 > 3 > 1c > 1d (entries 80–88, Table 6). Remarkably, the highest p,p'-DPA yield of 74 mol% was achieved with the ionic liquid 4b, which included a thiol group. Exploiting again the beneficial effect of the thiol group on this condensation reaction, the authors proposed the simultaneous use of mercaptoacetic acid and the most interesting ionic liquids (namely, 1a, 2, 3, 4a, 5, and 6) (entries 89–94, Table 6). The claimed results confirmed the positive role of thiol for the improvement of the p,p'-DPA yield, which increased up to 71–84 mol% depending on the employed ionic liquid. However, at the same time, a decrease in the p,p'-DPA/o,p'-DPA molar ratio was ascertained, which seems in disagreement with the findings reported in the literature. In any case, the thiol effect must be considered not only in terms of the thermodynamic aspects but also kinetic ones, and it should be better exploitable at lower LA conversions [31]. Due to the presence of the thiol group in the 1b and 4b ionic liquids, these were tested as additives together with the 4a ionic liquid as the main acid catalyst (entries 95 and 96, Table 6). The authors claimed further improvement of the p,p'-DPA yield and p,p'-DPA/o,p'-DPA molar ratio, proving that thiol-containing anions of ionic liquid play a key role in the optimization of p,p'-DPA synthesis.

According to the above data, the fine tunability of the catalytic properties of the ionic liquids certainly offers remarkable advantages for improving DPA synthesis by exploiting the Brønsted sulfonic groups to promote the LA conversion according to the general mechanism already discussed for the other Brønsted acids (Figure 2). Moreover, the inclusion of a thiol group within the anion of the ionic liquid remarkably improved both the yield and selectivity of the p,p'-DPA. Although mechanistic details have not been provided by the authors, the improved selectivity of the p,p'-DPA isomer could be attributed to the improved cooperation between the thiol group and its cationic counterpart in the ionic liquid. However, the main bottleneck limiting the use of ionic liquids for such industrial applications is their high cost in addition to their uneasy isolation/reuse, which requires an additional separation step of the reaction mixture using an appropriate solvent extraction given the high boiling points of the involved compounds.

Table 6. Ionic liquids applied to the synthesis of DPA.

Entry	LA/Phenol (mol/mol)	Catalyst (g_{LA}/g_{cat})	Additive ($mol_{LA}/mol_{additive}$)	T (°C)	t (h)	C_{LA} (mol%)	$Y_{p,p'\text{-DPA}}$ (mol%)	p,p'-DPA/o,p'-DPA Molar Ratio	Ref.
70	1/4	[BSMim]CF$_3$SO$_3$ [a] (0.6)	Ethanethiol (100)	60	24	81	79	100	[21]
71	1/4	[BSMim]OAc [b] (0.8)	Ethanethiol (100)	60	24	35	34	100	[21]
72	1/4	[BSMim]HSO$_4$ [c] (0.7)	Ethanethiol (100)	60	24	75	74	100	[21]
73	1/4	[BPy]HSO$_4$ [d] (1.0)	Ethanethiol (100)	60	24	68	67	90	[21]
74	1/4	[AMim]Br [e] (1.1)	Ethanethiol (100)	60	24	18	10	3.1	[21]
75	1/4	[BMim]Cl [f] (1.3)	Ethanethiol (100)	60	24	14	11	4.6	[21]
76	1/4	[BSMim]CF$_3$SO$_3$ [a] (0.6)	/	60	24	73	70	22.3	[21]
77	1/4.5	[BSMim]HSO$_4$ [c] (0.3)	Ethanethiol (100)	60	30	n.a. [g]	93	100	[21]
78	1/3.7	[EMIM][OTs] [h] (0.9)	/	75	6	n.a. [g]	59 [i]	n.a. [g]	[33]
79	1/3.7	[BMim]HSO$_4$ [j] (0.9)	/	75	6	n.a. [g]	68 [i]	n.a. [g]	[33]
80	1/4	1a (0.8) [k]	/	60	48	70	52	16.0	[22]
81	1/4	1c (0.7) [k]	/	60	48	n.a. [g]	34	32.0	[22]
82	1/4	1d (0.6) [k]	/	60	48	n.a. [g]	13	30.0	[22]
83	1/4	2 (0.6) [k]	/	60	48	n.a. [g]	48	21.0	[22]
84	1/4	3 (0.7) [k]	/	60	48	n.a. [g]	43	33.0	[22]
85	1/4	4a (0.8) [k]	/	60	48	n.a. [g]	48	32.0	[22]
86	1/4	4b (0.8) [k]	/	60	48	n.a. [g]	74	50.0	[22]
87	1/4	5 (0.7) [k]	/	60	48	n.a. [g]	47	33	[22]
88	1/4	6 (0.7) [k]	/	60	48	n.a. [g]	46	33	[22]
89	1/4	1a (0.8) [k]	Mercaptoacetic acid (20.0)	60	48	n.a. [g]	71	9.0	[22]
90	1/4	2 (0.6) [k]	Mercaptoacetic acid (20.0)	60	48	n.a. [g]	72	9.0	[22]
91	1/4	3 (0.7) [k]	Mercaptoacetic acid (20.0)	60	48	n.a. [g]	78	9.0	[22]
92	1/4	4a (0.8) [k]	Mercaptoacetic acid (20.0)	60	48	n.a. [g]	84	20.0	[22]
93	1/4	5 (0.7) [k]	Mercaptoacetic acid (20.0)	60	48	n.a. [g]	73	13.0	[22]
94	1/4	6 (0.7) [k]	Mercaptoacetic acid (20.0)	60	48	n.a. [g]	73	9.0	[22]
95	1/4	4a (0.9) [k]	1b (20.0)	60	48	n.a. [g]	85	100	[22]
96	1/4	4a (0.9) [k]	4b (20.0)	60	48	n.a. [g]	91	100	[22]

[a] 1-(4-sulphonic acid)butyl-3-methylimidazolium triflate; [b] 1-(4-sulphonic acid)butyl-3-methylimidazolium acetate; [c] 1-(4-sulphonic acid)butyl-3-methylimidazolium hydrogen sulphate; [d] 1-butylpyridinium hydrogen sulphate; [e] 1-allyl-3-methylimidazolium bromide; [f] 1-butyl-3-methylimidazolium chloride; [g] n.a. = not available; [h] 1-ethyl-3-methylimidazolium tosylate; [i] yield of p,p'-DPA + o,p'-DPA; [j] 1-butyl-3-methylimidazolium hydrogen sulphate; [k] chemical structure is reported in Appendix A.

3. Applications of DPA: Challenges and Opportunities

The presence of one carboxyl group and two phenolic ones makes DPA an interesting molecule to be exploited for the synthesis of a plethora of more added-value bio-products. On the other hand, the reactivity of DPA must be properly controlled to avoid unwanted reactions, including its self-condensation [29,38]. Moreover, it can undergo oxidation and degradation over time [39,40], which can limit its shelf life and usability in several applications, thus requiring careful attention to its stability during the synthesis of the DPA-derived polymers; for instance, when transforming the carboxyl group of DPA in esters, ethers, or amide derivatives [41,42]. To date, DPA has been exploited for the synthesis of epoxy and phenolic resins [6,43] as well as other polymers, such as polycarbonates [44,45], polyarylates [46], and polyesters [47]. Both DPA-based epoxy and phenolic resins find applications in the production of composites, adhesives, and coatings [48,49]. Some niche applications of DPA have been proposed, including the production of thermal paper for printers [50], coatings, and adhesives [51–53]. Moreover, DPA and its derivatives show good antioxidant, antiviral, and antibacterial properties, making them suitable for the development of valuable applications in the cosmetic, food, textile, and polymer sectors [27,33,54,55]. Such materials could be obtained from the diphenolate derivatives. In this context, alkyl levulinates (ALs) were proposed by Olson in 2001 [56] as a replacement for LA in the synthesis of new diphenolate esters (DPEs). Remarkably, the reaction between ethyl levulinate (EL) and phenol under conditions analogous to those used for LA enables the production of DPE, but only in low yields, thus highlighting a much more moderate reactivity of the ALs. However, when the reaction conditions were properly modified, very high yields of the DPE (95 mol%) were ascertained, which were even greater than those obtained when starting from LA (67 mol%). The most efficient method for the synthesis of DPE from EL was achieved by using concentrated sulfuric acid as the catalyst, followed by dilution with ethanol at the end of the reaction. The residual reactants were steam-distilled off the product, but no further details regarding synthesis or purification were reported. Pancrazzi et al. [57] performed the one-pot synthesis of DPEs, which involved the LA's esterification to EL and the cascade condensation of the latter with phenol and *ortho*-substituted phenols (R = 2-Me, 2-iso-Pr, 2-sec-Bu) without any intermediate purification step of ethyl levulinate. In principle, methyl and ethyl levulinate are the preferred ALs to be employed to exploit their highest esterification yields [58]. Therefore, two different approaches can be proposed for the synthesis of DPEs by employing ALs as starting materials (Figure 4).

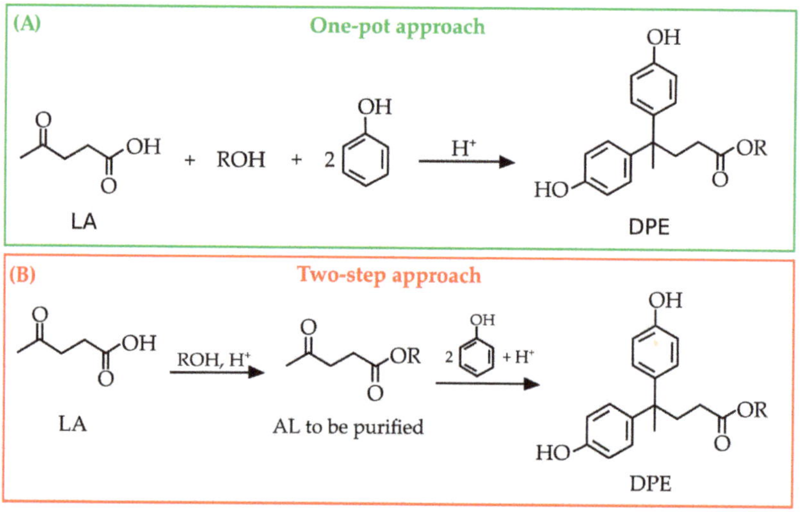

Figure 4. Comparison between the one-pot approach (**A**) and the two-step one (**B**) for the synthesis of DPE derivatives from ALs according to Pancrazzi et al. [57].

The one-pot approach should be preferred over the two-step one because the former allows for a significant reduction in the work-up costs. Certainly, ALs should be preferred over LA for reducing the formation of by-products, especially when working under heterogeneous catalysis. Remarkably, the use of esters can open the way to the production of a new class of DPE-based polymers, such as bis epoxy and polyurethane derivatives [56], thus expanding the possibilities for applications with respect to the traditional DPA-based polymers. In the following paragraphs, these main applications will be discussed, with an emphasis on their relevant advantages and limitations over the corresponding BPA-based materials.

3.1. Epoxy Resins

DPA-based epoxy resins are employed for the synthesis of thermosetting materials, including coatings, adhesives, matrix resins, and others [48,59]. The presence of DPA within epoxy resins confers remarkable valuable properties, including lightweight and high-strength composites [60]. Moreover, epoxy resins containing DPA could be used for encapsulating and protecting electronic devices (semiconductors, printed circuit boards, and sensors), thus valorizing their excellent electrical insulation properties, thermal stability, and protection against moisture and contaminants [48,61,62]. Furthermore, epoxy resins are used to create molds and patterns for casting various materials, including plastics and metals, with the latter showing promising dimensional stability and heat resistance [60].

In deepening the exploitation possibilities of DPA, it is well-known that the synthesis of BPA glycidyl ethers (DGEBA) represents valuable polymers due to their robust mechanical properties, favorable thermodynamic characteristics, and remarkable stability [63]. DGEBA and its oligomers account for 90% of the global production of epoxy resins [64]. A similar approach to that of BPA can also be proposed for DPA, thus obtaining the corresponding glycidyl ethers (DGEDPA). Both chemical structures of the corresponding glycidyl ether monomers are reported in Figure 5.

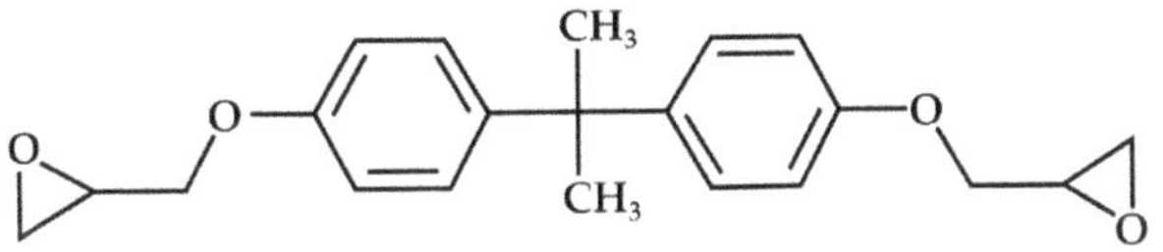

DGEBA

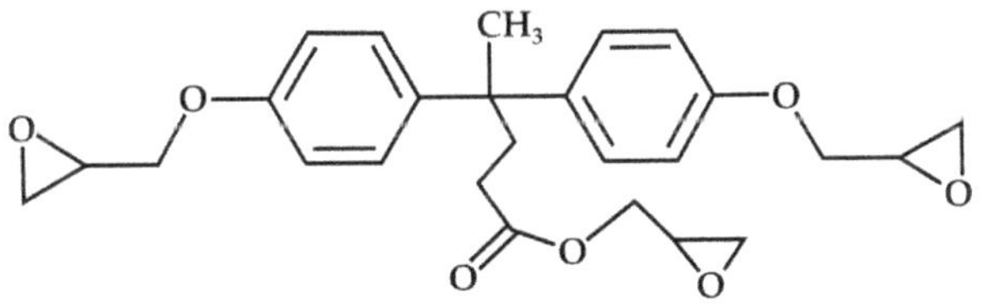

DGEDPA

Figure 5. Comparison between the chemical structures of DGEBA and DGEDPA.

The research towards the development of new DGEDPA-based materials is mainly aimed at solving some well-known DGEBA performance drawbacks, including high brittleness and poor impact resistance, due to its inherent rigid aromatic rings [65]. DGEDPA-based epoxy thermosets characterized by a hyperbranched structure may enhance mechanical strength and toughness simultaneously [66]. The hyperbranched structure generated through the reactivity of the carboxyl group of the DPA can significantly increase the crosslink density in thermosets, thereby enhancing their mechanical strength. This crosslinking could be obtained through the synthesis of ethers, esters, and amide derivatives of the DPA, thus exploiting the reactivity of its carboxylic group. In addition, hyperbranched polymers have a higher proportion of free volume than their linear counterparts, which helps to improve the toughness of DPA-based epoxy resins [67]. In this context, Wei et al. [42] synthesized a new DGEDPA-based polymer starting from DPA and epichlorohydrin to obtain a trifunctional epoxy resin. The obtained DGEDPA was cured with 3,3′-dimethyl-4,4′-diaminodicyclohexylmethane, and the final material exhibited better performance than that of the DGEBA-based one, including higher curing temperature, glass transition temperature, and thermal expansion characteristics [42]. Consequently, the authors claimed that this innovative epoxy resin shows promising potential for various critical applications, including aerospace and microelectronics. Regarding the electronic applications, McMaster et al. [68] tested a series of diglycidyl ethers of diphenolate esters (DGEDPE) as dielectric materials to be proposed in substitution of DGEBA. Methyl, ethyl, propyl, and butyl esters of DPA were synthesized and studied. Remarkably, the DGEDP-propyl showed the highest dielectric constant, with results comparable to those of the traditional DGEBA. The authors stated that variations in the dielectric characteristics of DGEDP-esters result from the complex interplay between segmental, local, and side-chain movements on one side and the influence of free volume and steric hindrance on the other. The research formed the groundwork for the development of new DGEDPE-based epoxy resins, with the intention to improve polarization and dielectric constants, which are valuable physical properties for such types of applications. Developing a different approach, Xiao et al. [65] synthesized different DPA-based hyperbranched epoxides (HBEPs) from DPA and dibromobutane to improve the mechanical properties of the final resin. A two-step synthetic approach was proposed by the authors (Figure 6). In the first step, hyperbranched polymers (HBPs) were synthesized through polymerization through a condensation mechanism starting from DPA and 1,4-dibromo butane in the presence of potassium carbonate and DMF. In the second step, the epoxidation reaction of HBPs to obtain HBEPs was carried out in the presence of an excess of epichlorohydrin. In this way, terminal functional groups were modified through epoxidation through epichlorohydrin, thus obtaining the epoxy resin. This methodology utilizes DPA as the branching component, thus improving the mechanical durability due to its aromatic ring composition and adding a well-tuned flexibility by including dibromo butane. Such hyperbranched thermosets exhibited exceptional mechanical strength and toughness, with much better performances than those of commercial DGEBA thermosets [65,69].

Qian et al. [41] proposed another innovative synthesis of a new class of epoxy resin with outstanding thermal performance, DPA-based polybenzoxazines, based on the amidation of DPA. First, the authors synthesized DPA hexanediammonium salt (DHDA) by reacting DPA and 1,6-hexanediamine in ethanol. Then, DHDA reacts with PEG-200, furfurylamine, and paraformaldehyde to obtain benzoxazine diphenolic hexanediamide (DHDA-fa) (Figure 7).

Polybenzoxazines offer distinct advantages compared to most conventional polymers. These advantages include minimal volumetric changes during curing, high glass transition temperatures, impressive thermal and mechanical characteristics, exceptional flame resistance, and a low dielectric constant [70]. The authors demonstrated that the incorporation of amide groups into a benzoxazine molecule can effectively improve the thermomechanical performances of these resins with respect to those of DPA-derived benzoxazines containing the free-ending carboxylic group in the DPA structure [41].

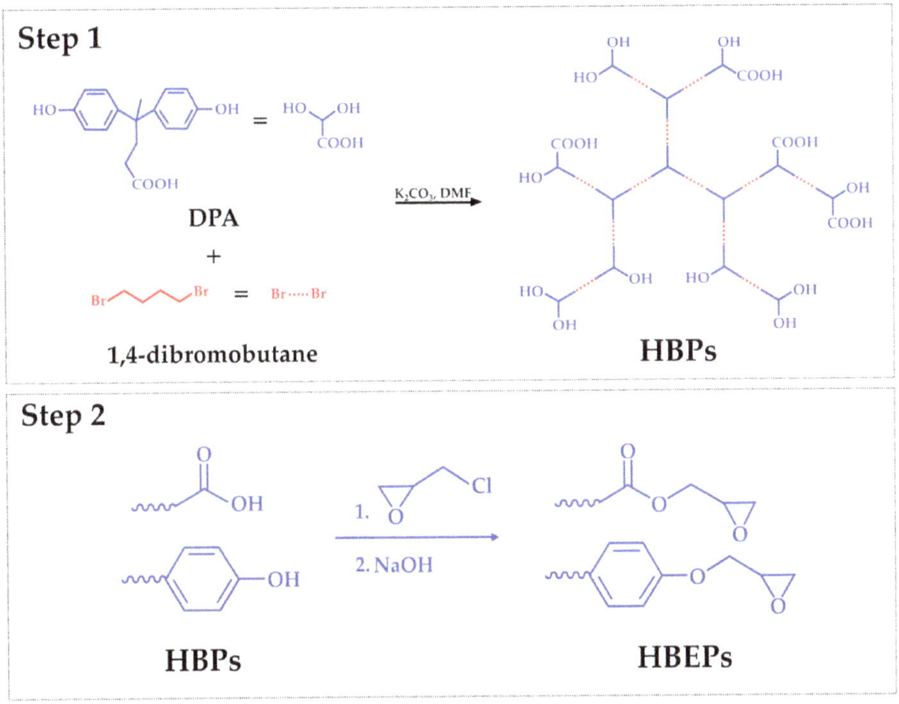

Figure 6. Synthesis of HBEPs, according to Xiao et al. [65].

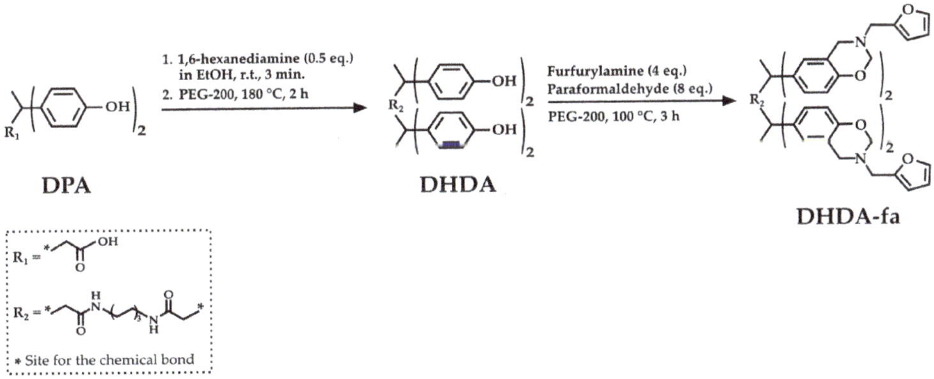

Figure 7. Chemical structure of DPA-derived benzoxazine diphenolic hexanediamide, according to Qian et al. [41].

3.2. Phenolic Resins

Phenolic resins are well-known for their heat resistance and durability, making them suitable for producing electrical and automotive components, as well as in the construction industry [71]. DPA has been employed for the synthesis of phenolic resins, including, in particular, novolaks [72,73]. In this context, Kiran et al. [74] synthesized a sulfonated poly(arylene ether sulfone) copolymer through direct copolymerization of *p,p′*-DPA, benzene 1,4-diol, and the synthesized sulfonated 4,4′-difluorodiphenylsulfone (Figure 8).

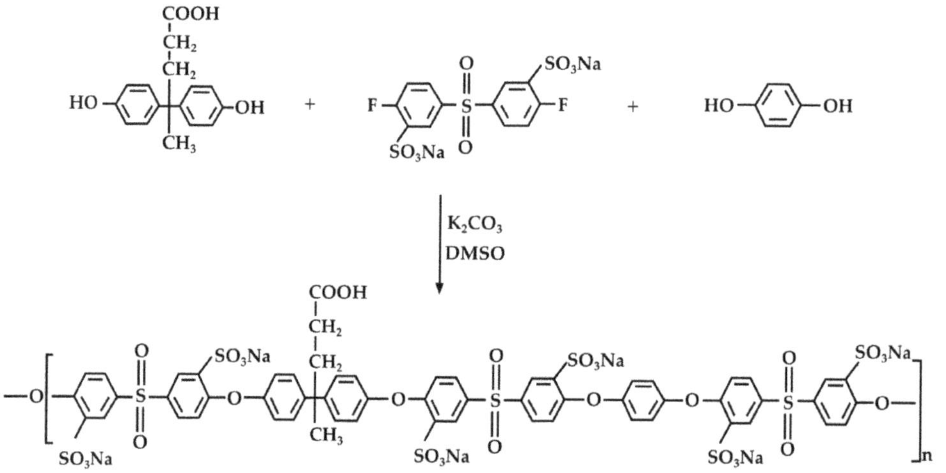

Figure 8. Synthesis of sulfonated poly(arylene ether sulfone) copolymer starting from DPA, according to Kiran et al. [74].

The copolymer was subsequently crosslinked with 4,4′-(hexafluoroisopropylidene) diphenol epoxy resin through a thermal curing reaction to synthesize crosslinked membranes for fuel cell applications. Zúñiga et al. [75] claimed the synthesis of a DPA-based porous polybenzoxazine as the phenolic resin. For this purpose, the foaming agent (CO_2) was generated in situ during the thermal curing. The DPA-benzoxazine monomer was thermally polymerized at different temperatures according to the reaction in Figure 9. According to the authors, the proposed synthesis is smart, as it uses the same chemical both as a crosslinking monomer and as a blowing agent. Moreover, the relative decarboxylation reaction can be controlled by varying the temperature, which gives an excellent and simple method for tuning foam properties. The obtained low-density foams exhibit a high glass transition temperature, with heterogeneous open-cell morphology.

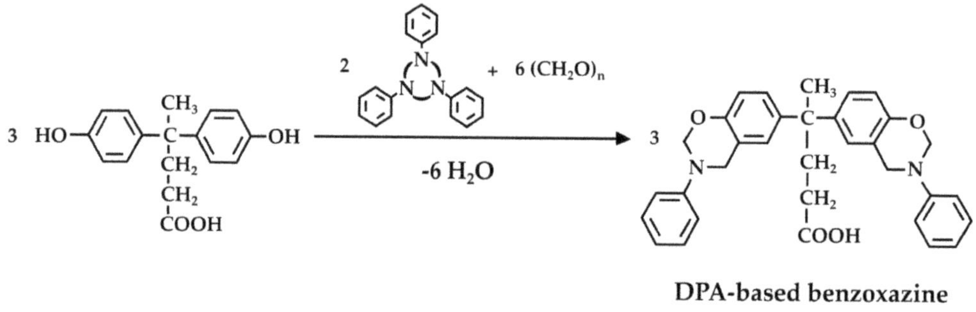

Figure 9. Synthesis of DPA-based polybenzoxazines according to Zúñiga et al. [75].

An alternative to DPA-based sulfonated poly(arylene ether sulfone) copolymers and porous polybenzoxazine is represented by DPE-based resins. To improve the thermal stability of aromatic polyesters and polycarbonates prepared from DPA, various DPEs have been proposed for the synthesis of new polymers, including methyl diphenolate (MDP), ethyl diphenolate (EDP), *n*-propyl diphenolate (PDP), *n*-butyl diphenolate (BDP), and *n*-hexyl diphenolate (HDP) [44,76]. For example, Ping et al. [76] synthesized two aromatic polyesters, poly(MDP–IPC) and poly(EDP–IPC), via interfacial polycondensation (IPC) of isophthaloyl chloride with methyl diphenolate (MDP) and ethyl diphenolate (EDP), respectively, and studied their thermal stability by considering the poly(DPA–IPC) as the reference. The authors found that the presence of carboxylic groups in poly(DPA–IPC) leads to lower thermal stability addressed to carboxyl group interactions, which can be weakened by working with the ester. In this context, poly(MDP–IPC) and poly(EDP–IPC) display single-stage thermal decomposition at higher temperatures, with slightly reduced glass transition temperatures due to increased side-chain flexibility. Overall, at the state of the art, DPE-based polymers exhibit better thermal performance and behavior than DPA-based ones.

3.3. Thermal Papers

DPA could also be used as a sustainable alternative to BPA to produce thermal papers, which are commonly used in point-of-sale (POS) receipt, fax, and label printers [50]. Thermal papers enable the production of images after exposure to a heat source, thus eliminating the need for the use of inks or toners. Generally, the thermosensitive layer contains a fluoran-type dye (leuco dye) serving as the thermal dye and a color developer, such as BPA, as the proton donor. These chemical components are responsible for color development after heating, according to Figure 10.

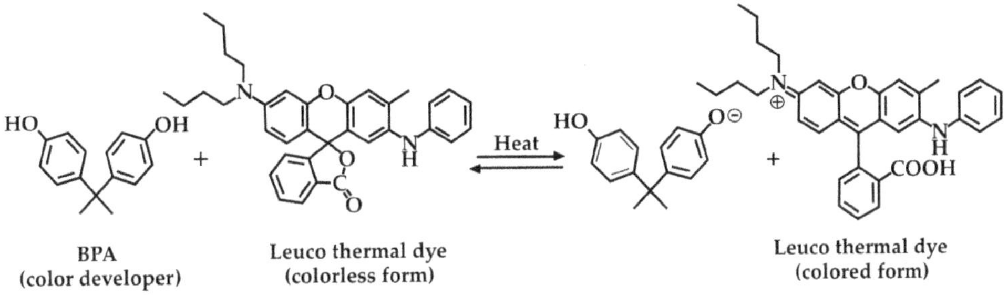

Figure 10. Reaction between BPA (color developer), and the leuco thermal dye (colorless) to give the colored open form of the dye [50].

The U.S. Environmental Protection Agency (EPA) has issued warnings about using BPA on thermal paper and is searching for possible substitutes, including DPA-based ones [77]. Moreover, DPA and its derivatives were proposed as drying agents for inkjet receptor media, as reported by Malhotra et al. [78]. In this context, when DPA is used in combination with a multivalent salt and a surfactant, it can efficiently dehydrate the medium, resulting in a better quality of the final printed product.

3.4. Coatings and Adhesives

DPA also finds applications for the formulation of coatings and adhesives [51–53]. To date, most of these applications of DPA are aimed at improving adhesion promotion, corrosion resistance, thermal stability, resin modification, adhesive strength, electrical insulation, and UV resistance [79]. Remarkably, Yan et al. [52] first synthesized the poly(diphenolic acid-phenyl phosphate) [poly(DPA-PDCP)] from DPA (Figure 11) and then used it in combination with a polyethylenimine (PEI) to produce, through layer-by-layer self-assembly, a flame-retardant surface coating for ramie fabric. Poly(DPA-PDCP)/PEI showed excellent flame-retardant properties, thus finding many potential applications in the automobile, train, and aerospace fields [53].

Furthermore, DPA is effective for improving the UV resistance of coatings by helping to prevent color fading and degradation when exposed to sunlight [79,80]. This is a valuable property for outdoor applications, such as architectural coatings and automotive finishes. For example, in the study by Zhang et al. [79], a novel, multifunctional, bioderived polyphosphate (PPD) for poly(lactic acid) (PLA) was synthesized through a three-step procedure (Figure 12). In the first step, 9,10-dihydro-9-oxa-10-phosphaphanthrene-10-

oxide (DOPO) reacted with polyformaldehyde (HCHO) to produce a DOPO-derived compound, 2-(6-oxido-6H-dibenz[*c,e*][1,2]oxaphosphorin-6-yl)-methanol (ODOPM). ODOPM reacted with DPA in the presence of N, N-dimethylformamide (DMF) and sulfuric acid (as the acid catalyst) to produce (6-oxidodibenzo[*c,e*]› [1,2]oxaphosphinin-6-yl)methyl 4,4-bis(4-hydroxyphenyl)pentanoate (DPOM). Lastly, DPOM reacted with phenylphosphonic dichloride (PPDC) in ethyl acetate (EAC) and triethylamine (Et$_3$N) to give the target PPD.

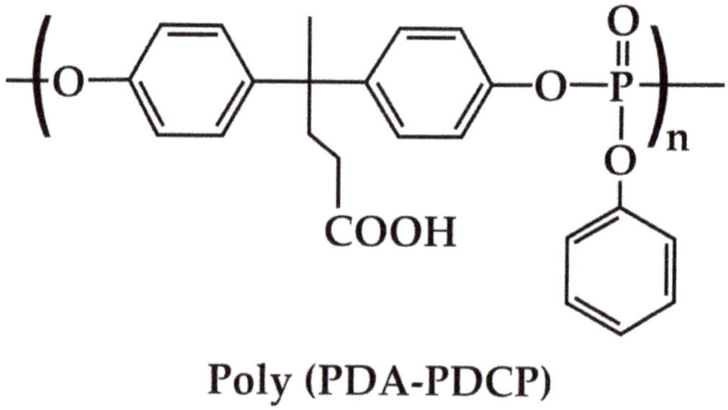

Figure 11. Structure of the [poly(DPA-PDCP)].

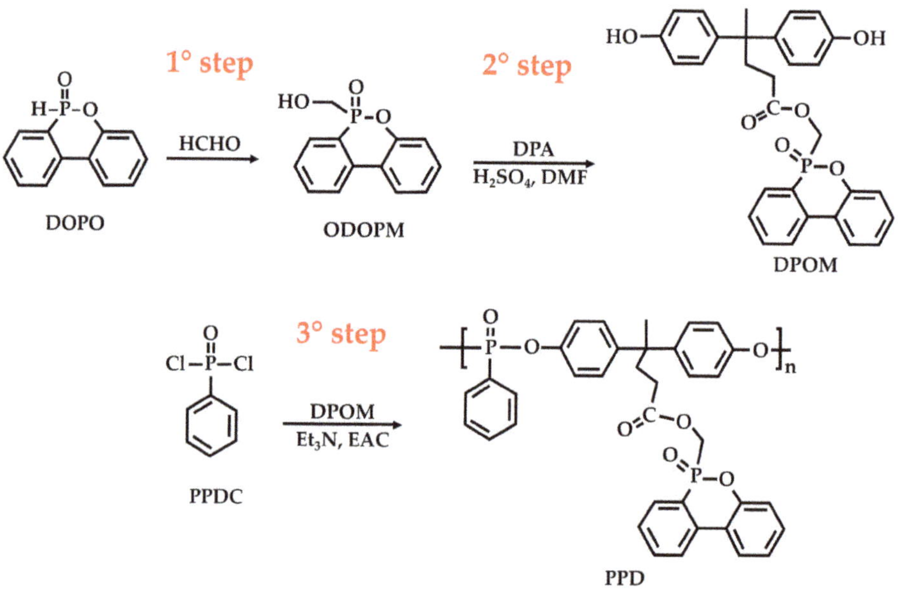

Figure 12. Synthesis of PPD, according to Zhang et al. [79].

3.5. Antioxidant Properties

DPA shows interesting antioxidant properties, which could potentially be exploited for various applications, such as in the packaging industry to extend the shelf life of products and in cosmetics for their anti-ageing effects. Moreover, DPA and its derivatives can be used in paint formulations, decorative surface coatings, additives for lubricating oil, plasticizers, surfactants, cosmetics, and the textile industry [27,54,55,81,82]. For example, in the rubber industry, DPA is employed as an antioxidant additive, thus helping to protect such products from the degradation of oxygen, heat, and other agents [83]. This can extend the lifespan of rubber components used in automotive and industrial applications. Similarly, in the paint industry, the addition of DPA derivatives can increase the durability of commercial paint formulations. As reported in the patent of Holmen et al. [82], such an improvement was achieved through the incorporation of a DPA-based polyamide. Even at the low amount of 1 wt% of DPA within the polyamide modifier, this formulation improves the longevity of an applied coating. In all of these applications, DPA serves as a valuable antioxidant additive by effectively scavenging free radicals and inhibiting the oxidation of materials, thereby extending the lifespan, maintaining performance, and ensuring the quality of various products and materials.

3.6. Other Possible Applications

Finally, DPA-coated surfaces show good antivirus and antibacterial properties. This valuable biological activity is attributed to the interaction between the DPA molecules and the membrane proteins of the pathogens, thus causing deformation of the cell membrane and the cell lysis [84]. In this context, Shen et al. [84] demonstrated the active biological activity of Co-DPA-1.0% fabric surface towards common pathogens, like *Escherichia coli* and *Staphylococcus aureus*. DPA can be employed as an antibacterial agent in textile finishes and coatings to enhance the resistance of textiles to biological factors, which can cause fading and degradation. For example, Shen et al. [84] employed DPA as a finishing reagent to impart antibacterial and antiviral functions to cotton fabric. After a simple pad–dry–cure process, DPA molecules were covalently linked onto cotton fiber surfaces via the esterification reaction between their carboxyl groups and the hydroxyl groups of cellulose on the fiber surfaces. The DPA phenolic groups on the cotton fibers enable the destruction of the pathogens through protein affinity interactions. Experimental results show that the DPA-modified fabrics realize not only a high bacteriostatic effect against both *Escherichia coli* and *Staphylococcus aureus* but also an excellent antivirus effect that allows for rapid virus inactivation in less than 20 min. It was also demonstrated that the modification process generated insignificant damage to the cotton fiber structure, and the resultant cotton fabrics are safe for human skin. Therefore, this work may open a new way to endow cotton textiles with antibacterial and antiviral functions. Moreover, some phenolic acids have been studied for their potential biological activity, including anti-inflammatory and anti-cancer properties [85,86], thus opening the way to the development of new therapeutic applications.

4. Conclusions

This review focused on the synthesis and potential applications of diphenolic acid, a valuable alternative to bisphenol A, and demonstrated that its synthesis from the condensation of levulinic acid and phenol still has many challenges and critical issues to be solved. The feasibility of this reaction has been variously investigated, requiring, similarly to bisphenol A, the use of a Brønsted acid catalyst, mild temperature, and long reaction time. Mineral acids give good results in terms of yield (about 70 mol%) at complete levulinic acid conversion, but the development of heterogeneous acid catalysts is highly desirable, thus minimizing the corrosion problems and simplifying the subsequent work-up operations. Commercial acid resins, such as Amberlyst-15 and zeolites, give promising catalytic performances similar to those of traditional homogeneous acids. Moreover, as discussed in detail, new ad hoc catalytic systems have been proposed by several authors, which, in some cases, maintain good catalytic performances after the reactivation step. In fact, catalyst deactivation was generally observed after its recycling mainly due to the adsorption of organic matter on the catalyst surface. On this basis, the thermal calcination of the spent catalyst is often effective to restore its pristine performances. However, this thermal treatment is feasible only for inorganic systems (sulfonated N-doped carbon nanotube, $H_3PW_{12}O_{40}$-silica materials, and zeolites), whilst it is inappropriate for sulfonic-acid-based organic resins (Amberlyst and Nafion), which would degrade at such high temperatures. Therefore, the inorganic catalysts are industrially more attractive from the perspective of a desirable scale-up of the reaction in the immediate future. The improvement of the regioselectivity of this reaction is specifically addressed because different isomers, including, mainly, the p,p'- and o,p'-ones, can be obtained starting from phenol, but only the first one is industrially attractive. The most studied solution to solve this issue involves the use of thiol additives, which mainly improve the $p,p'/o,p'$ molar ratio, in some cases up to 100, which allows for reaching a p,p'-DPA yield higher than 90 mol%. With these kinds of catalytic systems, the thiol effect should be properly tuned by preferring a kinetic control regime and therefore working at limited levulinic acid conversion rather than a thermodynamic one. In most cases, this last approach has been preferred by pursuing the maximum performance of the catalysts and thus mostly losing the advantage deriving from the thiol use. Remarkably, only batch-scale studies have been carried out up to now, whilst thiol use could be even better exploited through a continuous-scale approach. In addition to the catalytic synthesis of DPA, its main developed applications have also been discussed, which mainly deal with epoxy and phenolic resins and polycarbonates. Moreover, DPA shows many other potential valuable benefits in terms of antioxidant, flame-retardant, and UV protection properties and biological activity. In principle, the work-up step should be specifically evaluated depending on the use of the final material. In this context, some old patents for DPA synthesis reported some purification procedures, but the evaluation of DPA purity is limited by the technological techniques of those years in the field of characterization. Nowadays, the phase diagrams of the p,p'- and o,p'-isomers are not known, and the characterization of the heavier by-products has been little investigated. Therefore, further research on these issues is highly necessary to fill these gaps and improve the know-how mainly in the selective synthesis of the p,p'-isomer and the industrial separation/purification field. Lastly, other challenges to be further investigated include the use of substituted phenolic derivatives, which can be reasonably obtained from lignin sources, and/or the use of alkyl levulinates instead of levulinic acid to increase the selectivity to the p,p'-isomer and improve the physico-chemical properties of the final polymers. Very few promising results deriving from these approaches are already available, demonstrating that further improvement in this field is possible and thus favoring the growth of the production of DPA and its added-value, versatile derivatives.

Author Contributions: Conceptualization, A.M.R.G. and C.A.; writing—original draft preparation, D.L., S.F., N.D.F. and C.A.; writing—review and editing, D.L., S.F., N.D.F., C.A. and A.M.R.G.; supervision, A.M.R.G. and C.A. All authors have read and agreed to the published version of the manuscript.

Funding: The authors are grateful to the Italian Ministero dell'Istruzione, dell'Università e della Ricerca for the financial support provided through the PRIN 2020 LEVANTE project "LEvulinic acid Valorization through Advanced Novel Technologies" (Progetti di Ricerca di Rilevante Interesse Nazionale Bando 2020, 2020CZCJN7) and to the project NEST "Network 4 Energy Sustainable Transition" (code PE0000021), funded by the National Recovery and Resilience Plan (NRRP), Mission 4 Component 2 Investment 1.3–Call for tender No. 1561 of 11.10.2022 of MUR, funded by the European Union—Next Generation EU.

Data Availability Statement: Not applicable.

Acknowledgments: N.D.F. acknowledges the support of the RTD-A contract (no. 1112/2021, Prot. no. 0165823/2021) co-funded by the University of Pisa with respect to the PON "Ricerca e Innovazione" 2014–2020 (PON R&I FSE-REACT EU), Azione IV.6 "Contratti di ricerca su tematiche Green".

Conflicts of Interest: The authors declare no conflicts of interest.

Appendix A

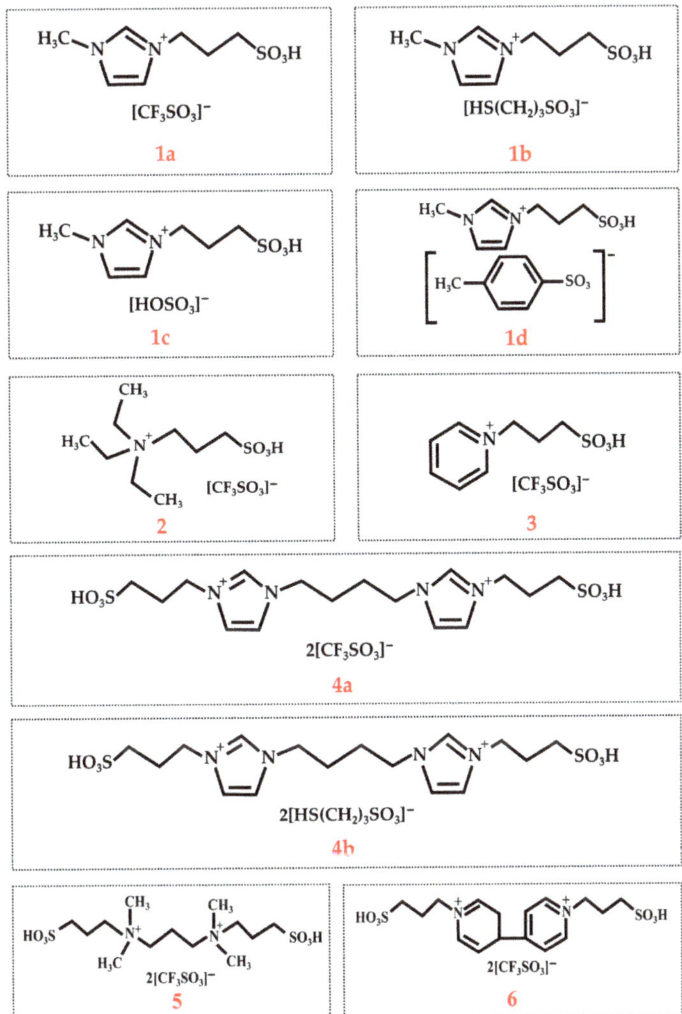

Figure A1. Ionic liquids employed for the synthesis of DPA.

References

1. Antonetti, C.; Licursi, D.; Fulignati, S.; Valentini, G.; Raspolli Galletti, A.M. New frontiers in the catalytic synthesis of levulinic acid: From sugars to raw and waste biomass as starting feedstock. *Catalysts* **2016**, *6*, 196. [CrossRef]
2. Licursi, D.; Antonetti, C.; Martinelli, M.; Ribechini, E.; Zanaboni, M.; Raspolli Galletti, A.M. Monitoring/characterization of stickies contaminants coming from a papermaking plant—Toward an innovative exploitation of the screen rejects to levulinic acid. *Waste Manag.* **2016**, *49*, 469–482. [CrossRef] [PubMed]
3. GF Biochemicals. Available online: https://gfbio.co/ (accessed on 25 October 2023).
4. Khan, M.A.; Dharmalingam, B.; Chuetor, S.; Cheng, Y.-S.; Sriariyanun, M. Comprehensive review on effective conversion of lignocellulosic biomass to levulinic acid. *Biomass Conv. Bioref.* **2023**. [CrossRef]
5. Liguori, F.; Marrodan, C.M.; Barbaro, P. Biomass-derived chemical substitutes for bisphenol A: Recent advancements in catalytic synthesis. *Chem. Soc. Rev.* **2020**, *49*, 6329–6363. [CrossRef] [PubMed]

6. Qian, Z.; Xiao, Y.; Zhang, X.; Li, Q.; Wang, L.; Fu, F.; Diao, H.; Liu, X. Bio-based epoxy resins derived from diphenolic acid via amidation showing enhanced performance and unexpected autocatalytic effect on curing. *J. Chem. Eng.* **2022**, *435 Pt 2*, 135022. [CrossRef]

7. Mohan, P. A critical review: The modification, properties, and applications of epoxy resins. *Polym. Plast. Technol.* **2013**, *52*, 107–125. [CrossRef]

8. Sharma, V.; Jain, D.; Rai, A.R.; Kumari, P.; Nagar, V.; Kaur, A.; Singh, A.; Verma, R.K.; Pandey, H.; Sankhla, M.S. Toxicological assessment and concentration analysis of Bisphenol A in food grade plastics: A systematic review. *Mater. Today Proc.* **2023**, *95*, 18–25. [CrossRef]

9. Russo, G.; Barbato, F.; Mita, D.G.; Grumetto, L. Occurrence of Bisphenol A and its analogues in some foodstuff marketed in Europe. *Food Chem. Toxicol.* **2019**, *131*, 110575. [CrossRef]

10. Rybczyńska-Tkaczyk, K.; Skóra, B.; Szychowski, K.A. Toxicity of bisphenol A (BPA) and its derivatives in divers biological models with the assessment of molecular mechanisms of toxicity. *Environ. Sci. Pollut. Res.* **2023**, *30*, 75126–75140. [CrossRef]

11. Heindel, J.J.; Blumberg, B.; Cave, M.; Machtinger, R.; Mantovani, A.; Mendez, M.A.; Nadal, A.; Palanza, P.; Panzica, G.; Sargis, R.; et al. Metabolism disrupting chemicals and metabolic disorders. *Reprod. Toxicol.* **2017**, *68*, 3–33. [CrossRef]

12. Pérez-Bermejo, M.; Mas-Pérez, I.; Murillo-Llorente, M.T. The role of the Bisphenol A in diabetes and obesity. *Biomedicines* **2021**, *9*, 666. [CrossRef] [PubMed]

13. EFSA, EFSA Scientific opinion on the risks to public health related to the presence of bisphenol A (BPA) in foodstuffs. *EFSA J.* **2015**, *13*, 3978. [CrossRef]

14. Hu, B.; Zhang, Z.-X.; Xie, W.-L.; Liu, J.; Li, Y.; Zhang, W.-M.; Fu, H.; Lu, Q. Advances on the fast pyrolysis of biomass for the selective preparation of phenolic compounds. *Fuel Process. Technol.* **2023**, *237*, 107465. [CrossRef]

15. Huang, X.; Ludenhoff, J.M.; Dirks, M.; Ouyang, X.; Boot, M.D.; Hensen, E.J.M. Selective production of biobased phenol from lignocellulose-derived alkylmethoxyphenols. *ACS Catal.* **2018**, *8*, 11184–11190. [CrossRef] [PubMed]

16. Rahaman, M.S.; Tulaphol, S.; Hossain, M.A.; Evrard, C.N.; Thompson, L.M.; Sathitsuksanoh, N. Kinetics of phosphotungstic acid-catalyzed condensation of levulinic acid with phenol to diphenolic acid: Temperature-controlled regioselectivity. *Mol. Catal.* **2021**, *514*, 111848. [CrossRef]

17. Bader, A.R. Resinous Material. Granted Patent US 2933472, 19 April 1960.

18. Yu, X.; Guo, Y.; Li, K.; Yang, X.; Xu, L.; Guo, Y.; Hu, J. Catalytic synthesis of diphenolic acid from levulinic acid over cesium partly substituted Wells-Dawson type heteropolyacid. *J. Mol. Catal. A Chem.* **2008**, *290*, 44–53. [CrossRef]

19. Zhu, R.; Peng, Y.; Deng, J. Sulfonated N-doped carbon nanotubes as magnetic solid acid catalysts for the synthesis of diphenolic acid. *ACS Appl. Nano Mater.* **2022**, *5*, 4214–4221. [CrossRef]

20. Li, K.; Hu, J.; Li, W.; Ma, F.; Xu, L.; Guo, Y. Design of mesostructured $H_3PW_{12}O_{40}$-silica materials with controllable ordered and disordered pore geometries and their application for the synthesis of diphenolic acid. *J. Mater. Chem.* **2009**, *19*, 8628–8638. [CrossRef]

21. Shen, Y.; Sun, J.; Wang, B.; Xu, F.; Sun, R. Catalytic synthesis of diphenolic acid from levulinic acid over Bronsted acidic ionic liquids. *BioResources* **2014**, *9*, 3264–3275. [CrossRef]

22. Liu, H.F.; Zeng, F.X.; Deng, L.; Liao, B.; Pang, H.; Guo, Q.X. Brønsted acidic ionic liquids catalyse the high-yield production of diphenolic acid/esters from renewable levulinic acid. *Green Chem.* **2013**, *15*, 81–84. [CrossRef]

23. Smith, R.V. Method of Preparing 4,4-bis (4-hydroxyaryl) Pentanoic Acids. Granted Patent US 3248421, 26 April 1966.

24. Das, D.; Lee, J.-F.; Cheng, S. Selective synthesis of Bisphenol-A over mesoporous MCM silica catalysts functionalized with sulfonic acid groups. *J. Catal.* **2004**, *223*, 152–160. [CrossRef]

25. Schutyser, W.; Koelewijn, S.-F.; Dusselier, M.; Vyver, S.V.D.; Thomas, J.; Yu, F.; Carbone, M.J.; Smet, M.; Puyvelde, P.V.; Dehaen, W.; et al. Regioselective synthesis of renewable bisphenols from 2,3-pentanedione and their application as plasticizers. *Green Chem.* **2014**, *16*, 1999–2007. [CrossRef]

26. Mendiratta, A.K. Process for Making Bisphenol A. Granted Patent EP0210366B1, 31 January 1990.

27. Morales, G.; Melero, J.A.; Paniagua, M.; López-Aguado, C.; Vidal, N. Beta zeolite as an efficient catalyst for the synthesis of diphenolic acid (DPA) from renewable levulinic acid. *Catal. Today* **2023**, *424*, 113801. [CrossRef]

28. Willems, G.J.; Liska, J. Method for Making Bis-Xylenols Containing Acid Moieties. Granted Patent US 5969180, 30 January 1998.

29. Ertl, J.; Cerri, E.; Rizzuto, M.; Caretti, D. Natural derivatives of diphenolic acid as substitutes for bisphenol-A. *AIP Conf. Proc.* **2014**, *1599*, 326. [CrossRef]

30. Van de Vyver, S.; Geboers, J.; Helsen, S.; Yu, F.; Thomas, J.; Smet, M.; Dehaen, W.; Sels, B.F. Thiol-promoted catalytic synthesis of diphenolic acid with sulfonated hyperbranched poly(arylene oxindole)s. *Chem. Commun.* **2021**, *48*, 3497–3499. [CrossRef] [PubMed]

31. Van de Vyver, S.; Helsen, S.; Geboers, J.; Yu, F.; Thomas, J.; Smet, M.; Dehaen, W.; Román-Leshkov, Y.; Hermans, I.; Sels, B.F. Mechanistic insights into the kinetic and regiochemical control of the thiol-promoted catalytic synthesis of diphenolic acid. *ACS Catal.* **2012**, *2*, 2700–2704. [CrossRef]

32. Guo, Y.; Li, K.; Yu, X.; Clark, J.H. Mesoporous $H_3PW_{12}O_{40}$-silica composite: Efficient and reusable solid acid catalyst for the synthesis of diphenolic acid from levulinic acid. *Appl. Catal. B Env.* **2008**, *81*, 182–191. [CrossRef]

33. Mthembu, L.D.; Lokhat, D.; Deenadayalu, N. Catalytic condensation of depithed sugarcane bagasse derived levulinic acid into diphenolic acid. *BioResources* **2021**, *16*, 2235–2248. [CrossRef]

34. Amarasekara, A.S.; Wiredu, B.; Grady, T.L.; Obregon, R.G.; Margetić, D. Solid acid catalyzed aldol dimerization of levulinic acid for the preparation of C10 renewable fuel and chemical feedstocks. *Catal. Commun.* **2019**, *124*, 6–11. [CrossRef]

35. Neumann, F.W.; Smith, W.E. By-product of the Bisphenol A reaction. Syntheses and structure assignments of 2,4′-isopropylidenediphenol and 4,4′-(4-hydroxy-m-phenylenediisopropylidene)-diphenol. *J. Org. Chem.* **1966**, *31*, 4318–4320. [CrossRef]

36. Margelefsky, E.L.; Bendjériou, A.; Zeidan, R.K.; Dufaud, V.; Davis, M.E. Nanoscale organization of thiol and arylsulfonic acid on silica leads to a highly active and selective bifunctional, heterogeneous catalyst. *J. Am. Chem. Soc.* **2008**, *130*, 13442–13449. [CrossRef] [PubMed]

37. Fulignati, S.; Licursi, D.; Di Fidio, N.; Antonetti, C.; Raspolli Galletti, A.M. Novel challenges on the catalytic synthesis of 5-hydroxymethylfurfural (HMF) from real feedstocks. *Catalysts* **2022**, *12*, 1664. [CrossRef]

38. Zúñiga, C.; Lligadas, G.; Ronda, J.C.; Galià, M.; Cádiz, V. Renewable polybenzoxazines based in diphenolic acid. *Polymer* **2012**, *53*, 1617–1623. [CrossRef]

39. Salomatova, V.A.; Pozdnyakov, I.P.; Yanshole, V.V.; Wu, F.; Grivin, V.P.; Bazhin, N.M.; Plyusnin, V.F. Photodegradation of 4,4-Bis(4-hydroxyphenyl)valeric acid and its inclusion complex with β-cyclodextrin in aqueous solution. *J. Photochem. Photobiol. A* **2014**, *274*, 27–32. [CrossRef]

40. Guo, L.; Huang, J.; Wu, J.; Luo, L.L. Effect of cyclodextrins on photodegradation of 4,4-Bis(4-hydroxyphenyl)pentanoic acid under UV irradiation. *Appl. Mech. Mater.* **2014**, *665*, 455–458. [CrossRef]

41. Qian, Z.; Zheng, Y.; Li, Q.; Wang, L.; Fu, F.; Liu, X. Amidation way of diphenolic acid for preparing biopolybenzoxazine resin with outstanding thermal performance. *ACS Sust. Chem. Eng.* **2021**, *9*, 4668–4680. [CrossRef]

42. Wei, J.; Duan, Y.; Wang, H.; Hui, J.; Qi, J. Bio-based trifunctional diphenolic acid epoxy resin with high Tg and low expansion coefficient: Synthesis and properties. *Polym. Bull.* **2023**, *80*, 10457–10471. [CrossRef]

43. Zúñiga, C.; Larrechi, M.S.; Lligadas, G.; Ronda, J.C.; Galià, M.; Cádiz, V. Polybenzoxazines from renewable diphenolic acid. *Polym. Chem.* **2011**, *49*, 1219–1227. [CrossRef]

44. Fischer, R.P.; Hartranft, G.R.; Heckles, J.S. Diphenolic acid ester polycarbonates. *J. Appl. Polym. Sci.* **1966**, *10*, 245–252. [CrossRef]

45. Moore, J.A.; Tannahill, T. Homo- and co-polycarbonates and blends derived from diphenolic acid. *High Perform. Polym.* **2001**, *13*, S305–S316. [CrossRef]

46. Trullemans, L.; Koelewijn, S.-F.; Boonen, I.; Cooreman, E.; Hendrickx, T.; Preegel, G.; Aelst, J.V.; Witters, H.; Elskens, M.; Puyvelde, P.V.; et al. Renewable and safer bisphenol A substitutes enabled by selective zeolite alkylation. *Nat. Sustain.* **2023**, *6*, 1693–1704. [CrossRef]

47. Kricheldorf, H.R.; Hobzova, R.; Schwarz, G. Cyclic hyperbranched polyesters derived from 4,4-bis(4′-hydroxyphenyl)valeric acid. *Polym.* **2003**, *44*, 7361–7368. [CrossRef]

48. Chen, J.; Chu, N.; Zhao, M.; Jin, F.-L.; Park, S.J. Synthesis and application of thermal latent initiators of epoxy resins: A review. *J. Appl. Polym. Sci.* **2020**, *137*, 49592. [CrossRef]

49. Zheng, Y.; Qian, Z.; Sun, H.; Jiang, J.; Fu, F.; Li, H.; Liu, X. A novel benzoxazine derived from diphenol amide for toughing commercial benzoxazine via copolymerization. *J. Polym. Sci.* **2022**, *60*, 2808–2816. [CrossRef]

50. Yun, J.; Kang, H.-S.; An, B.-K. Phenolic polymer-based color developers for thermal papers: Synthesis, characterization, and applications. *Ind. Eng. Chem. Res.* **2021**, *60*, 9456–9464. [CrossRef]

51. Ma, Z.; Li, C.; Fan, H.; Wan, J.; Luo, Y.; Li, B.-G. Polyhydroxyurethanes (PHUs) derived from diphenolic acid and carbon dioxide and their application in solvent- and water-borne PHU coatings. *Ind. Eng. Chem. Res.* **2017**, *56*, 14089–14100. [CrossRef]

52. Yan, H.; Li, N.; Fang, Z.; Wang, H. Application of poly(diphenolic acid-phenyl phosphate)-based layer by layer nanocoating in flame retardant ramie fabrics. *J. Appl. Polym. Sci.* **2017**, *134*. [CrossRef]

53. Yan, H.; Li, N.; Cheng, J.; Song, P.; Fang, Z.; Wang, H. Fabrication of flame retardant benzoxazine semi-biocomposites reinforced by ramie fabrics with bio-based flame retardant coating. *Polym. Compos.* **2018**, *39*, E480–E488. [CrossRef]

54. de Pinedo, A.T.; Peñalver, P.; Morales, J.C. Synthesis and evaluation of new phenolic-based antioxidants: Structure–activity relationship. *Food Chem.* **2007**, *103*, 55–61. [CrossRef]

55. Zhang, P.; Wu, L.; Bu, Z.; Li, B.-G. Interfacial polycondensation of diphenolic acid and isophthaloyl chloride. *J. Appl. Polym. Sci.* **2008**, *108*, 3586–3592. [CrossRef]

56. Olson, E.S. *Subtask 4.1—Conversion of Lignocellulosic Material to Chemicals and Fuels*; Final report 2001-EERC-06-09; U.S. Department of Energy: Pittsburgh, PA, USA, 2001.

57. Pancrazzi, F.; Castronuovo, G.; Maestri, G.; Constantin, A.M.; Voronov, A.; Maggi, R.; Mazzeo, P.P.; Motti, E.; Cauzzi, D.A.; Viscardi, R.; et al. Perfluorinated sulfonate resins as reusable heterogeneous catalysts for the one-pot synthesis of diPhenolic esters (DPEs). *Eur. J. Org. Chem.* **2022**, *2022*, e202201315. [CrossRef]

58. Raspolli Galletti, A.M.; Antonetti, C.; Fulignati, S.; Licursi, D. Direct alcoholysis of carbohydrate precursors and real cellulosic biomasses to alkyl levulinates: A critical review. *Catalysts* **2020**, *10*, 1221. [CrossRef]

59. Wang, P.; Xia, L.; Jian, R.; Ai, Y.; Zheng, X.; Chen, G.; Wang, J. Flame-retarding epoxy resin with an efficient P/N/S-containing flame retardant: Preparation, thermal stability, and flame retardance. *Polym. Degrad. Stab.* **2018**, *149*, 69–77. [CrossRef]

60. Maiorana, A.; Spinella, S.; Gross, R.A. Bio-based alternative to the diglycidyl ether of bisphenol A with controlled materials properties. *Biomacromolecules* **2015**, *16*, 1021–1031. [CrossRef] [PubMed]

61. Zhang, L.; Zhang, X.; Hua, W.; Xie, W.; Zhang, W.; Gao, L. Epoxy resin-hydrated halt shaped composite thermal control packaging material for thermal management of electronic components. *J. Clean. Prod.* **2022**, *363*, 132369. [CrossRef]
62. Gou, H.; Bao, Y.; Huang, J.; Fei, X.; Li, X.; Wei, W. Development of molding compounds based on epoxy resin/aromatic amine/benzoxazine for high-temperature electronic packaging applications. *Macromol. Mater. Eng.* **2022**, *307*, 2200351. [CrossRef]
63. Babayevsky, P.G.; Gillham, J.K. Epoxy thermosetting systems: Dynamic mechanical analysis of the reactions of aromatic diamines with the diglycidyl ether of bisphenol A. *J. Appl. Polym. Sci.* **1973**, *17*, 2067–2088. [CrossRef]
64. Auvergne, R.; Caillol, S.; David, G.; Boutevin, B.; Pascault, J.-P. Biobased thermosetting epoxy: Present and future. *Chem. Rev.* **2014**, *114*, 1082–1115. [CrossRef]
65. Xiao, L.; Li, W.; Li, S.; Chen, J.; Wang, Y.; Huang, J.; Nie, X. Diphenolic acid-derived hyperbranched epoxy thermosets with high mechanical strength and toughness. *ACS Omega* **2021**, *6*, 34142–34149. [CrossRef]
66. Chen, S.; Xu, Z.; Zhang, D. Synthesis and application of epoxy-ended hyperbranched polymers. *Chem. Eng. J.* **2018**, *343*, 283–302. [CrossRef]
67. Zhao, Y.; Huang, R.; Wu, Z.; Zhang, H.; Zhou, Z.; Li, L.; Dong, Y.; Luo, M.; Ye, B.; Zhang, H. Effect of free volume on cryogenic mechanical properties of epoxy resin reinforced by hyperbranched polymers. *Mater. Des.* **2021**, *202*, 109565. [CrossRef]
68. McMaster, M.S.; Yilmaz, T.E.; Patel, A.; Maiorana, A.; Manas-Zloczower, I.; Gross, R.; Singer, K.D. Dielectric properties of bio-based diphenolate ester epoxies. *ACS Appl. Mater. Interfaces* **2018**, *10*, 13924–13930. [CrossRef] [PubMed]
69. Chen, D.; Li, J.; Yuan, Y.; Gao, C.; Cui, Y.; Li, S.; Wang, H.; Peng, C.; Liu, X.; Wu, Z.; et al. A new strategy to improve the toughness of epoxy thermosets by introducing the thermoplastic epoxy. *Polymer* **2022**, *240*, 124518. [CrossRef]
70. Ishida, H. Overview and historical background of polybenzoxazine research. In *Handbook of Benzoxazine Resins*; Ishida, H., Agag, T., Eds.; Elsevier: Amsterdam, The Netherlands, 2011; Chapter 1; pp. 3–81. [CrossRef]
71. Asim, M.; Saba, N.; Jawaid, M.; Nasir, M.; Pervaiz, M. A Review on phenolic resin and its composites. *Curr. Anal. Chem.* **2018**, *14*, 185–197. [CrossRef]
72. Lee, Y.-K.; Kim, D.-J.; Kim, H.-J.; Hwang, T.-S.; Rafailovich, M.; Sokolov, J. Activation energy and curing behavior of resol- and novolac-type phenolic resins by differential scanning calorimetry and thermogravimetric analysis. *J. Appl. Polym. Sci.* **2003**, *89*, 2589–2596. [CrossRef]
73. De Medeiros, E.S.; Agnelli, J.A.M.; Joseph, K.; De Carvalho, L.H.; Mattoso, L.H.C. Curing behavior of a novolac-type phenolic resin analyzed by differential scanning calorimetry. *J. Appl. Polym. Sci.* **2003**, *90*, 1678–1682. [CrossRef]
74. Kiran, V.; Awasthi, S.; Gaur, B. Hydroquinone based sulfonated poly (arylene ether sulfone) copolymer as proton exchange membrane for fuel cell applications. *Express Polym. Lett.* **2015**, *9*, 1053–1067. [CrossRef]
75. Zúñiga, C.; Lligadas, G.; Ronda, J.C.; Galià, M.; Cádiz, V. Self-foaming diphenolic acid benzoxazine. *Polymer* **2012**, *53*, 3089–3095. [CrossRef]
76. Ping, Z.; Linbo, W.; Bo-Geng, L. Thermal stability of aromatic polyesters prepared from diphenolic acid and its esters. *Polym. Degrad. Stab.* **2009**, *94*, 1261–1266. [CrossRef]
77. Bisphenol A Alternatives in Thermal Paper. Available online: https://www.epa.gov/sites/default/files/2015-08/documents/bpa_final.pdf (accessed on 5 November 2023).
78. Malhotra, S.L.; Naik, K.N.; MacKinnon, D.N.; Jones, A.Y. Ink-Jet Printing Sheet for Transparency Preparation. Granted Patent US5683793, 4 November 1997.
79. Zhang, Y.; Jing, J.; Liu, T.; Xi, L.; Sai, T.; Ran, S.; Fang, Z.; Huo, S.; Song, P. A molecularly engineered bioderived polyphosphate for enhanced flame retardant, UV-blocking and mechanical properties of poly(lactic acid). *Chem. Eng. J.* **2021**, *411*, 128493. [CrossRef]
80. Bell, A.M.; Keltsch, N.; Schweyen, P.; Reifferscheid, G.; Ternes, T.; Buchinger, S. UV aged epoxy coatings Ecotoxicological effects and released compounds. *Water Res. X* **2021**, *12*, 100105. [CrossRef] [PubMed]
81. Chafetz, H.; Liu, C.S.; Papke, B.L.; Kennedy, T.A. Lubricating Oil Composition Containing the Reaction Product of an Alkenyl-succinimide with a Bis(hydroxyaromatic) Substituted Carboxylic Acid. Granted Patent US005445750A, 29 August 1995.
82. Holmen, R.E.; Olander, S.J. Paint composition for marking paved surfaces. Granted Patent US4031048, 21 June 1977.
83. Lu, Y.; Wang, J.; Wang, L.; Song, S. Diphenolic acid-modified PAMAM/chlorinated butyl rubber nanocomposites with superior mechanical, damping, and self-healing properties. *Sci. Technol. Adv. Mater.* **2021**, *22*, 14–25. [CrossRef] [PubMed]
84. Shen, L.; Jiang, J.; Liu, J.; Fu, F.; Diao, H.; Liu, X. Cotton fabrics with antibacterial and antiviral properties produced by a simple pad-dry-cure process using diphenolic acid. *Appl. Surf. Sci.* **2022**, *600*, 154152. [CrossRef]
85. El-Demerdash, F.M.; Tousson, R.M.; Kurzepa, J.; Habib, S.L. Xenobiotics, oxidative stress, and antioxidants. *Oxid. Med. Cell. Longev.* **2018**, *2018*, 9758951. [CrossRef]
86. Abotaleb, M.; Liskova, A.; Kubatka, P.; Büsselberg, D. Therapeutic potential of plant phenolic acids in the treatment of cancer. *Biomolecules* **2020**, *10*, 221. [CrossRef]

MDPI AG
Grosspeteranlage 5
4052 Basel
Switzerland
Tel.: +41 61 683 77 34

Molecules Editorial Office
E-mail: molecules@mdpi.com
www.mdpi.com/journal/molecules